AP® PHYSICS 1 PREMIUM PREP

10th Edition

The Staff of The Princeton Review

PrincetonReview.com

Penguin Random House

The Princeton Review
110 East 42nd St., 7th Floor
New York, NY 10017

Published in the United States by Penguin Random House LLC, New York.

Terms of Service: The Princeton Review Online Companion Tools ("Student Tools") for retail books are available for only the two most recent editions of that book. Student Tools may be activated only once per eligible book purchased for a total of 24 months of access. Activation of Student Tools more than once per book is in direct violation of these Terms of Service and may result in discontinuation of access to Student Tools Services.

ISBN: 978-0-593-51680-5
ISSN: 2690-5582

The material in this book is up-to-date at the time of publication. However, changes may have been instituted by the testing body in the test after this book was published.

If there are any important late-breaking developments, changes, or corrections to the materials in this book, we will post that information online in the Student Tools. Register your book and check your Student Tools to see if there are any updates posted there.

Editor: Selena Coppock
Production Editors: Emma Parker, Liz Dacey
Production Artist: Deborah Weber
Content Contributor: Felicia Tam

Printed in the United States of America.

10 9 8 7 6 5 4 3 2 1

10th Edition

For customer service, please contact **editorialsupport@review.com**, and be sure to include:

- full title of the book
- ISBN
- page number

Acknowledgments

Special thanks to Felicia Tam for her fantastic work on the 10th Edition of this book. Additionally, the staff of The Princeton Review would like to thank Deborah Weber, Liz Dacey, and Emma Parker for their time and attention to each page.

Contents

Get More (Free) Content at PrincetonReview.com/prep

As easy as 1·2·3

1 Go to PrincetonReview.com/prep or scan the **QR code** and enter the following ISBN for your book: **9780593516805**

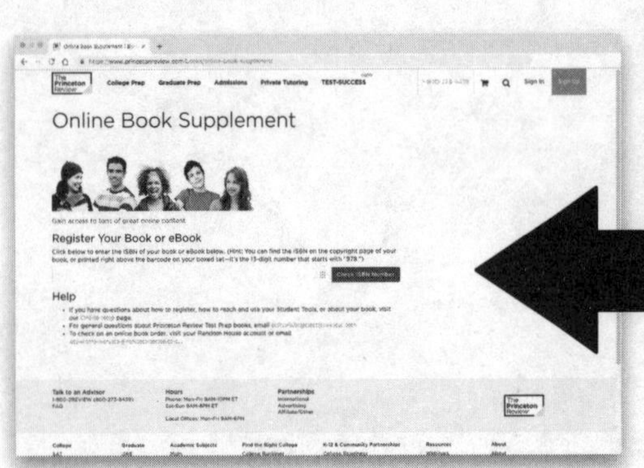

2 Answer a few simple questions to set up an exclusive Princeton Review account. *(If you already have one, you can just log in.)*

3 Enjoy access to your **FREE** content!

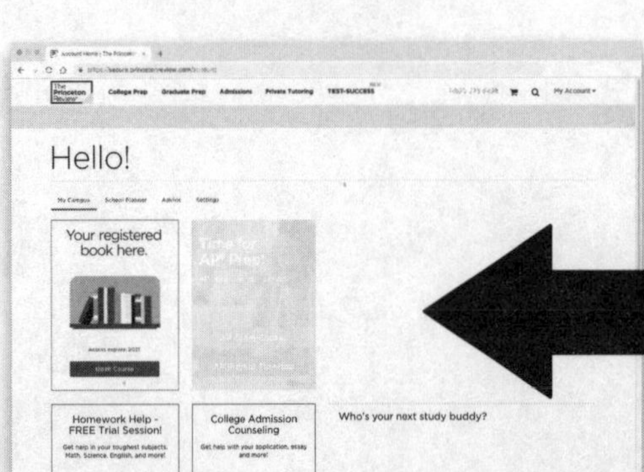

Once you've registered, you can...

- Get any updates or new information about assorted AP courses and exams from the College Board
- Access your fifth AP Physics 1 practice test (there are 4 right here in your book and 1 online), plus an answer key and complete answers and explanations
- Access comprehensive study guides and a variety of printable resources, including bubble sheets for practice tests, AP score conversion charts, and important equations and formulas
- Take a full-length practice SAT and ACT
- Get valuable advice about the college application process, including tips for writing a great essay and where to apply for financial aid
- Check to see if there have been any corrections or updates to this edition
- Use our searchable rankings of *The Best 389 Colleges* to find out more information about your dream school

LET'S GO MOBILE! Access all of these free, additional resources by downloading the new Princeton Review app at www.princetonreview.com/mobile/apps/highschool or scan the QR Code to the right.

Need to report a potential **content** issue?

Contact **EditorialSupport@review.com** and include:

- full title of the book
- ISBN
- page number

Need to report a **technical** issue?

Contact **TPRStudentTech@review.com** and provide:

- your full name
- email address used to register the book
- full book title and ISBN
- Operating system (Mac/PC) and browser (Chrome, Firefox, Safari, etc.)

Look For These Icons Throughout The Book

 **ONLINE ARTICLES**

 **PROVEN TECHNIQUES**

 **APPLIED STRATEGIES**

 **OTHER REFERENCES**

 **STUDY BREAK**

 **ONLINE VIDEO**

Part I
Using This Book to Improve Your AP Score

- Preview: Your Knowledge, Your Expectations
- Your Guide to Using This Book
- How to Begin

PREVIEW: YOUR KNOWLEDGE, YOUR EXPECTATIONS

Your route to a high score on the AP Physics 1 Exam depends a lot on how you plan to use this book. Respond to the following questions.

Welcome to AP Physics 1 Prep! Be sure to check out great AP online content that comes with this book Follow the instructions on the "Get More (Free) Content" page at the beginning of the book to access these materials!

1. Rate your level of confidence about your knowledge of the content tested by the AP Physics 1 Exam:

   A. Very confident—I know it all
   B. I'm pretty confident, but there are topics for which I could use help.
   C. Not confident—I need quite a bit of support
   D. I'm not sure.

2. If you have a goal score in mind, circle your goal score for the AP Physics 1 Exam:

   5 4 3 2 1 I'm not sure yet.

3. What do you expect to learn from this book? Circle all that apply to you.

   A. A general overview of the test and what to expect
   B. Strategies for how to approach the test
   C. The content tested by this exam
   D. I'm not sure yet.

YOUR GUIDE TO USING THIS BOOK

This book is organized to provide as much—or as little—support as you need, so you can use this book in whatever way will be most helpful for improving your score on the AP Physics 1 Exam.

- The remainder of **Part I** will provide guidance on how to use this book and help you determine your strengths and weaknesses.

- **Part II** of this book contains your first practice test, a Diagnostic Answer Key, detailed answers and explanations for each question, and a scoring guide. (Bubble sheets can be found online in your Student Tools.) This is where you should begin your test preparation in order to realistically determine:
  - your starting point right now
  - which question types you're ready for and which you might need to practice
  - which content topics you are familiar with and which you will want to carefully review

  Once you have nailed down your strengths and weaknesses with regard to this exam, you can focus your test preparation, build a study plan, and be efficient with your time. Our Diagnostic Answer Key will assist you with this process.

- **Part III** of this book will:
  - provide information about the structure, scoring, and content of the AP Physics 1 Exam
  - help you to make a study plan
  - point you toward additional resources

- **Part IV** of this book will explore the following strategies:
  - how to approach multiple-choice questions
  - how to approach free-response questions
  - how to manage your time to maximize the number of points available to you

- **Part V** of this book covers the content you need to know for the AP Physics 1 Exam.

- **Part VI** of this book contains Practice Tests 2, 3, and 4, their answers and explanations, and a score conversion chart. (Bubble sheets can be found online in your Student Tools.) If you skipped Practice Test 1, we recommend that you do all four (with at least a day or two between them) so that you can compare your progress between them. Additionally, this will help to identify any external issues: if you get a certain type of question wrong more than once, you probably need to review it. If you got it wrong only once, you may have run out of time or been distracted by something. In either case, comparing the results of your practice tests will allow you to focus on the factors that caused the discrepancy in scores and to be as prepared as possible on the day of the test.

You may choose to use some parts of this book over others, or you may work through the entire book. This will depend on your needs and how much time you have.

HOW TO BEGIN

1. **Take a Test**

   Before you can decide how to use this book, you need to take a practice test. Doing so will give you insight into your strengths and weaknesses, and the test will also help you make an effective study plan. If you're feeling test-phobic, remind yourself that a practice test is a tool for diagnosing yourself—it's not how well you do that matters but how you use information gleaned from your performance to guide your preparation.

   So, before you read further, take AP Physics 1 Practice Test 1 starting at page 9 of this book. Be sure to do so in one sitting, following the instructions that appear before the test.

2. **Check Your Answers**

   Using the Diagnostic Answer Key on page 32, follow our three-step process to identify your strengths and weaknesses with regard to the tested topics. This will help you determine which content review chapters to prioritize when studying this book. Don't worry about the explanations for now, and don't worry about missed questions. We'll get to that soon.

3. Reflect on the Test

After you take your first test, respond to the following questions:

- How much time did you spend on the multiple-choice questions?
- How much time did you spend on each free-response question?
- How many multiple-choice questions did you miss?
- Do you feel you had the conceptual understanding to address the subject matter of the free-response questions?

4. Read Part III of This Book and Make Your Study Plan

As discussed in the Guide section, Part III will provide information on how the test is structured and scored. It will also set out areas of content that are tested.

As you read Part III, re-evaluate your answers to item 3 above (Reflect on the Test). At the end of Part III, you will revisit and refine the answers. You will then be able to make a study plan, based on your needs and the time available, that will allow you to use this book most effectively.

5. Engage with Parts IV and V as Needed

Notice the word **engage**. You'll get more out of this book if you use it intentionally than if you read it passively, hoping for an improved score through osmosis.

Strategy chapters will help you think about your approach to the question types on this exam. Part IV opens with a reminder to think about how you approach questions now and then closes with a reflection section asking you to think about how/whether you will change your approach in the future.

Part V is designed to provide a review of the content tested on the AP Physics 1 Exam, including the level of detail you need to know and how the content is tested. You will have the opportunity to assess your command of the content of each chapter through test-appropriate questions and a reflection section.

6. **Take Practice Test 2 and Assess Your Performance**

Once you feel you have developed the strategies you need and gained the knowledge you lacked, you should take Practice Test 2, which begins on page 291. You should do so in one sitting, following the instructions at the beginning of the test.

When you are done, check your answers to the multiple-choice sections. Approach a teacher to read your essays and provide feedback.

Once you have taken the test, reflect on those areas you still need to improve upon, and revisit the respective chapters. Through this reflective and engaging approach, you will continue to improve.

7. **Keep Working**

After you have revisited certain content chapters, continue the process of testing, reflection, and engaging with the next practice test in this book. Remember that you have 5 practice tests total (4 in this book, 1 online), so be sure to use them all! Consider what additional work you need to do and how you will change your strategic approach to adapt to different parts of the test.

As we will discuss in Part III, there are other resources available to you, including a wealth of information on AP Students, the official homepage for all things AP. You should continue to explore areas that can stand to improve right up to the day of the test.

Stay Up to Date!
For late-breaking information about test dates, exam formats, and any other changes pertaining to AP Physics 1, make sure to check the College Board's website at apstudents.collegeboard.org/courses/ap-physics-1-algebra-based

Part II
Practice Test 1

- Practice Test 1
- Practice Test 1: Diagnostic Answer Key and Explanations
- How to Score Practice Test 1

Practice Test 1

AP® Physics 1 Exam

SECTION I: Multiple-Choice Questions

DO NOT OPEN THIS BOOKLET UNTIL YOU ARE TOLD TO DO SO.

At a Glance

Total Time
90 minutes
Number of Questions
50
Percent of Total Grade
50%
Writing Instrument
Pen required

Instructions

Section I of this examination contains 50 multiple-choice questions. Fill in only the ovals for numbers 1 through 50 on your answer sheet.

CALCULATORS MAY BE USED ON BOTH SECTIONS OF THE AP PHYSICS 1 EXAM.

Indicate all of your answers to the multiple-choice questions on the answer sheet. No credit will be given for anything written in this exam booklet, but you may use the booklet for notes or scratch work. Please note that there are two types of multiple-choice questions: single-select and multi-select questions. After you have decided which of the suggested answers is best, completely fill in the corresponding oval(s) on the answer sheet. For single-select, you must give only one answer; for multi-select you must give BOTH answers in order to earn credit. If you change an answer, be sure that the previous mark is erased completely. Here is a sample question and answer.

Sample Question	Sample Answer
Chicago is a (A) state (B) city (C) country (D) continent	Ⓐ ● Ⓒ Ⓓ

Use your time effectively, working as quickly as you can without losing accuracy. Do not spend too much time on any one question. Go on to other questions and come back to the ones you have not answered if you have time. It is not expected that everyone will know the answers to all the multiple-choice questions.

About Guessing

Many candidates wonder whether or not to guess the answers to questions about which they are not certain. Multiple-choice scores are based on the number of questions answered correctly. Points are not deducted for incorrect answers, and no points are awarded for unanswered questions. Because points are not deducted for incorrect answers, you are encouraged to answer all multiple-choice questions. On any questions you do not know the answer to, you should eliminate as many choices as you can, and then select the best answer among the remaining choices.

GO ON TO THE NEXT PAGE.

ADVANCED PLACEMENT PHYSICS 1 TABLE OF INFORMATION

CONSTANTS AND CONVERSION FACTORS	
Proton mass, $m_p = 1.67 \times 10^{-27}$ kg	Electron charge magnitude, $e = 1.60 \times 10^{-19}$ C
Neutron mass, $m_n = 1.67 \times 10^{-27}$ kg	Coulomb's law constant, $k = 1/4\pi\varepsilon_0 = 9.0 \times 10^9\ \text{N}\cdot\text{m}^2/\text{C}^2$
Electron mass, $m_e = 9.11 \times 10^{-31}$ kg	Universal gravitational constant, $G = 6.67 \times 10^{-11}\ \text{m}^3/\text{kg}\cdot\text{s}^2$
Speed of light, $c = 3.00 \times 10^8$ m/s	Acceleration due to gravity at Earth's surface, $g = 9.8\ \text{m/s}^2$

UNIT SYMBOLS								
	meter,	m	kelvin,	K	watt,	W	degree Celsius,	°C
	kilogram,	kg	hertz,	Hz	coulomb,	C		
	second,	s	newton,	N	volt,	V		
	ampere,	A	joule,	J	ohm,	Ω		

PREFIXES		
Factor	Prefix	Symbol
10^{12}	tera	T
10^9	giga	G
10^6	mega	M
10^3	kilo	k
10^{-2}	centi	c
10^{-3}	milli	m
10^{-6}	micro	μ
10^{-9}	nano	n
10^{-12}	pico	p

VALUES OF TRIGONOMETRIC FUNCTIONS FOR COMMON ANGLES							
θ	$0°$	$30°$	$37°$	$45°$	$53°$	$60°$	$90°$
$\sin\theta$	0	1/2	3/5	$\sqrt{2}/2$	4/5	$\sqrt{3}/2$	1
$\cos\theta$	1	$\sqrt{3}/2$	4/5	$\sqrt{2}/2$	3/5	1/2	0
$\tan\theta$	0	$\sqrt{3}/3$	3/4	1	4/3	$\sqrt{3}$	∞

The following conventions are used in this exam.

I. The frame of reference of any problem is assumed to be inertial unless otherwise stated.

II. Assume air resistance is negligible unless otherwise stated.

III. In all situations, positive work is defined as work done on a system.

IV. The direction of current is conventional current: the direction in which positive charge would drift.

V. Assume all batteries and meters are ideal unless otherwise stated.

GO ON TO THE NEXT PAGE.

ADVANCED PLACEMENT PHYSICS 1 EQUATIONS

MECHANICS

$v_x = v_{x0} + a_x t$

$x = x_0 + v_{x0}t + \frac{1}{2}a_x t^2$

$v_x^2 = v_{x0}^2 + 2a_x(x - x_0)$

$\vec{a} = \frac{\sum \vec{F}}{m} = \frac{\vec{F}_{net}}{m}$

$|\vec{F}_f| \le \mu |\vec{F}_n|$

$a_c = \frac{v^2}{r}$

$\vec{p} = m\vec{v}$

$\Delta\vec{p} = \vec{F}\,\Delta t$

$K = \frac{1}{2}mv^2$

$\Delta E = W = F_{\parallel}d = Fd\cos\theta$

$P = \frac{\Delta E}{\Delta t}$

$\theta = \theta_0 + \omega_0 t + \frac{1}{2}\alpha t^2$

$\omega = \omega_0 + \alpha t$

$x = A\cos(2\pi ft)$

$\vec{\alpha} = \frac{\sum \vec{\tau}}{I} = \frac{\vec{\tau}_{net}}{I}$

$\tau = r_{\perp}F = rF\sin\theta$

$L = I\omega$

$\Delta L = \tau\,\Delta t$

$K = \frac{1}{2}I\omega^2$

$|\vec{F}_s| = k|\vec{x}|$

$U_s = \frac{1}{2}kx^2$

$\rho = \frac{m}{V}$

$\Delta U_g = mg\,\Delta y$

$T = \frac{2\pi}{\omega} = \frac{1}{f}$

$T_s = 2\pi\sqrt{\frac{m}{k}}$

$T_p = 2\pi\sqrt{\frac{\ell}{g}}$

$|\vec{F}_g| = G\frac{m_1 m_2}{r^2}$

$\vec{g} = \frac{\vec{F}_g}{m}$

$U_G = -\frac{Gm_1 m_2}{r}$

a = acceleration
A = amplitude
d = distance
E = energy
f = frequency
F = force
I = rotational inertia
K = kinetic energy
k = spring constant
L = angular momentum
ℓ = length
m = mass
P = power
p = momentum
r = radius or separation
T = period
t = time
U = potential energy
V = volume
v = speed
W = work done on a system
x = position
y = height
α = angular acceleration
μ = coefficient of friction
θ = angle
ρ = density
τ = torque
ω = angular speed

ELECTRICITY

$|\vec{F}_E| = k\left|\frac{q_1 q_2}{r^2}\right|$

$I = \frac{\Delta q}{\Delta t}$

$R = \frac{\rho\ell}{A}$

$I = \frac{\Delta V}{R}$

$P = I\,\Delta V$

$R_s = \sum_i R_i$

$\frac{1}{R_p} = \sum_i \frac{1}{R_i}$

A = area
F = force
I = current
ℓ = length
P = power
q = charge
R = resistance
r = separation
t = time
V = electric potential
ρ = resistivity

WAVES

$\lambda = \frac{v}{f}$

f = frequency
v = speed
λ = wavelength

GEOMETRY AND TRIGONOMETRY

Rectangle
$A = bh$

Triangle
$A = \frac{1}{2}bh$

Circle
$A = \pi r^2$
$C = 2\pi r$

Rectangular solid
$V = \ell wh$

Cylinder
$V = \pi r^2\ell$
$S = 2\pi r\ell + 2\pi r^2$

Sphere
$V = \frac{4}{3}\pi r^3$
$S = 4\pi r^2$

A = area
C = circumference
V = volume
S = surface area
b = base
h = height
ℓ = length
w = width
r = radius

Right triangle
$c^2 = a^2 + b^2$
$\sin\theta = \frac{a}{c}$
$\cos\theta = \frac{b}{c}$
$\tan\theta = \frac{a}{b}$

c
a
θ
90°
b

GO ON TO THE NEXT PAGE.

THIS PAGE IS LEFT INTENTIONALLY BLANK.

GO ON TO THE NEXT PAGE.

AP PHYSICS 1

SECTION I

Note: To simplify calculations, you may use $g = 10 \text{ m/s}^2$ in all problems.

Directions: Each of the questions or incomplete statements is followed by four suggested answers or completions. Select the one that is best in each case and then fill in the corresponding circle on the answer sheet.

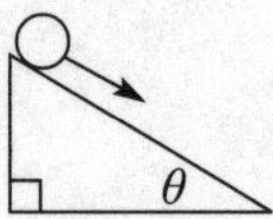

1. If a ball is rolling down an inclined plane without slipping, which force is responsible for exerting the torque that causes its rotation?

   (A) Normal force
   (B) Gravity
   (C) Kinetic friction
   (D) Static friction

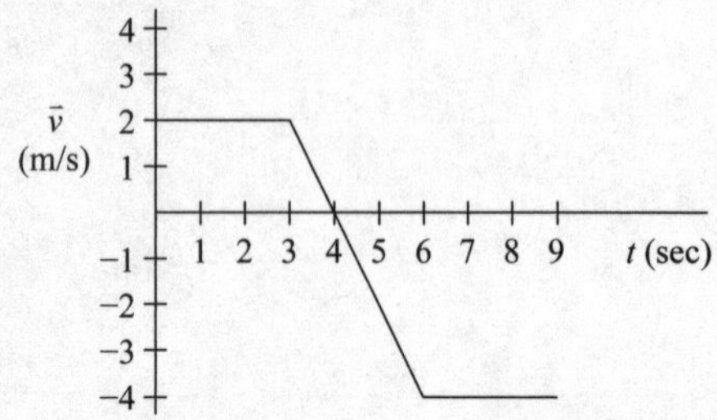

2. The graph above shows the velocity of an object as a function of time. What is the net displacement of the object over the time shown?

   (A) −23 m
   (B) −9 m
   (C) 9 m
   (D) 23 m

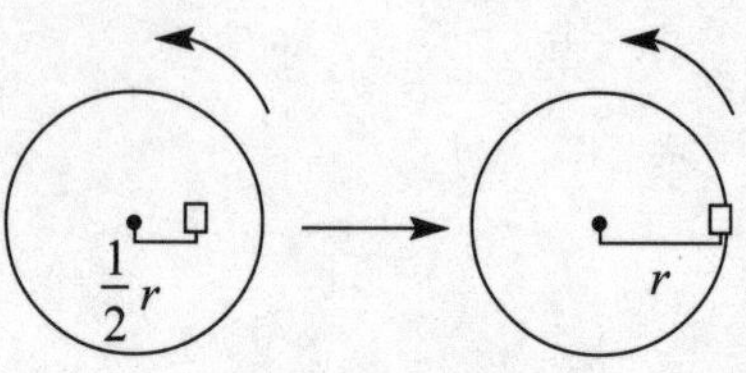

Top view

3. An object is resting on a platform that rotates at a constant speed. At first, it is a distance of half the platform's radius from the center. If the object is moved to the edge of the platform, what happens to the centripetal force that it experiences? Assume the platform continues rotating at the same speed.

   (A) Increases by a factor of 4
   (B) Increases by a factor of 2
   (C) Decreases by a factor of 2
   (D) Decreases by a factor of 4

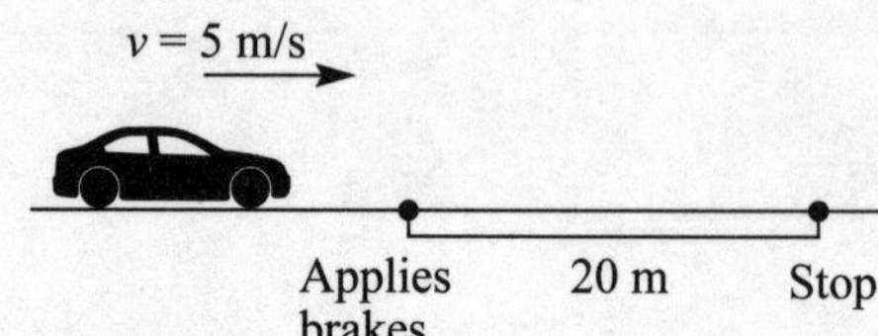

4. A car of mass 1000 kg is traveling at a speed of 5 m/s. The driver applies the brakes, generating a constant friction force, and skids for a distance of 20 m before coming to a complete stop. Given this information, what is the coefficient of friction between the car's tires and the ground?

   (A) 0.25
   (B) 0.2
   (C) 0.125
   (D) 0.0625

GO ON TO THE NEXT PAGE.

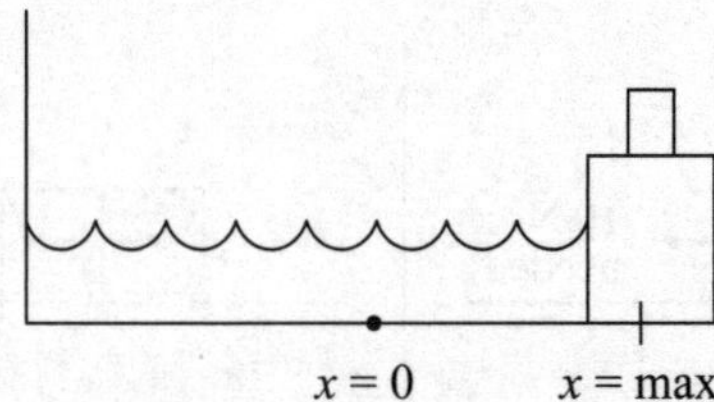

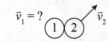

5. A spring-block system is oscillating without friction on a horizontal surface. If a second block of equal mass were placed on top of the original block at a time when the spring is at maximum compression, which of the following quantities would NOT be affected? Assume that the top block stays on the bottom block.
   (A) Frequency
   (B) Maximum speed
   (C) Amplitude
   (D) All of the above quantities would be affected.

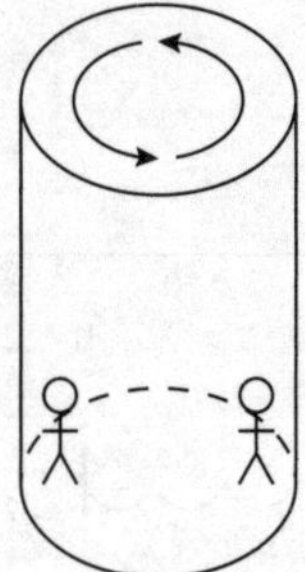

6. A certain theme park ride involves people standing against the walls of a cylindrical room that rotates at a rapid pace, making them stick to the walls without needing support from the ground. Once the ride achieves its maximum speed, the floor drops out from under the riders, but the circular motion holds them in place. Which of the following factors could make this ride dangerous for some riders but not others?
   (A) The mass of the individuals
   (B) The coefficient of friction of their clothing in contact with the walls
   (C) Both of the above
   (D) None of the above

Post-Collision

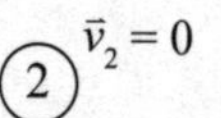

$\vec{v}_1$ $\vec{v}_2 = 0$ | $\vec{v}_1 = ?$ $\vec{v}_2$

7. Two balls collide as shown above. Given the final direction of the second ball's motion after the collision, which of the following is a possible direction for the first ball to move after the collision?
   (A) ↖
   (B) ←
   (C) →
   (D) ↘

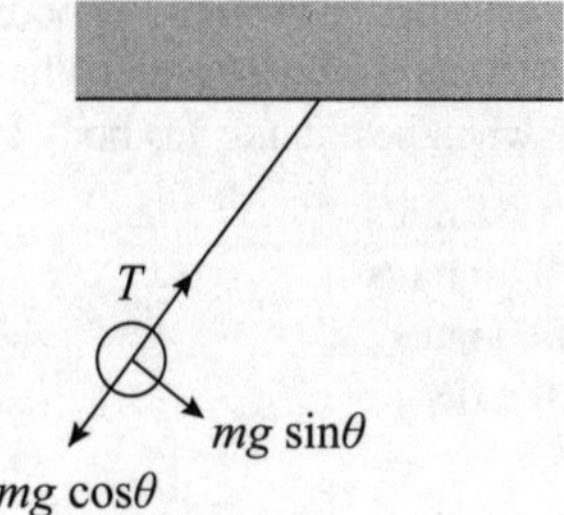

8. As a pendulum swings back and forth, it is affected by two forces: gravity and tension in the string. Splitting gravity into component vectors, as shown above, produces $mg\sin\theta$ (the restoring force) and $mg\cos\theta$. Which of the following correctly describes the relationship between the magnitudes of tension in the string and $mg\cos\theta$?
   (A) Tension > $mg\cos\theta$
   (B) Tension = $mg\cos\theta$
   (C) Tension < $mg\cos\theta$
   (D) The relationship depends on the position of the ball.

GO ON TO THE NEXT PAGE.

9. A car traveling at a velocity of v_0 has a minimum stopping distance of d. What is the minimum stopping distance of the car when it travels at a velocity of $2v_0$? Assume that the acceleration is the same in both cases.
   (A) $d/2$
   (B) $2d$
   (C) $4d$
   (D) $8d$

10. A car accelerates from 0 to 25 m/s in 5 s. If the car's tires have a diameter of 70 cm, how many revolutions does a tire make during this acceleration?
    (A) 14.2 revolutions
    (B) 28.4 revolutions
    (C) 89 revolutions
    (D) 179 revolutions

11. A diver rotates at a rate of 18 rad/s when his body is tucked and has a moment of inertia of 4.5 kg·m^2. Before he hits the water, he extends his body so that it has a moment of inertia of 15 kg·m^2. What is his body's rotation rate when he extends his body?
    (A) 1.62 rad/s
    (B) 5.40 rad/s
    (C) 60.0 rad/s
    (D) 200 rad/s

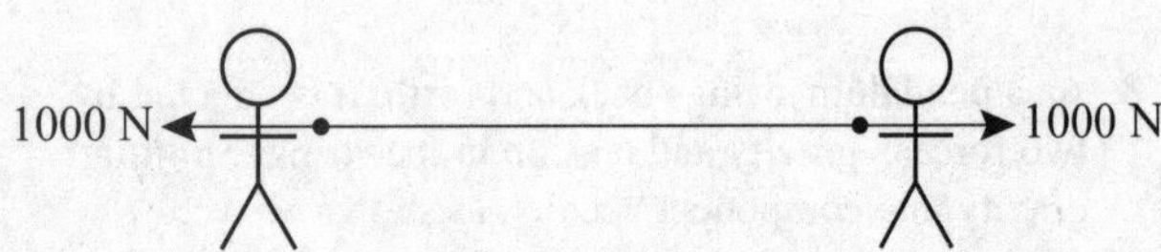

12. If two people pull with a force of 1000 N each on opposite ends of a rope and neither person moves, what is the magnitude of tension in the rope?
    (A) 0 N
    (B) 500 N
    (C) 1000 N
    (D) 2000 N

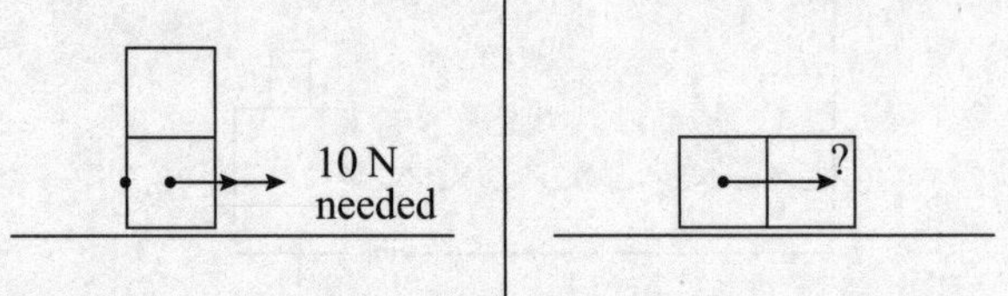

13. Two identical blocks are stacked on top of each other and placed on a table. To overcome the force of static friction, a force of 10 N is required. If the blocks were placed side by side and pushed as shown in the figure above, how much force would be required to move them?
    (A) $\frac{10\sqrt{2}}{2}$ N
    (B) 10 N
    (C) $10\sqrt{2}$ N
    (D) 20 N

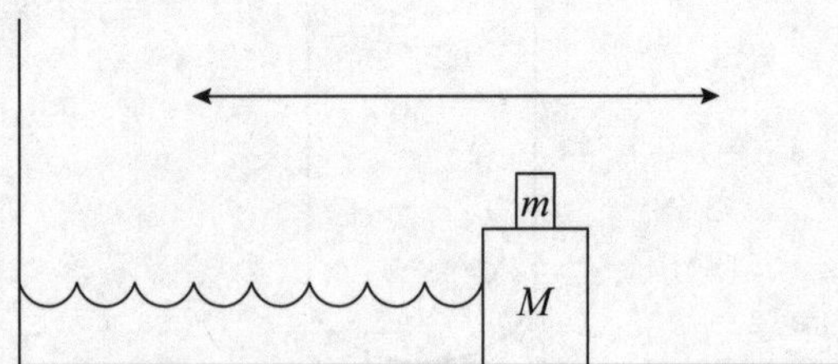

14. A block of known mass M is connected to a horizontal spring that is sliding along a flat, frictionless surface. There is an additional block of known mass m resting on top of the first block. Which of the following quantities would NOT be needed to determine whether the top block will slide off the bottom block?
    (A) The maximum coefficient of static friction between the blocks
    (B) The amplitude of the system's motion
    (C) The spring constant
    (D) The period of the motion

GO ON TO THE NEXT PAGE.

Questions 15–17 all refer to the following scenario.

A 35 kg child ziplines from rest on a platform 12 m high to the ground with negligible friction. In the last 8 m of the zipline, which is approximately horizontal, a braking mechanism applies a constant force opposing the motion to slow the child down so that she's traveling at 2.5 m/s when she finally lands on the safety padding at the end. The padding compresses for 1.5 s when the child lands on it and stops.

15. What is the magnitude of the force applied by the braking mechanism?

    (A) 350 N
    (B) 511 N
    (C) 525 N
    (D) 4090 N

16. What is the average force acting on the child as she comes to a stop on the safety padding?

    (A) 1.7 N
    (B) 58 N
    (C) 73 N
    (D) 150 N

17. If the height of the platform had been 15 m, by what factor would her maximum speed have increased?

    (A) $\frac{\sqrt{5}}{2}$
    (B) $\frac{5}{4}$
    (C) $\sqrt{\frac{5}{2}}$
    (D) $\frac{5}{2}$

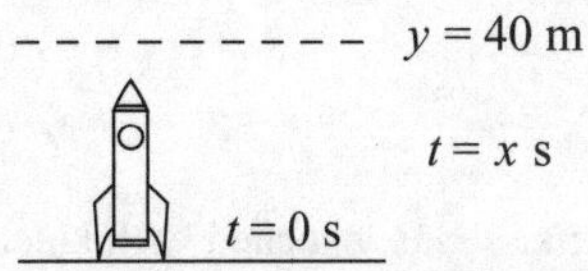

18. A rocket is launched into the air. A few moments after liftoff, the rocket is 40 m above the ground. After another 5 s, the rocket is now 200 m off the ground. What is the average velocity of the rocket during this 5 s interval of the flight?

    (A) 16 m/s
    (B) 32 m/s
    (C) 48 m/s
    (D) 64 m/s

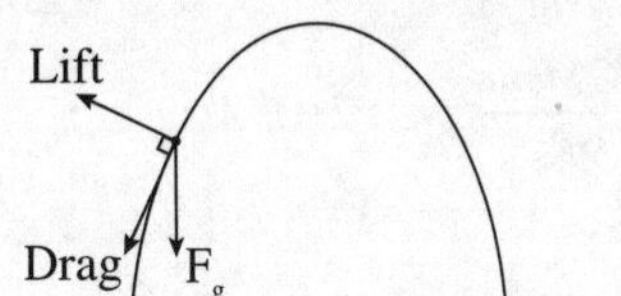

19. In projectile motion, an object experiences three forces: gravity, drag, and lift. These are depicted in the picture above. Given this information, how does lift affect the speed of a projectile?

    (A) It increases the speed.
    (B) It decreases the speed.
    (C) Its effect on speed varies.
    (D) It has no effect.

20. An object is launched from a cliff with a speed of 20 m/s at an angle of 30° above the horizontal. It lands on the ground below 5 s later. How high is the cliff? Assume air resistance is negligible.

    (A) 25 m
    (B) 50 m
    (C) 75 m
    (D) 125 m

GO ON TO THE NEXT PAGE.

21. A bicycle wheel with a diameter of 62.2 cm and a rotational inertia of 0.40 kg·m^2 is rotating at a rate of 8 rad/s when the brakes are applied. The brakes apply a normal force of 85 N on the brake pad at the edge of the wheel. The coefficient of kinetic friction between the brake pad and the wheel is 0.32. How long does it take for the bicycle to stop?

(A) 0.19 s
(B) 0.29 s
(C) 0.38 s
(D) 2.6 s

22. A rocket of mass m is launched with kinetic energy K_0 from the surface of the Earth. How much less kinetic energy does the rocket have at an altitude of two Earth radii in terms of the gravitational constant, G; the mass of the Earth, m_E; the radius of the Earth, R_E; and the mass of the rocket?

(A) $\frac{GM_E m}{2R_E}$

(B) $\frac{2GM_E m}{3R_E}$

(C) $\frac{3GM_E m}{4R_E}$

(D) $\frac{8GM_E m}{9R_E}$

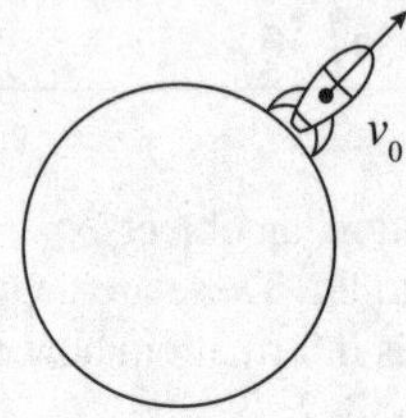

23. Escape velocity is defined as the minimum speed at which an object must be launched to "break free" from a massive body's gravitational pull. Which of the following principles could be used to derive this speed for a given planet?

(A) Conservation of Linear Momentum
(B) Newton's Third Law
(C) Conservation of Angular Momentum
(D) Conservation of Energy

24. What is the average power used to pitch a 0.142 kg baseball at 42 m/s, if it takes 0.3 s to accelerate it from rest?

(A) 251 W
(B) 417 W
(C) 501 W
(D) 835 W

25. The muzzle velocity of a 0.003-kg paintball pellet is 90 m/s. If it takes 0.002 s for the pellet to be shot from the paintball gun, what is the average force of the recoil on the paintball gun?

(A) 24 N
(B) 60 N
(C) 68 N
(D) 135 N

26. A 40-kg box slides down an incline that is angled 30° above the horizontal. The box starts from rest at a height of 8 m, and the coefficient of kinetic friction is 0.2. What is the total work done on the box as it slides to the bottom of the incline?

(A) 1109 J
(B) 2091 J
(C) 3200 J
(D) 4309 J

27. A runner on a relay team starts running as her teammate approaches with the baton. Once she takes the baton, she accelerates at 2.4 m/s^2 for 2 s to reach a speed of 10 m/s. How far does she travel from the time she takes the baton to the time she reaches her top speed?

(A) 4.8 m
(B) 15.2 m
(C) 20.0 m
(D) 24.8 m

GO ON TO THE NEXT PAGE.

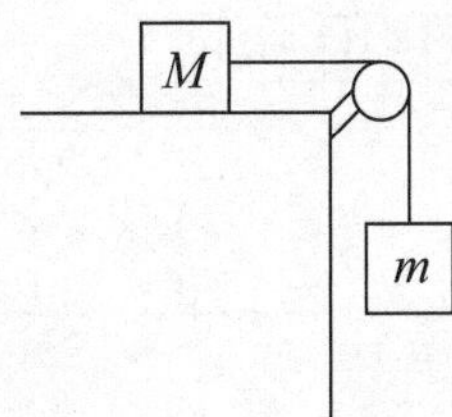

28. A block of mass M is at rest on a table. It is connected by a string and pulley system to a block of mass m hanging off the edge of the table. Assume the hanging mass is heavy enough to make the resting block move. Knowing the acceleration of the system and the mass of each block is sufficient to calculate all of the following quantities EXCEPT which one?
    (A) Net force on each block
    (B) Tension in the string
    (C) Coefficient of kinetic friction between the table and the block of mass M
    (D) The speed of the block of mass M when it reaches the edge of the table

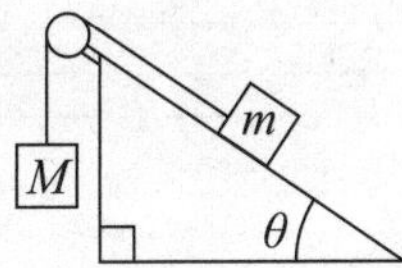

29. A block of mass m is connected by a string that runs over a frictionless pulley to a heavier block of mass M. The smaller block rests on an inclined plane of angle θ, and the larger block hangs over the edge, as shown above. In order to prevent the blocks from moving, the coefficient of static friction must be

    (A) $\dfrac{mg\sin\theta}{Mg - mg\cos\theta}$

    (B) $\dfrac{Mg - mg\sin\theta}{Mg\cos\theta}$

    (C) $\dfrac{Mg - mg\sin\theta}{mg\cos\theta}$

    (D) $\dfrac{Mg - mg\cos\theta}{mg\sin\theta}$

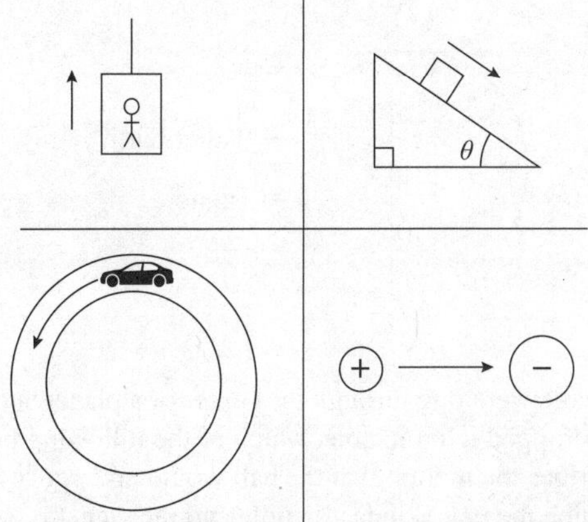

30. Which of the following forces does not do work in its given situation?
    (A) Normal force, as a person goes up in an elevator
    (B) Frictional force, as a box slides down a ramp
    (C) Centripetal force, as a car drives around a circular track
    (D) Gravitational force, as a positive charge moves toward a negative charge

0.2 m

$m = 1$ kg $m = 1$ kg

31. Two identical spheres of mass 1 kg are placed near each other. To keep the spheres apart, a light spring is placed between them. The spring has a natural length of 1.0 m. If equilibrium is achieved when the spheres are 0.2 m apart, what is the spring constant of the spring?
    (A) 4.17×10^{-10} N/m
    (B) 1.67×10^{-9} N/m
    (C) 2.09×10^{-9} N/m
    (D) 8.34×10^{-9} N/m

GO ON TO THE NEXT PAGE.

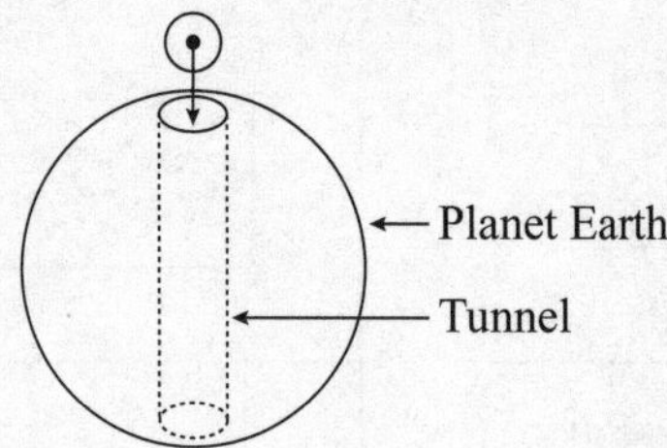

32. If a hole were dug through the center of a planet and a ball dropped into the hole, which of the following best describes the motion that the ball would undergo? Assume the ball is indestructible and the planet is a perfect sphere.
    (A) It would continuously gain speed and eventually escape the gravitational pull of the planet.
    (B) It would fall to the center of the planet and get stuck there because gravity is always pulling things toward the center of the planet.
    (C) It would fall to the other end of the hole, come to a momentary stop, fall back to the starting location, and then repeat this back-and-forth motion indefinitely.
    (D) None of the above is correct.

33. An empty mine car of mass m starts at rest at the top of a hill of height h above the ground, then rolls down the hill and into a horizontal semicircular banked turn of radius r. Ignoring rolling friction so that the only forces acting on the mine car are the normal force from the track and gravity, what is the magnitude of centripetal force on the car as it rounds the banked curve?
    (A) mgh
    (B) $2mgh/r$
    (C) mgh/r
    (D) $mgh/(2r)$

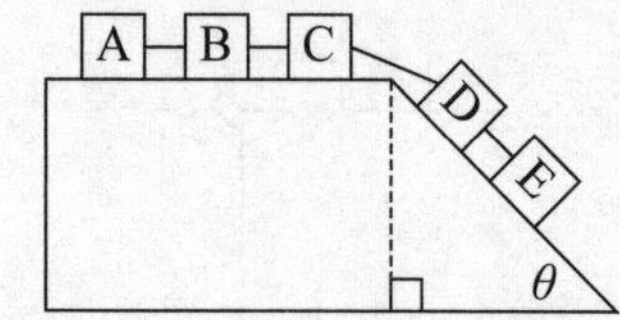

34. Five boxes are linked together, as shown above. If both the flat and slanted portions of the surface are frictionless, what will be the acceleration of the box marked B?
    (A) $\frac{1}{5}g\sin\theta$
    (B) $\frac{2}{5}g\sin\theta$
    (C) $\frac{5}{7}g\sin\theta$
    (D) $\frac{2}{5}g\cos\theta$

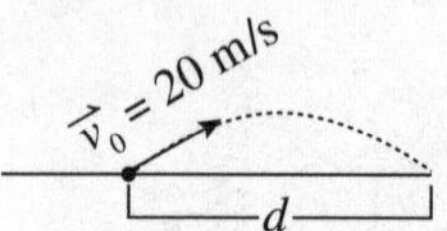

35. A soccer ball on a level field is kicked at an angle of 30 degrees above the horizontal with an initial speed of 20 m/s, as shown above. What is d, the distance from the spot of the kick to the spot where the ball lands? Ignore air resistance.
    (A) $20\sqrt{3}$ m
    (B) 40 m
    (C) $40\sqrt{3}$ m
    (D) 80 m

GO ON TO THE NEXT PAGE.

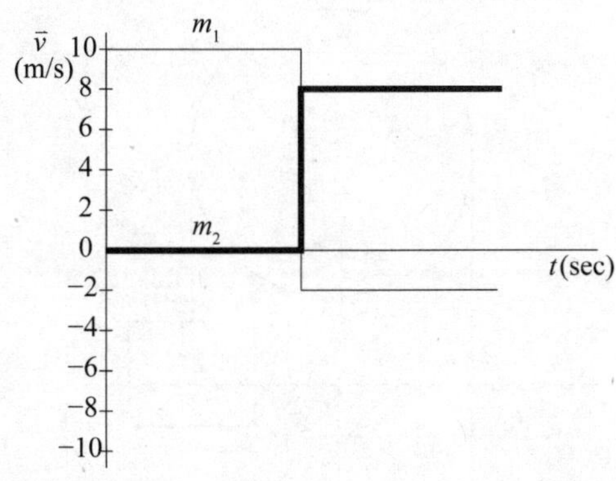

Questions 36–37 refer to the graph above.

36. Two objects of masses m_1 and m_2 undergo a collision. The graph above shows their velocities with respect to time both before and after the collision. If m_1 = 10 kg, then m_2 must be

    (A) 5 kg
    (B) 10 kg
    (C) 15 kg
    (D) 20 kg

37. If the two objects have masses of m_1 = 4 kg and m_2 = 6 kg, what type of collision does the graph represent?

    (A) Perfectly elastic
    (B) Perfectly inelastic
    (C) Neither of the above
    (D) Cannot be determined

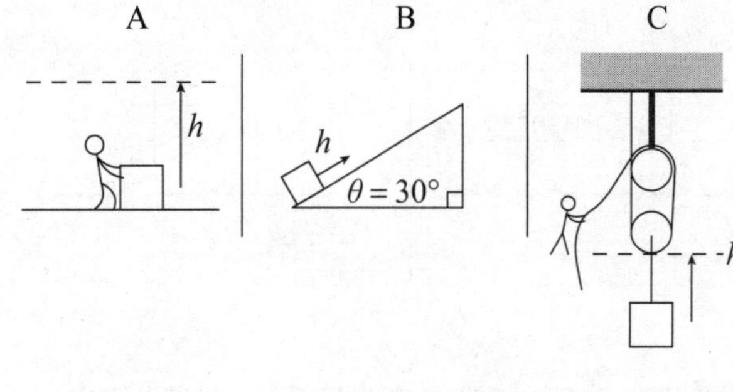

38. The diagrams above show a box of mass m being lifted from the ground up to a height of h via three different methods. In situation A, the box is simply lifted by a person. In B, it is pushed up a ramp with an incline angle of 30 degrees. In C, it is lifted by a pulley system. Assuming ideal conditions (no friction) for all of these situations, which of the following correctly ranks the amount of work required to lift the box in each case?

    (A) A > B > C
    (B) A > B = C
    (C) C > B > A
    (D) A = B = C

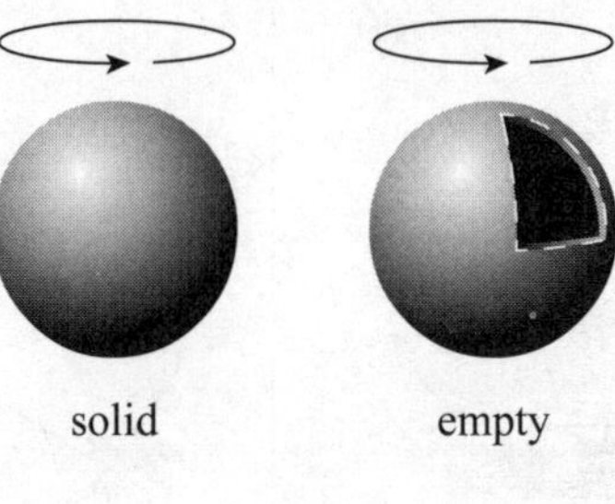

$m_1 = m_2$

$\tau_1 = \tau_2$

39. Two spheres of equal size and equal mass are rotated with an equal amount of torque. One of the spheres is solid with its mass evenly distributed throughout its volume, and the other is hollow with all of its mass concentrated at the edges. If both spheres start at rest, which one is rotating faster after 1 second?

    (A) Solid sphere
    (B) Hollow sphere
    (C) They are rotating with the same angular velocity.
    (D) Additional information is required to determine the relative rates of rotation.

GO ON TO THE NEXT PAGE.

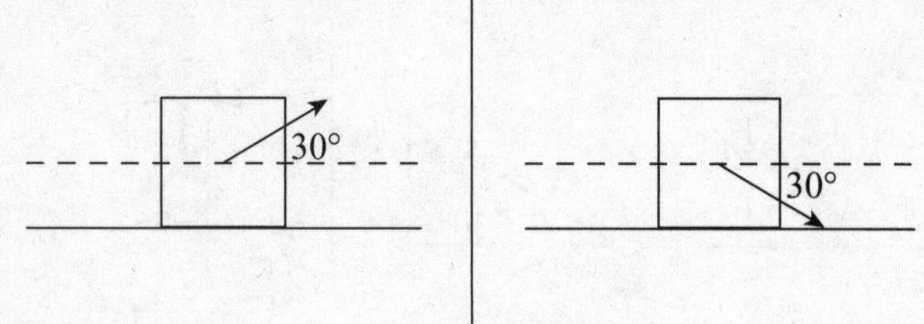

40. Pulling a block of mass m to the right by a connected string at an angle of 30° above the horizontal (as shown in the left picture) with a force equal to the block's weight produces a friction force F_{kf}. If the same block were to be pulled at an angle of 30° beneath the horizontal (as shown in the right picture), what would be the friction force? Assume that the applied force is enough to make the block move in both cases.

(A) $3F_{kf}$
(B) $2F_{kf}$
(C) $F_{kf}/2$
(D) $F_{kf}/3$

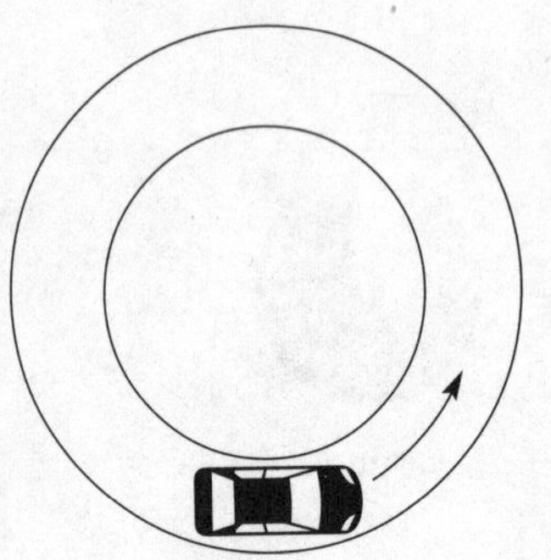

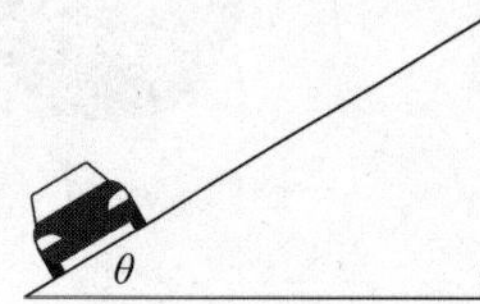

top view side view

41. A car is traveling in a horizontal circle of radius r, on a frictionless banked turn, at a constant speed v. How does the magnitude of the normal force on the car compare with the car's weight?

(A) $F_N < F_g$
(B) $F_N = F_g$
(C) $F_N > F_g$
(D) The answer depends on the speed of the car, the radius of the turn, and the angle of the embankment.

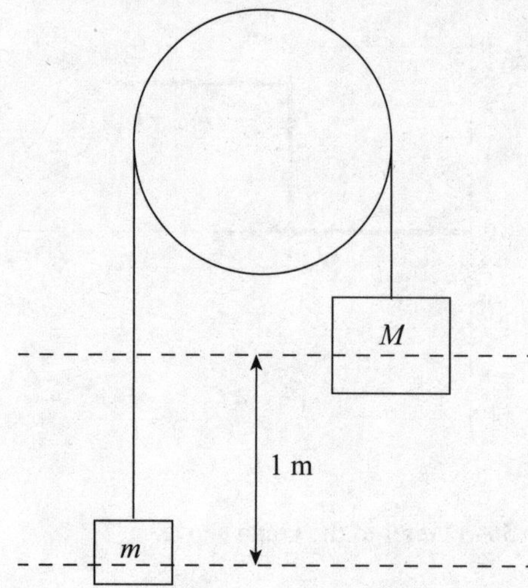

42. The pulley above is both frictionless and massless, and the blocks are initially at rest. How long will it take for the two blocks to reach an equal height if block M of mass 15 kg starts 1 m above block m of mass 5 kg?

(A) $\sqrt{\frac{2}{5}}$ s
(B) $\sqrt{\frac{1}{5}}$ s
(C) $\sqrt{\frac{1}{20}}$ s
(D) $\frac{1}{5}$ s

43. A person pushes a 60 kg grocery cart, initially at rest, across a parking lot. He exerts a pushing force directed 20° below the horizontal. If the person pushes the cart with a force of 300 N for 5 m across horizontal ground and then releases the cart, the cart has a speed of 3 m/s. What is the work done by friction during this motion?

(A) −1230 J
(B) −1140 J
(C) 1140 J
(D) 1230 J

GO ON TO THE NEXT PAGE.

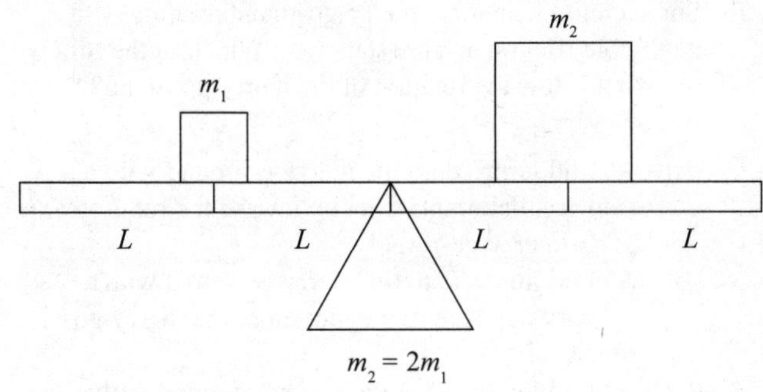

44. The system above is currently not balanced. Which of the following proposed changes would keep the system out of balance? Assume the plank has no mass of its own.
    (A) Adding a mass equal to m_2 on the far left side and a mass equal to m_1 and on the far right side
    (B) Stacking both masses directly on top of the fulcrum
    (C) Moving the fulcrum a distance $L/3$ to the right
    (D) Moving both masses a distance $L/3$ to the left

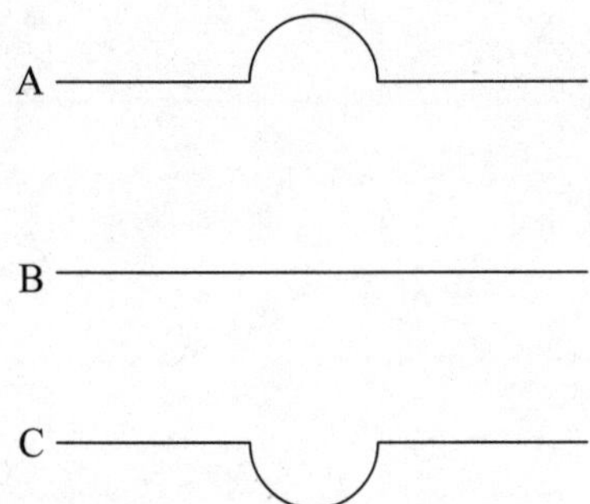

45. Three identical balls are rolled from left to right across the three tracks above with the same initial speed. Assuming the tracks all have negligible friction and the balls have enough initial speed to reach the ends of each track, which set correctly orders the average speed of the balls on the three tracks?
    (A) A = B = C
    (B) B > C > A
    (C) C > B > A
    (D) Cannot be determined

Directions: For each of the questions 46–50, <u>two</u> of the suggested answers will be correct. Select the two answers that are best in each case, and then fill in both of the corresponding circles on the answer sheet.

46. Which two velocity-versus-time graphs depict situations of uniformly accelerated motion? Select two answers.

(A) 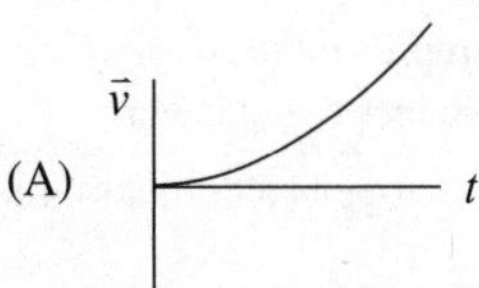

(B) 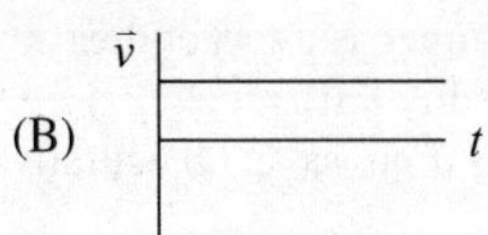

(C) 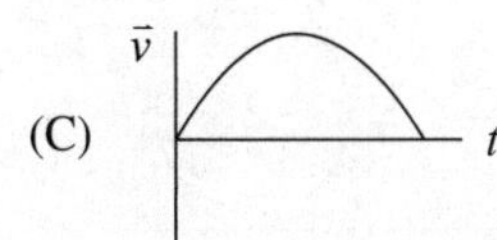

(D) 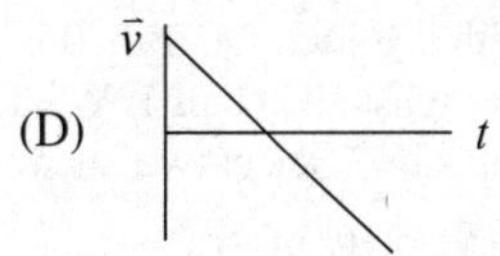

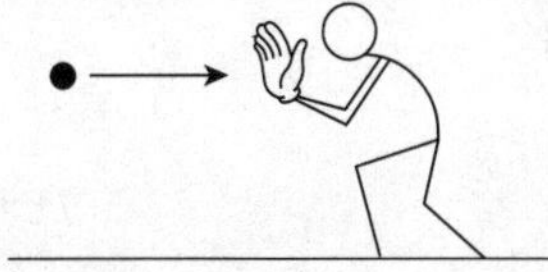

47. A man is standing on a frictionless surface. A ball is thrown horizontally to him, and he catches it with his outstretched hand, as shown above. Which two of the following values will remain the same after the catch as they were before the catch? Select two answers.
    (A) Angular momentum of the man-and-ball system
    (B) Angular momentum of the ball about the man's center of mass
    (C) Mechanical energy of the man-and-ball system
    (D) Momentum of the man-and-ball system

GO ON TO THE NEXT PAGE.

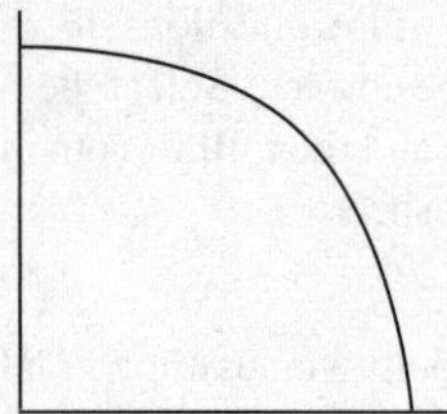

48. The graph above could be a representation of which two of the following situations? Select two answers.
    (A) Vertical position versus time, as an object falls with no air resistance
    (B) Kinetic energy versus time, as an object falls with no air resistance
    (C) Kinetic energy versus time, as a rocket flies at constant speed but steadily burns fuel
    (D) Potential energy versus time, as an object falls with no air resistance

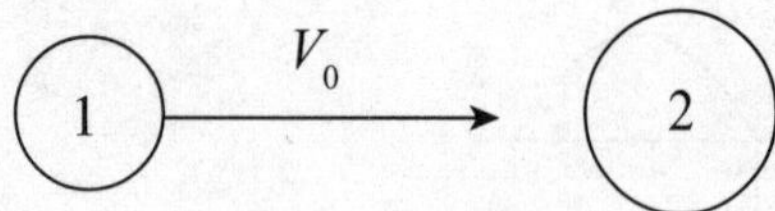

49. Ball 1 rolls toward ball 2 with a velocity v_0. Ball 2 is initially at rest and has a greater mass than ball 1. What are possible outcomes of the collision? Select two answers.
    (A) Ball 1 rebounds with a velocity of $-v_0$.
    (B) Ball 1 and ball 2 have the same final velocity.
    (C) Ball 2 has a final velocity of v_0.
    (D) The final velocities of ball 1 and ball 2 are not parallel.

50. Some children are on a merry-go-round rotating with negligible friction at a constant rate. Which of the following would slow the rotation of the merry-go round? Select two answers.
    (A) A child jumps onto the merry-go-round with a velocity that points directly toward the rotation axis of the merry-go-round
    (B) A child jumps onto the merry-go-round with a velocity opposite to the motion of the merry-go-round
    (C) A child jumps off of the merry-go round with a velocity that points directly away from the rotation axis of the merry-go-round
    (D) A child jumps onto off of the merry-go-round with a velocity opposite to their instantaneous velocity before jumping

END OF SECTION I

DO NOT CONTINUE UNTIL INSTRUCTED TO DO SO.

AP PHYSICS 1
SECTION II
Free-Response Questions
Time—90 minutes
Percent of total grade—50

General Instructions

Use a separate piece of paper to answer these questions. Show your work. Be sure to write CLEARLY and LEGIBLY. If you make an error, you may save time by crossing it out rather than trying to erase it.

GO ON TO THE NEXT PAGE.

AP PHYSICS 1

SECTION II

Directions: Questions 1–5 here are as follows: one experimental design question (worth 12 points), one quantitative/qualitative translation question (worth 12 points), one paragraph argument short-answer question (worth 7 points), and two additional short-answer questions (worth 7 points each). You have a total of 90 minutes to complete this section. Show your work for each part in the space provided after that part.

1. A car of known mass m_1 will collide with a second car of known mass m_2. The collision will be head on, and both cars will only move linearly both before and after the collision. In a clear, coherent, paragraph-length response, explain a method for determining whether the collision is perfectly elastic, perfectly inelastic, or neither. If the collision is perfectly inelastic, include at least one possible cause of energy loss.

GO ON TO THE NEXT PAGE.

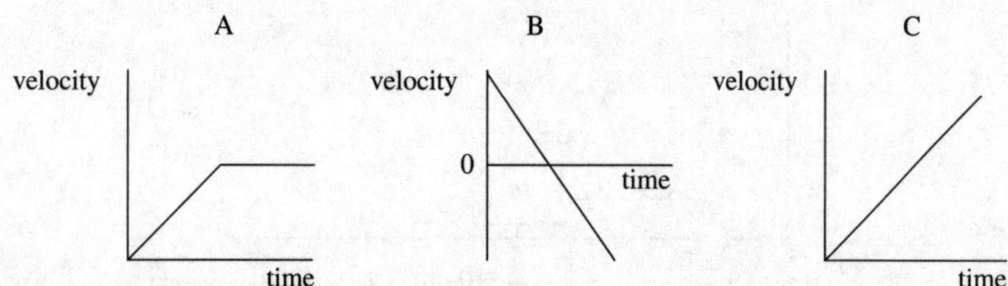

2. Each of the three graphs above depicts the velocity of an object as a function of time. Each graph is of a different object, as labeled above. For each velocity graph, draw a displacement versus time graph and an acceleration versus time graph. Additionally, give an example of a situation that could produce each of the three graphs above.

GO ON TO THE NEXT PAGE.

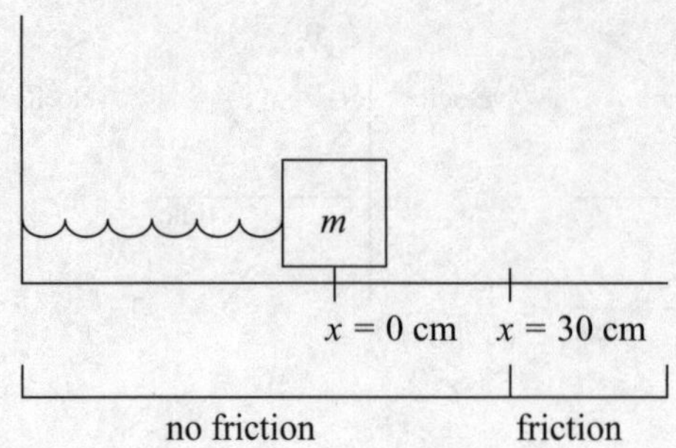

3. In the diagram above, a spring-block system is oscillating on a flat horizontal surface. Part of the surface is frictionless, and part of the surface is frictional. The block starts at rest at position $x = 0$ cm. The block is then pushed to the left, compressing the spring, until it reaches the position $x = -50$ cm. After being pushed in, the block is released and allowed to move naturally.

   (A) Draw a graph of the system's position as a function of time starting from the moment it is released, and explain your reasoning behind the graph you draw. Be sure to label important values on the graph.

   (B) Given that the mass $m = 2$ kg and the spring constant $k = 100$ N/m, what is the magnitude of work done by the frictional surface?

GO ON TO THE NEXT PAGE.

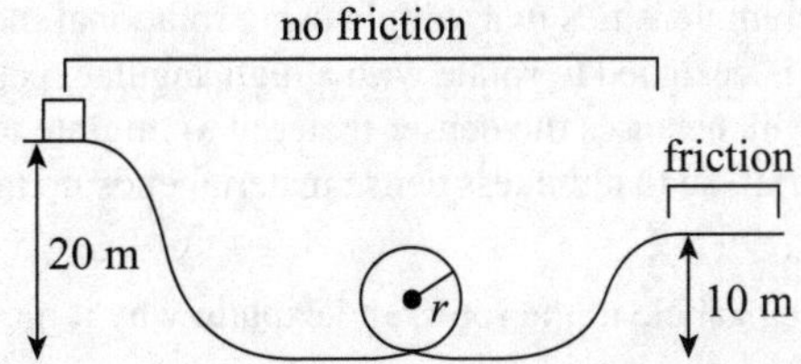

4. A roller coaster cart rides along the track shown above. The initial drop, the loop, and the ascending ramp are all frictionless. The flat portion of the track after the ascending ramp is frictional.

(A) How fast would the cart be moving just before it enters the loop?

(B) Would the normal force on the cart be greater just after entering the loop or at the peak of the loop? Explain why using relevant equations.

(C) What is the greatest possible radius for the loop that would allow the cart to still make it through?

(D) If the coefficient of static friction for the final segment of the track is 0.2, how long does the segment need to be to allow the cart to come to a complete stop due to friction alone?

GO ON TO THE NEXT PAGE.

5. Centrifuges separate materials of different densities in a sample using rotational motion. The samples are placed in centrifuge tubes that are placed in a rotor, which is designed to rotate with a high angular speed. As the rotor spins, the samples experience a large centripetal acceleration, which causes the denser material to migrate toward the outer part of the rotor. The denser material displaces the less dense material, so that the less dense material ends up toward the center of the rotor.

(A) Describe the motion of a dense particle in the rotor, and explain why it moves toward the outer part of the rotor.

(B) Although many centrifuges are designed to rotate at set angular speeds, it is the centripetal acceleration that determines the efficacy of the separation. If a rotor of radius r_1 rotating at an angular speed ω_1 results in a centripetal acceleration a_c for material at the outer portion of the rotor, what angular speed should be used to achieve the same acceleration with a rotor of radius r_2?

(C) The acceleration of a centrifuge is often reported as a multiple of the acceleration due to gravity. What angular speed would produce an acceleration of $1000g$ in a 20-cm radius rotor?

STOP

END OF EXAM

Practice Test 1: Diagnostic Answer Key and Explanations

PRACTICE TEST 1: DIAGNOSTIC ANSWER KEY

Let's take a look at how you did on Practice Test 1. Follow the three-step process in the diagnostic answer key below and go read the explanations for any questions you got wrong, or you struggled with but got correct. Once you finish working through the answer key and the explanations, go to the next chapter to make your study plan.

Check your answers and mark any correct answers with a ✔ in the appropriate column.

Section I: Multiple Choice							
Q #	**Ans.**	**✔**	**Chapter #, Title**	**Q #**	**Ans.**	**✔**	**Chapter #, Title**
1	D		**10,** Torque	21	C		**10,** Rotational Inertia
2	B		**4,** Acceleration	22	B		**7,** Conservation of Mechanical Energy
3	B		**6,** Uniform Circular Motion	23	D		**7,** Conservation of Mechanical Energy
4	D		**5,** Friction	24	B		**7,** Power
5	C		**9,** Simple Harmonic Motion	25	D		**8,** Impulse
6	B		**6,** Uniform Circular Motion	26	B		**7,** Conservation of Energy with Nonconservative Forces
7	D		**8,** Conservation of Linear Momentum	27	B		**4,** Uniformly Accelerated Motion and the Big Five
8	D		**5,** Friction **6,** Uniform Circular Motion	28	D		**5,** Pulleys
9	C		**4,** Uniformly Accelerated Motion and the Big Five	29	C		**5,** Friction **5,** Pulleys
10	B		**10,** Rotational Motion	30	C		**7,** Work at an Angle
11	B		**10,** Rotational Motion	31	C		**6,** Newton's Law of Gravitation **9,** Simple Harmonic Motion
12	C		**5,** Newton's 1st Law	32	C		**6,** Newton's Law of Gravitation
13	B		**5,** Friction	33	B		**6,** Uniform Circular Motion **7,** Conservation of Mechanical Energy
14	D		**5,** Friction **9,** Simple Harmonic Motion	34	B		**5,** Pulleys
15	B		**7,** The Work–Energy Theorem	35	A		**4,** Projectile Motion
16	B		**8,** Impulse	36	C		**8,** Conservation of Linear Momentum
17	A		**7,** Conservation of Mechanical Energy	37	A		**8,** Collisions
18	B		**4,** Uniformly Accelerated Motion and the Big Five	38	D		**7,** Conservation of Energy with Nonconservative Forces
19	D		**4,** Projectile Motion	39	A		**10,** Rotational Inertia
20	C		**4,** Projectile Motion	40	A		**5,** Friction

Section I: Multiple Choice (Continued)							
Q #	**Ans.**	**✔**	**Chapter #, Title**	**Q #**	**Ans.**	**✔**	**Chapter #, Title**
41	C		**6,** Uniform Circular Motion	46	B, D		**4,** Acceleration
42	B		**5,** Pulleys	47	A, D		**8,** Conservation of Linear Momentum **10,** Angular Momentum
43	B		**7,** The Work–Energy Theorem	48	A, D		**4,** Free Fall **7,** Conservation of Mechanical Energy
44	A		**10,** Equilibrium	49	B, D		**8,** Conservation of Linear Momentum
45	C		**7,** Conservation of Mechanical Energy	50	A, B		**10,** Angular Momentum

Section II: Free Response			
Q #	**Ans.**	**✔**	**Chapter #, Chapter Title, Section Title**
1	See Explanation		**8,** Momentum, Collisions
2	See Explanation		**4,** Kinematics, Acceleration
3(A)	See Explanation		**9,** Simple Harmonic Motion
3(B)	See Explanation		**7,** Energy, Conservation of Energy with Nonconservative Forces
4(A)	See Explanation		**7,** Energy, Conservation of Mechanical Energy
4(B)	See Explanation		**6,** Circular Motion and Gravitations, Uniform Circular Motion
4(C)	See Explanation		**7,** Energy, Conservation of Mechanical Energy
4(D)	See Explanation		**5,** Dynamics Friction **7,** Conservation of Energy with Nonconservative Forces
5(A)	See Explanation		**6,** Circular Motion and Gravitations, Uniform Circular Motion
5(B)	See Explanation		**6,** Circular Motion and Gravitations, Uniform Circular Motion **10,** Torque and Rotational Motion What is Angular Velocity?
5(C)	See Explanation		**6,** Circular Motion and Gravitations, Uniform Circular Motion **10,** Torque and Rotational Motion, What is Angular Velocity?

Tally your correct answers from Step 1 by chapter. For each chapter, write the number of correct answers in the appropriate box. Then, divide your correct answers by the number of total questions (which we've provided) to get your percent correct.

CHAPTER 4 TEST SELF-EVALUATION

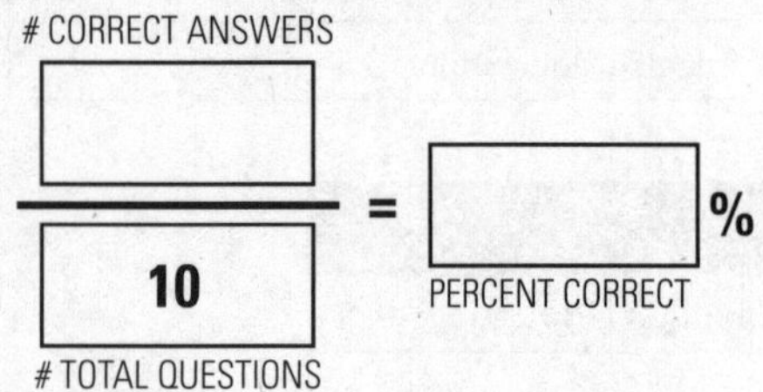

CHAPTER 5 TEST SELF-EVALUATION

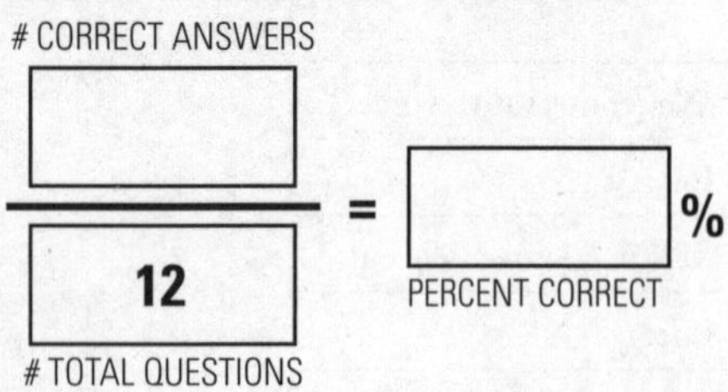

CHAPTER 6 TEST SELF-EVALUATION

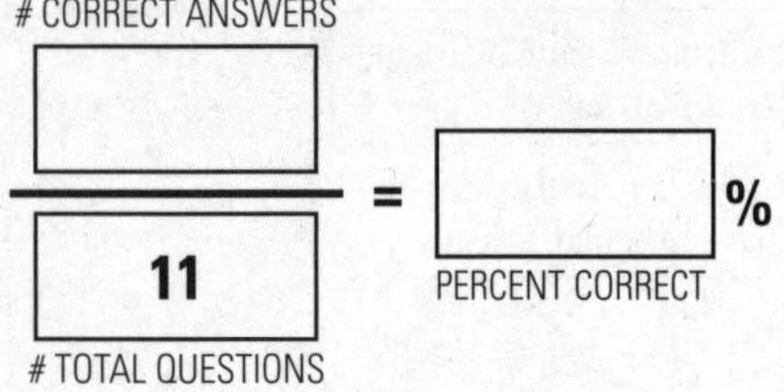

CHAPTER 7 TEST SELF-EVALUATION

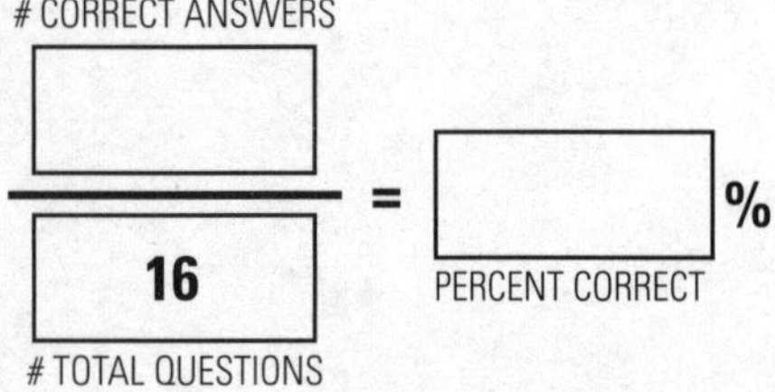

CHAPTER 8 TEST SELF-EVALUATION

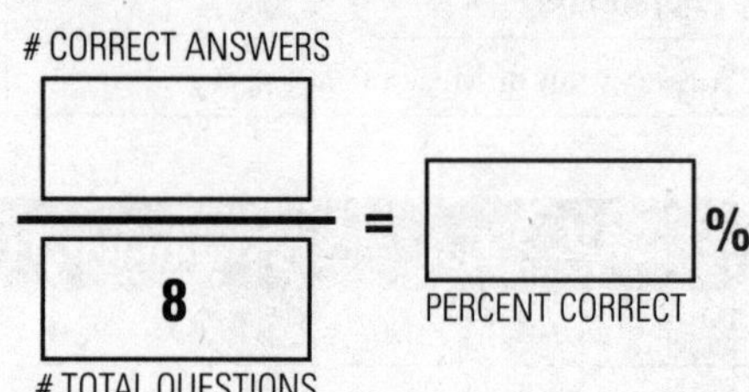

CHAPTER 9 TEST SELF-EVALUATION

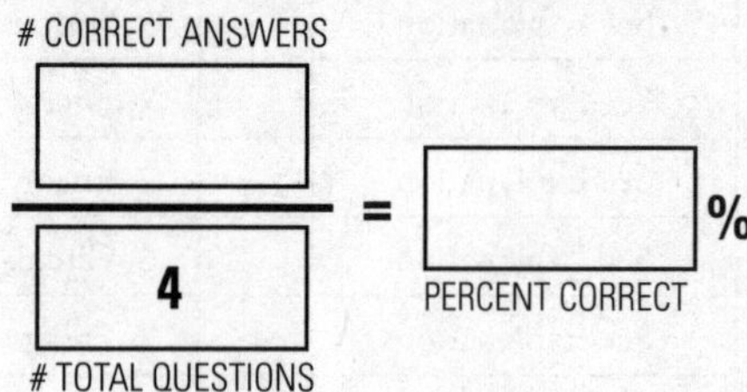

CHAPTER 10 TEST SELF-EVALUATION

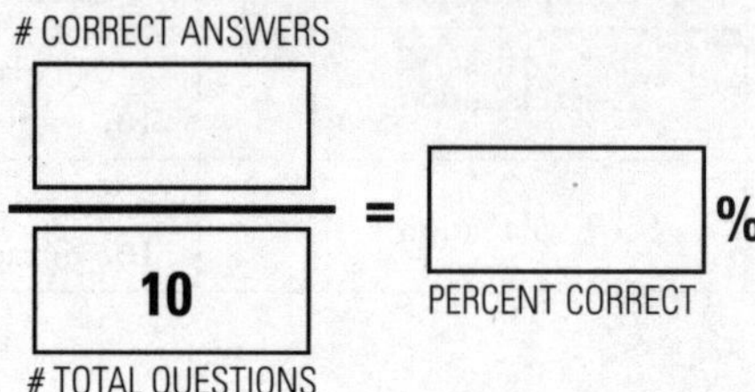

Use the results above to customize your study plan. You may want to start with, or give more attention to, the chapters with the lowest percents correct.

PRACTICE TEST 1: ANSWERS AND EXPLANATIONS

Section I: Multiple-Choice Questions

1. **D** As a ball rolls down an inclined plane, the three forces act on the ball, as shown in the diagram below.

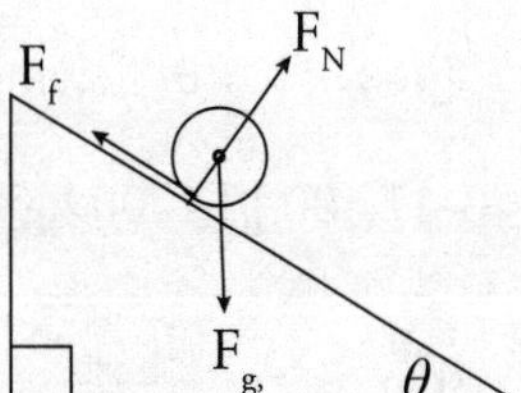

In order for a force to exert a torque, the force must be located off-center, and must not point directly at or away from the object's center of mass. An object's weight always acts at its center of mass, so (B) is wrong. The normal force does act at the edge of the ball, but it points directly out of the ramp, which is straight at the center of the ball. Choice (A) is wrong. Friction acts where the ball meets the inclined plane, *and* it points parallel to the surface, which is tangent to the ball's surface. So friction is what causes rotation. What kind of friction do we have? The phrase "without slipping" tells you that even though the ball as a whole is moving, the spot on the ball that touches the ramp is at rest with respect to the ramp. That means it is *static* friction, (D).

2. **B** In a velocity-versus-time (also referred to as a *v*-versus-*t*) graph, the displacement of an object is the area between the curve and the horizontal axis; the area below the *t*-axis counts as negative. In this case, the area can be broken into multiple rectangles and triangles.

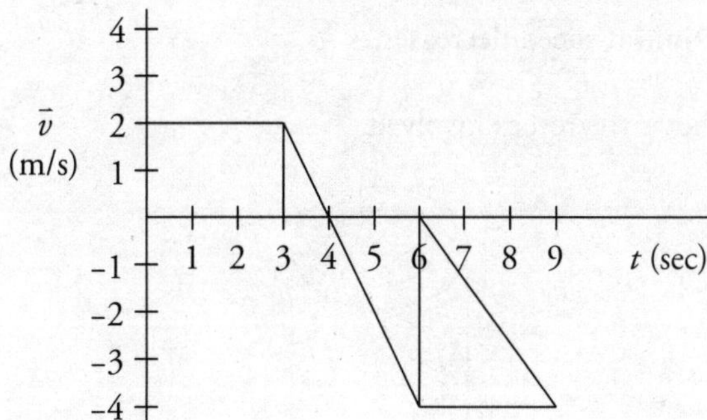

In order of the number labels, each shape has an area of 6, 1, 4, and 12, respectively. The last two, however, are below the *t*-axis, so their area represents negative displacement. Therefore, the net displacement is 6 m + 1 m – 4 m – 12 m = –9 m. As a check, it looks like most of the area is below the *t*-axis, so the answer should be negative.

3. **B** The formula for centripetal force is $F_c = mv^2/r$, which initially seems to indicate the force is inversely proportional to the radius. However, in the case of circular motion, an object's linear speed is $v = \omega r$. Substituting this value into the equation gives $F_c = m(\omega r)^2/r = m\omega^2 r$. So it turns out that the force is directly proportional to r, which means doubling the radius will double the force on the object.

4. **D** First, you know $F_f = \mu F_N = \mu(mg)$. Second, you know that $W = \Delta KE = KE_f - KE_0 = -KE_0$ (since $KE_f = 0$ in this case) $= -\frac{1}{2}mv_0^2$. Plugging in numbers, you get $W = -\frac{1}{2}(1000 \text{ kg})(5 \text{ m/s})^2 = -12{,}500 \text{ J}$. You also know that $W = Fd\cos\theta = [\mu(mg)]d\cos\theta$. Solving for μ, you get $\mu = W/[(mg)d\cos\theta]$. Plugging in the numbers then gives $\mu = -12{,}500 \text{ J}/[(1000 \text{ kg})(10 \text{ m/s}^2)(20 \text{ m})(\cos 180°)] = 0.0625$.

5. **C** The frequency of a spring-block system is $f = \frac{1}{2\pi}\sqrt{\frac{k}{m}}$, so it would be affected by the change in mass. Furthermore, when a spring is at maximum compression or extension, all of its energy is potential energy, which is given by $U = \frac{1}{2}kx^2$. In adding this block, none of the relevant values are changed, so the spring will still extend to the same length, which means the amplitude is unchanged. Finally, maximum speed will be limited by the maximum K of the system (which will be unchanged since the maximum U was unchanged). $K = \frac{1}{2}mv^2$, so the increased mass would have to be balanced by a decrease in speed to leave the K unaltered.

If you want a physical, rather than mathematical, explanation, think about the acceleration of the pair of blocks. The spring exerts the same force as before, but we effectively raised the block's mass. Thus, the acceleration is less than before. Since the distance from the starting point (maximum compression) to equilibrium is unchanged, the blocks will pick up less speed than the single block did before. Thus, maximum speed decreases.

6. **B** The diagram below shows the forces involved.

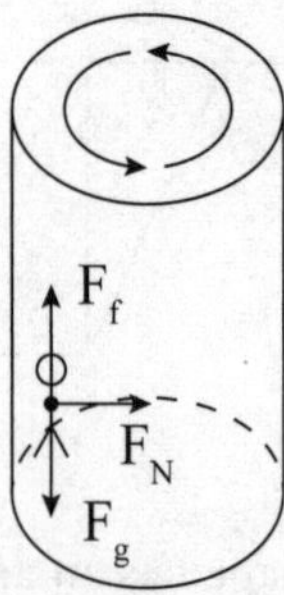

In this case, the person will remain suspended in the air as long as $F_f = F_g$. Furthermore, because this is an example of uniform circular motion, you know $F_N = F_C = mv^2/r$. So you can rewrite the first equation as $\mu(mv^2/r) = mg$. Thus, the coefficient of friction is an important factor, but the mass of the person is not since it exists on both sides of the equation and will cancel out.

7. **D** In any collision, the total momentum of the system must be conserved. Prior to the collision, you see that all of the momentum is horizontal. Therefore, the net momentum after the collision must also be purely horizontal. If one ball is moving up after the collision, the other must be moving down to cancel out that vertical momentum.

8. **D** Whenever the pendulum is in motion, it will be experiencing circular motion, which means there must be a centripetal force. Centripetal force is a net force toward the center, so that means tension must be the greater force whenever the pendulum is moving. However, the pendulum is not always moving. At the two extreme edges of its motion, the pendulum is motionless for an instant as it changes directions. At those times, the net centripetal force is zero, meaning the two given forces are equal.

9. **C** Apply the kinematic equation $v^2 = v_0^2 + 2a(x - x_0)$. For a stopping distance of *d*, $x - x_0 = d$ and $v = 0$, so $0 = v_0^2 + 2ad \rightarrow d = \frac{-v_0^2}{2a}$. For an initial velocity of $2v_0$, the stopping distance d' is

$$d' = \frac{-(2v_0)^2}{2a} = \frac{-4v_0^2}{2a} = 4\left(\frac{-v_0^2}{2a}\right) = 4d$$

10. **B** Use rotational kinematics to determine the angular displacement of the tire.

$$\theta = \theta_0 + \omega_0 t + \frac{1}{2}\alpha t^2 \rightarrow \Delta\theta = \frac{1}{2}\alpha t^2$$

Convert the linear velocity to angular velocity using $\omega = \frac{v}{r}$ to determine the angular acceleration. Remember to use the radius of the tire in the equation, not the diameter.

$$\alpha = \frac{\omega - \omega_0}{t} = \frac{\frac{v}{r} - 0}{t} = \frac{v}{rt}$$

$$\Delta\theta = \frac{1}{2}\left(\frac{v}{rt}\right)t^2 = \frac{vt}{2r} = \frac{(25\text{ m/s})(5\text{ s})}{2(0.35\text{ m})} = 178.6\text{ rad}$$

Finally, convert to revolutions: $178.6\text{ rad} \times \frac{1\text{ rev}}{2\pi\text{ rad}} = 28.4\text{ rev}$

11. **B** In the absence of external torque, the diver's angular momentum is conserved,

$$L_0 = L \rightarrow I_0\omega_0 = I\omega \rightarrow \omega = \frac{I_0\omega_0}{I} = \frac{(4.5\text{ kg}\cdot\text{m}^2)(18\text{ rad/s})}{15\text{ kg}\cdot\text{m}^2} = 5.40\text{ rad/s}$$

12. **C** Just focus on one end of the rope. Nothing is moving, so the net force must be 0. If a person pulls with a force of 1000 N and nothing moves, then the resulting tension must also be 1000 N. Additionally, tension is a constant magnitude throughout a string. The direction of the tension can change (as in the case of a pulley system), but the magnitude will remain the same.

13. **B** The force of static friction will be $F_f = \mu F_N = \mu(mg)$. Changing the arrangement of the blocks does not change any of these three quantities, so the force will remain the same. Thus, 10 N will again be required to move them.

14. **D** It's easiest to consider both blocks as one system: calculate the maximum acceleration of the pair of blocks caused by the spring (which occurs at maximum displacement from equilibrium), and determine whether the friction between the two blocks is strong enough to give the top block the necessary acceleration to keep up with the bottom block. Applying Newton's Second Law to the system of blocks, F_{spring} = (combined mass) * $a \rightarrow kA = (M + m)a$. In order for the top block to accelerate at this rate, static friction must be strong enough that $F_{sf} = ma \rightarrow \mu sFN = ma$. You don't need to do any math to see that μs (coefficient of static friction), A (amplitude), and k (spring constant) appear, but the period doesn't.

15. **B** The work done by a nonconservative force is equal to the change in total mechanical energy. $W_{nc} = \Delta E = \Delta K + \Delta U$. The child starts at rest, so $K_i = 0$, and set $h = 0$ at the ground so that $U_f = 0$. $W_{nc} = K_f - K_i + U_f - U_i = K_f - U_i$. Apply the definition of work to solve for the force,

$$W_{nc} = F_{nc}d = \frac{1}{2}mv_f^2 - mgh_i$$

$$F_{nc} = \frac{1}{d}\left(\frac{1}{2}mv_f^2 - mgh_i\right) = \frac{1}{8\text{ m}}\left(\frac{1}{2}(35\text{ kg})(2.5\text{ m/s})^2 - (35\text{ kg})(10\text{ m/s}^2)(12\text{ m})\right) = -511\text{ N}$$

16. **B** The average force acting on the child is her change in momentum divided by the stopping time.

$$F_{avg} = \frac{\Delta p}{\Delta t} = \frac{mv_f - mv_i}{\Delta t} = \frac{(35\text{kg})(2.5\text{ m/s}) - 0}{1.5\text{ s}} = 58\text{ N}$$

17. **A** Since the breaking mechanism is horizontal, the child's maximum speed is achieved right before the breaking mechanism, when all her potential energy had converted to kinetic energy according to the Conservation of Mechanical Energy. $K_i + U_i = K_f + U_f \Rightarrow U_i = K_f \Rightarrow \frac{1}{2}mv_f^2 = mgh_i \Rightarrow$ $v_f = \sqrt{2gh_i}$. Since $v_f \propto \sqrt{h_i}$, an increase in the height by $\frac{15}{12} = \frac{5}{4}$ corresponds to an increase in the maximum speed by $\sqrt{\frac{5}{4}} = \frac{\sqrt{5}}{2}$.

18. **B** The average velocity is given by the equation:

$$\bar{v} = \frac{\Delta x}{\Delta t} = \frac{200\text{ m} - 40\text{ m}}{5\text{ s}} = 32\text{ m/s}$$

19. **D** Because the lift force is perpendicular to the velocity of the object, lift can have no effect on the magnitude of velocity. The magnitude of velocity is also known as speed, so (D) is correct.

20. **C** The flight time of a projectile depends on the motion in the *y*-direction. Use kinematics to determine the displacement of the object, using the *y*-component of the initial velocity, $v_{0y} = v_0 \sin\theta$.

$$y = y_0 + v_{0y}t - \frac{1}{2}gt^2$$

$$\Delta y = v_0 \sin\theta t - \frac{1}{2}gt^2 = (20\text{ m/s})\sin 30°(5\text{ s}) - \frac{1}{2}(10\text{ m/s}^2)(5\text{ s})^2 = -75\text{ m}$$

The object ends up 75 m below its launch point, so the cliff is 75 m high.

21. **C** The brakes exert a friction force that causes torque against the motion of the wheel. The friction force is directed tangent to wheel, which is perpendicular to the lever arm, so $\tau = rF_f$. Angular acceleration can be calculated from the torque $\tau = I\alpha = rF_f \Rightarrow \alpha = \frac{r\mu F_N}{I} = \frac{(0.311\text{ m})(0.32)(85\text{ N})}{0.40\text{ kg}\cdot\text{m}^2} =$ 21 rad/s^2. Use the Big 5 #2 for rotational kinematics to solve for the time.

$$\omega = \omega_0 - \alpha t \Rightarrow 0 = 8\text{ rad/s} - (21\text{ rad/s}^2)t \Rightarrow t = 0.38\text{ s}$$

22. **B** Conservation of Energy applies to the rocket, $K_0 + U_0 = K + U$. The gravitational potential energy far from the Earth's surface is $U_G = -\frac{Gm_1m_2}{r}$. At an altitude of $2R_E$, the rocket is a distance of $3R_E$ from the center of the Earth. Thus, $K = K_0 + U_0 - U = K_0 - \frac{GM_Em}{R_E} + \frac{GM_Em}{3R_E} = K_0 - \frac{2GM_Em}{3R_E}$.

Therefore, the rocket's kinetic energy at an altitude of $2R_E$ is $\frac{2GM_Em}{3R_E}$ less than it was on the Earth's surface.

23. **D** If an object leaves a planet at the exact escape velocity, its speed will approach zero as its distance from the planet approaches infinity. Zero speed implies zero *KE*, and infinite separation implies zero *Ug* if the reference separation is infinity. Thus, the total mechanical energy of an object traveling at escape velocity is 0 J. That's why Conservation of Energy is the relevant concept.

24. **B** The average power used by the pitcher is the work he does on the baseball per unit time. By the Work–Energy Theorem, the work done on the baseball is its change in kinetic energy.

$$P = \frac{W}{\Delta t} = \frac{\Delta K}{\Delta t} = \frac{\frac{1}{2}mv^2 - 0}{\Delta t} = \frac{\frac{1}{2}(0.142\text{ kg})(42\text{ m/s})^2}{0.3\text{ s}} = 417\text{ W}$$

25. **D** According to Newton's Third Law, the force the pellet exerts on the gun is equal and opposite to the force the gun exerts on the pellet. The average force exerted on the pellet times the time it is exerted is the impulse the pellet experiences. Impulse is also equal to the change in momentum:

$$F\Delta t = \Delta p \rightarrow F = \frac{m\Delta v}{\Delta t} = \frac{(0.003 \text{ kg})(90 \text{ m/s} - 0 \text{ m/s})}{0.002 \text{ s}} = 135 \text{ N}$$

26. **B** The total work done on the box is the sum of the work done by each of the forces acting on the box. The work done by gravity is related to the change in gravitational potential energy of the box.

$$W_g = -\Delta U_g = -mg\Delta h = -(40 \text{ kg})(10 \text{ m/s}^2)(-8 \text{ m}) = 3200 \text{ J}$$

The normal force is perpendicular to the displacement of the box, so it does zero work on the box.

Friction acts in the opposite direction of the displacement of the box. The length of its displacement is $d = \dfrac{h}{\sin\theta}$.

$$W_f = F_f d \cos 180° = -\mu_k F_N d = -\mu_k mg\cos\theta\left(\frac{h}{\sin\theta}\right)$$

$$= -(0.2)(40 \text{ kg})(10 \text{ m/s}^2)\cos 30°\left(\frac{8 \text{ m}}{\sin 30°}\right) = -1109 \text{ J}$$

Therefore, the total work done on the box is $W_{total} = W_g + W_f = 3200 \text{ J} - 1109 \text{ J} = 2091 \text{ J}$.

27. **B** Apply the equation $x = x_0 + vt - \frac{1}{2}at^2$ with the moment she takes the baton as the initial point, where $x_0 = 0$. Substitute for the final velocity, acceleration, and time to get $x = 0 + (10 \text{ m/s})(2 \text{ s}) - \frac{1}{2}(2.4 \text{ m/s}^2)(2 \text{ s})^2 = 15.2 \text{ m}$.

28. **D** The net force on each block can be found by using Newton's Second Law, $F_{Net} = ma$. The tension in the string can be found by focusing on the hanging mass. You know the net force on it will be $F_{Net} = ma = F_g - T = (mg) - T$, which means $T = (mg) - (ma)$. Finally, the coefficient of kinetic friction can be found by looking at the top block. For that block, the net force in the horizontal direction will be the same as the overall net force since the two vertical forces (normal and gravity) will cancel out. Thus, you get $F_{Net} = ma = T - F_f = T - \mu F_N = T - \mu mg$. The only value in the answers you cannot calculate is the speed of the block when it reaches the edge. In order to compute this value, you would need to know how far from the edge the block is when it begins moving.

29. **C** If nothing is moving, then you know that the net force will be 0. Looking first at the forces perpendicular to the plane, you get $F_N = F_g\cos\theta = mg\cos\theta$. Next, using Newton's Second Law and defining "up the ramp" as positive, you can say $F_{Net} = T - F_g\sin\theta - F_f = 0$. Solving for tension and plugging in all the variables gives you $T = mg\sin\theta + \mu mg\cos\theta$.

Next, looking at the hanging block, you can again use Newton's Second Law to determine that $F_{Net} = F_g - T = Mg - (mg\sin\theta + \mu mg\cos\theta) = 0$. Thus, solving for μ gives you $\mu = \dfrac{Mg - mg\sin\theta}{mg\cos\theta}$.

30. **C** The basic formula for work is $W = Fd\cos\theta$, where θ is the angle between the force and the direction of motion. Centripetal force, by definition, is always perpendicular to motion. Therefore, θ will always be 90°, and cos 90° = 0, which means this force cannot do work.

31. **C** At equilibrium, the force of gravity pulling the spheres together is equal in magnitude to the spring force keeping the spheres apart, $F_g = F_s$. When the spheres are 0.2 m apart, the spring is compressed 0.8 m from its natural length. Therefore,

$$G\frac{m_1m_2}{r^2} = kx \rightarrow k = G\frac{m_1m_2}{xr^2} = \left(6.67\times10^{-11}\ \text{N}\cdot\text{m}^2/\text{kg}^2\right)\frac{(1\ \text{kg})^2}{(0.8\ \text{m})(0.2\ \text{m})^2} = 2.09\times10^{-9}\ \text{N/m}$$

32. **C** Consider the situation from the perspective of Conservation of Energy. When it is initially dropped from one end of the planet, the ball has some amount of potential energy and no kinetic energy. When it reaches the center of the Earth, it will have no more potential energy, so all of that energy is now kinetic. Because of inertia, the ball will continue in that direction until the energy has again been converted entirely to potential energy. This would happen just as it reaches the other end of the hole. At that point, it would fall again, and the process would repeat infinitely. This is exactly the same motion as an ideal spring system.

33. **B** At the top of the ramp, the car will have potential energy $PE = mgh$. At the bottom, this same amount of energy has been converted into kinetic energy $KE = \frac{1}{2}mv^2 = mgh$. Multiplying both sides by 2 produces $mv^2 = 2mgh$. Finally, divide both sides by r to get $mv^2/r = 2mgh/r$. The first term is the general formula for centripetal force, so the second term must also be equivalent to that force in this situation.

34. **B** First, find the net force on the system. Each of the top three blocks will contribute nothing to the net force. The forces on each of the two blocks on the ramp will be the following:

The normal force and the perpendicular component of gravity will cancel out. This leaves just the parallel component of gravity to be a net force. Using Newton's Second Law, you get $F_{Net} = F_{g,parallel}$ $2mg\sin\theta = 5ma$. The 2 comes in because there are two blocks on the ramp and each one will contribute the net force discussed earlier. The 5 is necessary because the blocks cannot accelerate individually. Either they all move or none do. Solving for a then gives $a = \frac{2}{5}g\sin\theta$.

35. **A** First, split up v_0 into horizontal and vertical components.

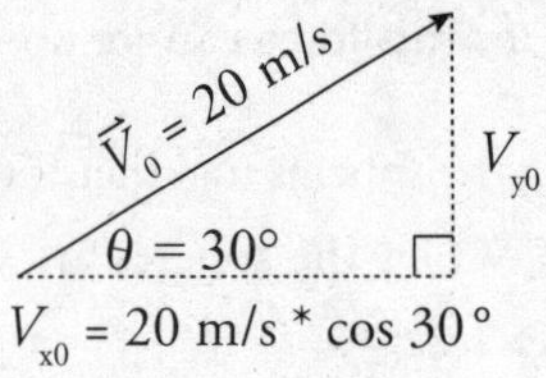

$\vec{V}_0 = 20$ m/s

$V_{y0} = 20$ m/s * sin 30°

$\theta = 30°$

$V_{x0} = 20$ m/s * cos 30°

$v_{x0} = 20$ m/s * $\sqrt{3}/2 = 10\sqrt{3}$ m/s, and $v_{y0} = 20$ m/s * 1/2 = 10 m/s. Then, make a column of vertical values. Take the positive direction to be up.

Vertical

$y - y_0 = 0$ (since it starts and finishes on the ground)
$v_{y0} = 10$ m/s
v_{yf} = don't care
$a_y = -10$ m/s^2
t = ?

Plugging these into Big Five #3 gives t = 0 s or 2 s, meaning there are two times at which the elevation of the ball is 0 m. (What is Big Five #3? Or, better yet, what are the 5 Big Five equations that we love so much? We cover them all in Chapter 4: Kinematics. Flip there now to learn more, or come back to this page later for further clarification.) Zero seconds corresponds to the instant the ball leaves the ground; 2 seconds corresponds to the time the ball lands on the ground. (Alternatively, you could use Big Five #2 to calculate the time the ball spends traveling up, and then double it to get the total time in flight, as done in Examples 25 and 26 of Chapter 4.)

Finally, make your column of horizontal values.

Horizontal

$x - x_0$ = ?
$v_x = 10\sqrt{3}$ m/s (Remember, forward speed is constant for an ideal projectile.)
t = 2 s (You just solved for this.)

Solve for the horizontal displacement by using the definition of velocity, $x - x_0 = v_x t$, or just $d = v_x t$. This gives $d = (10\sqrt{3}$ m/s) * 2 s = $20\sqrt{3}$ m/s.

36. **C** For any collision, momentum must be conserved. That means $m_1v_{1,0} + m_2v_{2,0} = m_1v_{1,f} + m_2v_{2,f}$. Plugging in known values, that gives (10)(10) + m_2(0) = (10)(–2) + m_2(8). Solving for m_2 gives m_2 = 120/8 = 15 kg.

37. **A** In a perfectly inelastic collision, the two objects stick together. That is not true in the graph because the objects have different velocities post-collision.

For a perfectly elastic collision, kinetic energy must be conserved. Check this using $\frac{1}{2}m_1v_{1,0}^2 + \frac{1}{2}m_2v_{2,0}^2 = \frac{1}{2}m_1v_{1,f}^2 + \frac{1}{2}m_2v_{2,f}^2$. Plugging in the values gives $\frac{1}{2}(4)(10)^2 + \frac{1}{2}(6)(0)^2 = \frac{1}{2}(4)(-2)^2 + \frac{1}{2}(6)(8)^2$. Calculating each side gives 200 + 0 = 8 + 192, which is true. Therefore, kinetic energy is conserved, which makes this a perfectly elastic collision.

38. **D** In all of these cases, mechanical energy must be conserved. Thus, the energy required will be *mgh* for all of them.

39. **A** You know $\mathrm{T}_{\text{Net}} = I\alpha$. The torque applied to each is equal, so the sphere with a smaller moment of inertia (I) will experience a greater angular acceleration. Moment of inertia is smaller for an object with its mass concentrated closer to the center, so the solid sphere will have a smaller moment of inertia and thus increase its angular speed more quickly.

40. **A** You know $F_f = \mu F_N$. In both cases, μ will be the same, so you need only concern yourself with F_N. There are three vertical forces in this problem: gravity, the normal force, and the vertical component of the applied force. In the first case, that means $F_N = F_g - mg\sin\theta = mg - mg\sin 30 = mg - mg/2 = mg/2$.

In the second case, $F_N = F_g + mg\sin\theta = mg + mg\sin 30° = mg + mg/2 = 3mg/2$. Therefore, the force in this case will be three times what it was in the first case.

41. **C** First, the force diagram for this situation would look like the following picture:

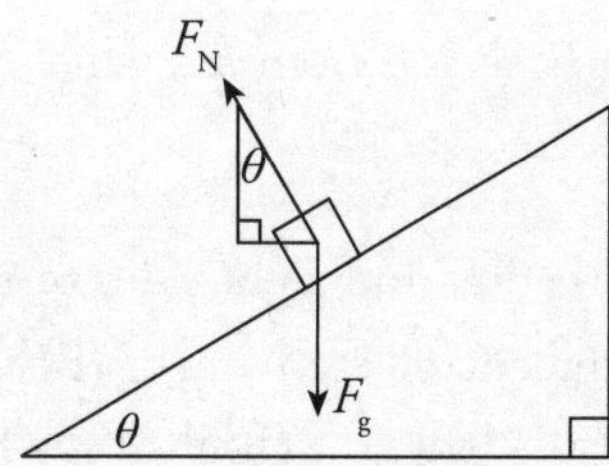

The car's path is horizontal, which means the net force in the vertical direction is zero. Since the turn is frictionless, the only forces acting on the car are normal force and weight. Thus, the vertical component of the normal force must be large enough to cancel out weight. This can happen only if the magnitude of the normal force is greater than the gravitational force on the car.

42. **B** Using Newton's Second Law and taking up as the positive direction, you get $F_{\text{Net}} = ma = T - mg$ for the first block and $F_{\text{Net}} = M(-a) = T - Mg$ for the second block. Subtracting the second equation from the first gives $ma + Ma = Mg - mg$. Solving for a then gives $a = \frac{g(M-m)}{m+M} = \frac{10(15-5)}{5+15} = \frac{100}{20} = 5 \text{ m/s}^2$.

From there, the problem is simply a situation of uniform accelerated motion. The blocks need to move a distance of d = 0.5 m each, the starting speed is 0 m/s, and the acceleration is 5 m/s^2. Using $d = v_0t + \frac{1}{2}at^2$ gives $(0.5) = (0)t + \frac{1}{2}(5)t^2$, so $t = \sqrt{\frac{1}{5}}$ s.

43. **B** The work done by kinetic friction is always negative, so (C) and (D) are eliminated. The total work done on the cart is the sum of the work done by each of the forces acting on the cart, $W_{total} = W_{person} + W_f + W_g + W_N$. The normal force and gravity don't do any work on the cart because it is perpendicular to the displacement, so $W_{total} = W_{person} + W_f$. By the Work–Energy Theorem, total work is equal to the change in kinetic energy. $W_{person} + W_f = \Delta K \Rightarrow W_f = \Delta K - W_{person} = K_f - K_i - W_{person} = \frac{1}{2}mv_f^2 - 0 - F_{person}d\cos\theta = \frac{1}{2}(60\text{ kg})(3\text{ m/s})^2 - (300\text{ N})(5\text{ m})\cos 20° = -1140\text{ J}$.

44. **A** In this situation, the left-hand block will provide counterclockwise torque, and the right-hand block will provide clockwise torque. Therefore, the two must be equal in magnitude for the system to be balanced. You know that the formula for torque is $\tau = Fr\sin\theta$. In this problem, θ will always be 90 degrees, and sin 90° = 1, so that term will be ignored for the rest of the explanation. Furthermore, the only forces involved in this problem are the forces of gravity on the blocks, and you know $F_g = mg$.

For (A), the net counterclockwise torque (left side of the system) would be $\tau = (m_2g)(2L) + (m_1g)(L) = (2m_1g)(2L) + (m_1g)(L) = 5m_1gL$. The clockwise torque (right side of the system) would be $\tau = (m_2g)(L) + (m_1g)(2L) = (2m_1g)(L) + (m_1g)(2L) = 4m_1gL$. Thus, it would not be balanced. Choice (B) would make both torques 0, so that would be balanced. Choices (C) and (D) both result in the fulcrum being twice as far from m_1 as it is from m_2, which would counteract the difference in their weights.

45. **C** The correct answer is (C). For the left and right sides of the tracks, all three tracks are identical. Looking only at the middle segment of each track, (B) would hold steady throughout. Choice (A) would decrease the speed of the ball as it climbed the hill and then increase the speed back to the original amount as it descended the hill. That means it spends the entire middle segment at a speed less than the initial speed. Choice (C) would be the opposite of (A). The ball gains speed as it descends, and then loses speed as it climbs back up, eventually ending at the original speed. Therefore, it spends the entire middle segment at a speed greater than the initial amount.

46. **B, D** On a velocity-versus-time graph, the acceleration at any instant is the slope of the graph at that point. *Uniformly* accelerated motion means the acceleration is *constant,* and only straight lines have constant slope. Thus, the correct answers are the two graphs with lines, (B) and (D).

47. **A, D** First, the momentum of a system will always be conserved, regardless of the type of collision that takes place. Thus, (D) must be true. You can eliminate (C) because that is true only for perfectly elastic collisions, and this collision is not. Much like linear momentum, angular momentum is always conserved for an entire system. Therefore, (A) is correct, but (B) is not because it does not encompass the entire system.

48. **A, D** Choice (B) can be eliminated because *KE* would increase in this situation because of the object's increasing speed. Choice (C) is also incorrect. In this situation, *KE* would decrease, but $KE = \frac{1}{2}mv^2$, meaning it would be a linear decrease. Choices (A) and (D) are correct because a graph of a falling object loses more altitude each second than it did the previous second, and gravitational potential energy is proportional to altitude.

49. **B, D** In a collision of freely-moving objects, momentum is conserved, and kinetic energy is either conserved or lost. In order for ball 1 to rebound with a velocity of $-v_0$, ball 2 would have to have a positive final velocity to conserve momentum. However, this scenario would result in an increase in the kinetic energy of the system, so eliminate (A). Similarly, kinetic energy is proportional to mass, and ball 2 is more massive than ball 1. Therefore, ball 2 cannot have a final velocity of v_0, as it would have more kinetic energy than the system started with, eliminating (C). The two balls can have the same final velocity if they undergo a perfectly inelastic collision. The balls could also end up with equal and opposite *y*-components of velocity, but different *x*-components of velocity, such that their final velocities are not parallel.

50. **A, B** In the absence of torque, the angular momentum of the merry-go-round is conserved. When a child jumps on or off the merry-go-round in the radial direction, as in (A) and (C), they do not exert a torque on the merry-go-round. Since angular momentum $L = I\omega$, the rotation rate slows if the moment of inertia increases. A child jumping onto the merry-go-round increases the moment of inertia of the system, making (A) correct, whereas a child jumping off decreases the moment of inertia of the system, making (C) incorrect. A change in angular momentum is caused by a torque on the system. A child jumping onto the merry-go-round with a velocity opposite to the motion of the merry-go-round imparts a torque to decrease the moment of inertia of the merry-go-round when they land. The decrease in L and increase in I both contribute to decreasing ω, making (B) correct. In contrast, a child jumping off of the merry-go-round with a velocity opposite to their instantaneous velocity before jumping imparts a torque that increases L. The increase in L and decrease in I both increase ω, making (D) incorrect.

Section II: Free-Response Questions

1. The defining characteristic of a perfectly elastic collision is that no kinetic energy is lost. Therefore, you must record the speeds of each car before and after the collision. Then, the equation $\frac{1}{2}m_1v_{1,0}^2 + \frac{1}{2}m_2v_{2,0}^2 = \frac{1}{2}m_1v_{1,f}^2 + \frac{1}{2}m_2v_{2,f}^2$ can be used to check whether or not it is elastic.

   The defining feature of a perfectly inelastic collision is that the objects stick together after the collision. This can be simply observed. If the two cars remain intact post-collision, then the collision is perfectly inelastic.

   If neither of the above is true, then it is neither perfectly elastic nor perfectly inelastic.

   Finally, a few possible sources of energy loss would be the heat generated as metal is warped, the sound made by the crash, and even the light of any sparks generated.

2. Read the graphs column by column. Object A is on the left, Object B in the middle, and Object C on the right. (See more in Chapter 4: the slope of an x versus t graph gives velocity.) The slope of a v versus t graph gives acceleration. Object A could be made from a car or plane accelerating up to a certain speed and then maintaining it. Object B could be made from any case of ideal projectile motion where the initial velocity is up. Object C could be anything that uniformly accelerates from a stop.

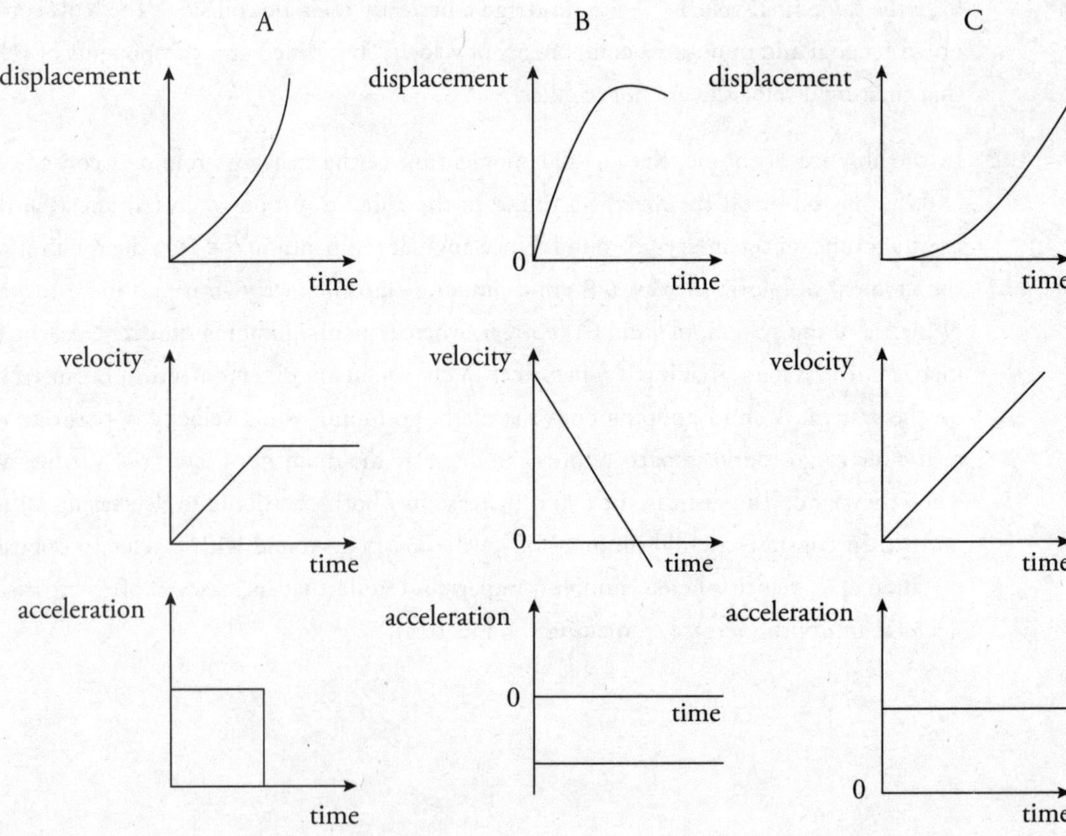

3. (A) Your graph should look like this:

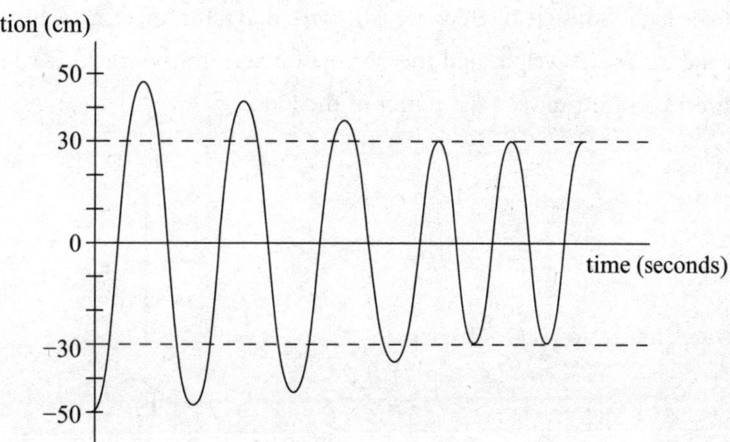

First, it must start at position –50 as the problem indicates. Second, it must go up to a point that is nearly 50 but not quite. This is because the frictional surface will remove some of the energy from the system as the block slides across it. Each time the block swings back to the frictional surface, it will drain a little more of the energy, making the amplitude slowly taper off until it eventually is only on the frictionless surface. Because the frictionless surface extends to 30 cm, it will oscillate between that point and –30 cm indefinitely.

(B) The energy of a spring system can be found using $PE = \frac{1}{2}kA^2$, where A is the amplitude of the system, so the initial energy of the system will be $PE = \frac{1}{2}(100)(0.5)^2 = 50(0.25) = 12.5$ J. After the oscillations have tapered off, it will have a new energy of $PE = \frac{1}{2}(100)(0.3)^2 = 50(0.09) = 4.5$ J. Therefore, the frictional surface has caused 8 J of energy to be lost.

4. (A) Using Conservation of Energy, set the potential energy at the top of the ramp equal to the kinetic energy at the bottom to get $mgh = \frac{1}{2}mv^2$. Solving for v gives $v = \sqrt{2gh} = \sqrt{(2)(10)(20)} = \sqrt{400} = 20$ m/s.

(B) You're asked to compare a force when the car's at the bottom of the loop to a force when the car's at the top of the loop. So, start by drawing two force diagrams, one for each situation. The only forces acting on the car are its weight and the normal force from the track. It will be easiest to make the positive direction point toward the center of the loop.

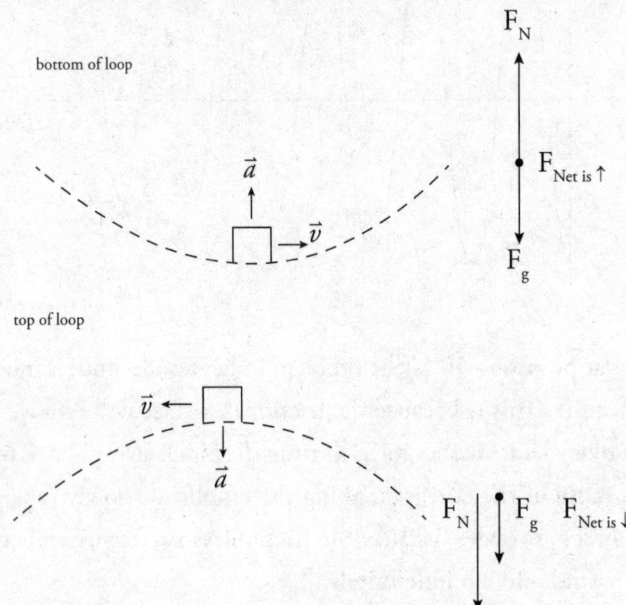

At the bottom, the net force points upward. In terms of vectors, $\mathbf{F}_{\text{net, bottom}} = \mathbf{F}_{\text{N, bottom}} + \mathbf{F}_g$, but $\mathbf{F}_g$ points in the negative direction. So in terms of magnitudes, $F_{\text{net, bottom}} = F_{\text{N, bottom}} - F_g$. Solving for normal force gives $F_{\text{net, bottom}} + F_g = F_{\text{N, bottom}}$, or

$$F_{\text{N, bottom}} = F_{\text{net, bottom}} + mg \text{ (equation 1)}$$

Meanwhile, at the top, net force points downward, and both the normal force and gravitational force point in the positive direction. Thus, $F_{\text{net, top}} = F_{\text{N, top}} + F_g \rightarrow F_{\text{net, top}} - F_g = F_{\text{N, top}} \rightarrow$

$$F_{\text{N, top}} = F_{\text{net, top}} - mg \text{ (equation 2)}$$

Since mg is positive, this says that at the bottom, the normal force is *greater* than the net force, while at the top, the normal force is *less* than the net force. It seems like the normal force is greater just after entering the loop than it is at the peak of the loop. However, the net force isn't the same at both locations, so you don't know for sure yet.

To prove that your hunch is correct, you need to know the net force at each location. In general, the motion of the car about the loop is *not* uniform circular motion because the car's speed changes. However, at the very bottom and at the very top, the path is momentarily horizontal, so the speed is momentarily constant. Therefore, it's okay to apply UCM at these two locations only.

At the top, $F_{net, top} = mv_{top}^2/r$; at the bottom, $F_{net, bottom} = mv_{bottom}^2/r$. Obviously, the car is going faster at the bottom of the loop than at the top, so $v_{bottom} > v_{top}$, so $F_{net, bottom} > F_{net, top}$. Therefore, at the bottom, equation (1) says that $F_{N, bottom}$ = (large F_{net}) + (positive value). At the top, equation (2) says that $F_{N, top}$ = (small F_{net}) – (positive value). Thus, the normal force must be greater at the bottom than at the top.

(C) Mechanical energy will be constant throughout since the track is frictionless, so set the energy at the beginning equal to the energy at the top of the loop. This gives $mgh = mg(2r) + \frac{1}{2}mv^2$, where r is the radius of the circle and v is the speed at the top of the loop.

Next, recall that in order for the car to complete the loop, it must be going fast enough to create some normal force at the top. In general, this equation would look like $\frac{mv^2}{r} = F_N + F_g$. If the loop is at its maximum height, F_N approaches 0 at the top, so $\frac{mv^2}{r} = mg$. Moving the r to the other side gives $mv^2 = mgr$.

Substituting this value into the first equation gives $mgh = mg(2r) + \frac{1}{2}(mgr)$. You can cancel out m and g since they appear in every term, so $h = 2r + \frac{1}{2}r = 2.5r$. The problem shows $h = 20$, so solving for r gives $r = h/(2.5) = (20)/(2.5) = 8$ m.

(D) Conservation of Energy can be used to find the speed of the car as it enters the final segment. Set the potential energy of the initial position equal to the potential and kinetic energies as the cart enters the final stretch. This gives $mgh_1 = mgh_2 + \frac{1}{2}mv^2$. First, realize m can be dropped since it appears in every term. Solving for v then gives $v = \sqrt{2g(h_1 - h_2)} = \sqrt{(2)(10)(20-10)} = \sqrt{200} = 10\sqrt{2}$ m/s.

In order for the friction to bring the cart to a stop, it must do work to the cart. You know $W = \Delta KE$, so $Fd\cos\theta = \frac{1}{2}mv_f^2 - \frac{1}{2}mv_0^2$. The first KE term can be dropped since the final speed will be 0. Substitute $F_f = \mu F_N = \mu mg$ for F term since friction is the force doing the work. Additionally, $\theta = 180°$ because the frictional force will act in a direction opposite the cart's motion. All of this gives $(\mu mg)d\cos 180° = -\frac{1}{2}mv_0^2$. Canceling the m on each side and solving for d then gives $d = -\frac{1}{2}v_0^2/(\mu g\cos 180°) = -\frac{\frac{1}{2}(10\sqrt{2})^2}{(0.2)(10)(-1)} = 50$ m.

5. (A) As the rotor spins, the particle moves in a curved path with its velocity tangent to the path. Inertia therefore causes the particle to move in the direction of this tangential velocity, which is toward the outside of its curved path. The outward motion will continue until the particle encounters a force from the outer walls of the centrifuge tube that prevents further outward motion.

(B) Centripetal acceleration depends on the angular speed and radius of rotation, $a_c = \omega^2 r$. The two rotors have the centripetal acceleration, so $\omega_1^2 r_1 = \omega_2^2 r_2 \rightarrow \omega_2^2 = \frac{\omega_1^2 r_1}{r_2} \rightarrow \omega_2 = \omega_1\sqrt{\frac{r_1}{r_2}}$.

(C) Apply the equation for centripetal acceleration:

$$a_c = \omega^2 r \rightarrow \omega = \sqrt{\frac{a_c}{r}} = \sqrt{\frac{1000(10 \text{ m/s}^2)}{0.2 \text{ m}}} = 224 \text{ rad/s}$$

HOW TO SCORE PRACTICE TEST 1

Section I: Multiple Choice

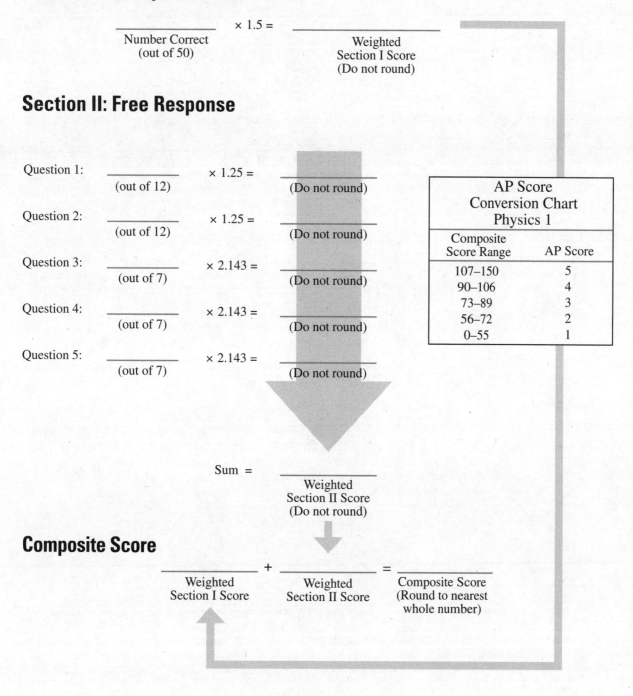

______ × 1.5 = ______

Number Correct (out of 50)

Weighted Section I Score (Do not round)

Section II: Free Response

Question 1:	______ (out of 12)	× 1.25 =	______ (Do not round)
Question 2:	______ (out of 12)	× 1.25 =	______ (Do not round)
Question 3:	______ (out of 7)	× 2.143 =	______ (Do not round)
Question 4:	______ (out of 7)	× 2.143 =	______ (Do not round)
Question 5:	______ (out of 7)	× 2.143 =	______ (Do not round)

Sum = ______

Weighted Section II Score (Do not round)

AP Score Conversion Chart Physics 1	
Composite Score Range	AP Score
107–150	5
90–106	4
73–89	3
56–72	2
0–55	1

Composite Score

______ + ______ = ______

Weighted Section I Score

Weighted Section II Score

Composite Score (Round to nearest whole number)

Part III
About the AP Physics 1 Exam

- The Structure of the AP Physics 1 Exam
- A Quick Word About Equations
- How AP Exams Are Used
- Other Resources
- Designing Your Study Plan

THE STRUCTURE OF THE AP PHYSICS 1 EXAM

The AP Physics 1 Exam consists of two sections: a multiple-choice section and a free-response section. The multiple-choice section consists of two question types. Single-select questions are each followed by four possible responses, only one of which is correct. Multi-select questions are a new addition to the AP Physics 1 Exam, and require two of the listed answer choices to be selected to answer the question correctly. There are five multi-select questions that always appear at the end of the multiple-choice section.

Section	Question Type	Number of Questions	Weight	Time
IA	Multiple-Choice questions: single-select	45	50%	90 minutes
IB	Multiple-Choice questions: multi-select	5		
II	Free-Response Questions	5	50%	90 minutes
II	Question 1: Experimental Design (12 points)			
II	Question 2: Qualitative/Quantitative Translation (12 points)			
II	Question 3: Paragraph Argument Short Answer (7 points)			
II	Questions 4 and 5: Short Answer (7 points each)			

The free-response section consists of five multi-part questions, which require you to write out your solutions, showing your work. The total amount of time for this section is 90 minutes. Unlike the multiple-choice section, which is scored by a computer, the free-response section is graded by high school and college teachers. They have guidelines for awarding partial credit, so you may still receive partial points should you not correctly respond to every part of the question.

You are allowed to use a calculator on the entire AP Physics 1 Exam—including both the multiple-choice and free-response sections. Four-function, scientific, or graphing calculators may be used, provided that they don't have any unapproved features or capabilities (a list of approved graphing calculators is available on the College Board's website). In addition, a table of equations commonly used in physics will be provided to you at the exam site. This can be found online and we've included it with all practice tests.

Grades on the AP Physics 1 Exam are reported as a number: either 1, 2, 3, 4, or 5. Here is a description of each of these five numerical scores plus data on how students scored on the May 2022 AP Physics 1 Exam:

Score	2022 Percentage	Credit Recommendation	College Grade Equivalent
5	7.9%	Extremely Well Qualified	A
4	17.0%	Well qualified	A–, B+, B
3	18.3%	Qualified	B–, C+, C
2	27.1%	Possibly Qualified	–
1	29.6%	No Recommendation	–

Scores from May 2022 AP administration. Data taken from the College Board website.

Colleges are generally looking for a 4 or 5, but some may grant credit for a 3. How well do you have to do to earn each grade? Each test is curved, so scores vary from year to year, but as you can see above, in May 2022 nearly 60% of test-takers earned scores of 1 or 2, so you'll want to study hard and prepare for this very difficult exam. We'll get to that soon, we promise!

So, what is on the exam and how do you prepare? Take a look at the following list of the major topics covered on the AP Physics 1 Exam.

AP Physics 1: Algebra-Based Course Content

You may be using this book as a supplementary text as you take an AP Physics 1 course at your high school, or you may be using it on your own. The College Board is very detailed in what they require your AP teacher to cover in their AP Physics 1 course. They explain that you should be familiar with the following topics:

- kinematics
- dynamics: Newton's laws
- circular motion and universal law of gravitation
- simple harmonic motion: simple pendulum and mass-spring systems
- impulse, linear momentum, and conservation of linear momentum: collisions
- work, energy, and conservation of energy
- rotational motion: torque, rotational kinematics and energy, rotational dynamics, and conservation of angular momentum, kinetic energy

In addition, the College Board says that you should be familiar with the following Big Ideas:

- Big Idea 1: Systems
- Big Idea 2: Fields
- Big Idea 3: Force Interactions
- Big Idea 4: Change
- Big Idea 5: Conservation

Updates!

In February 2021, the College Board announced that 3 units would no longer be tested on the AP Physics 1 Exam: Electric Charge and Electric Force, DC Circuits, Mechanical Waves, and Sound. These were formerly Units 8, 9, and 10 in the College Board's AP Physics 1 Course Description. Now, units 1–7 will be represented on the AP Physics 1 Exam in approximately similar proportion to their relative weight from the Course Description. More on that on the next page.

These are the Big Ideas that should be covered in your AP Physics 1 course, but the exam gets a bit more detailed as to what you should know. As we mentioned in a sidebar on the previous page, in early 2021 the College Board announced that starting with the 2021 exam, Units 8–10 from their AP Physics 1 Course Description (available on the College Board website) would no longer be tested. So we have eliminated coverage of those topics (electric charge and electric force, DC Circuits, and mechanical waves and sound) and now we are all about units 1–7. Knowing that, here is a run-down of the units that the College Board will be testing and an estimate of their exam weighting:

More Updates!
In 2022, the College Board announced that some topics in AP Physics 1 and AP Physics 2 (fluids, waves) will realign in the next year or two. At the time of this printing, the realignment seems to have been pushed off to the 2023–2024 school year. To learn more about the specific information, visit the College Board's website for breaking news.

- Unit 1: Kinematics 12–18%
- Unit 2: Dynamics 16–20%
- Unit 3: Circular Motion and Gravitation 6–8%
- Unit 4: Energy 20–28%
- Unit 5: Momentum 12–18%
- Unit 6: Simple Harmonic Motion 4–6%
- Unit 7: Torque and Rotational Motion 12–18%

A QUICK WORD ABOUT EQUATIONS

As you know, you will be given an Equation Sheet on the day of your exam and that Equation Sheet is yours for the duration of the test. You will see many equations in different forms and it can be hard to recall what is equivalent to what, so we have made a handy chart for you. Here is a tally of a few equations that are equivalent to other equations and where you can find them in this book and on your Equation Sheet.

Equation(s)	Page	Location on AP Physics 1 Equations Sheet
Big Five #2, #3, #5	93	First three equations in top left
Newton's Second Law	117	Fourth equation below the first three
Force of Friction	124, 125	Fifth equation from top
Power	172	$P = \Delta E/\Delta t$ is the equation on the sheet. Note specifically that the change in energy is the work done, so these two equations are equivalent.
Gravitational Force	144	Third equation from the bottom of the second column. Don't be thrown by the use of absolute value bars; they mean the same thing.

HOW AP EXAMS ARE USED

Different colleges use AP Exams in different ways, so it is important that you go to a particular college's website to determine how it uses AP Exams. The three categories below represent the main ways in which AP Exam scores can be used.

- **College Credit.** Some colleges will award you college credit if you score well on an AP Exam. These credits count toward your graduation requirements, meaning that you can take fewer courses while in college. Given the cost of college, this could be quite a benefit, indeed.
- **Satisfy Requirements.** Some colleges will allow you to "place out" of certain requirements if you do well on an AP Exam, even if they do not give you actual college credits. For example, you might not need to take an introductory-level course, or perhaps you might not need to take a class in a certain discipline at all.
- **Admissions Plus.** Even if your AP Exam will not result in college credit or even allow you to place out of certain courses, most colleges will respect your decision to push yourself by taking an AP course or even an AP Exam outside of a course. A high score on an AP Exam shows understanding of more difficult content than is taught in many high school courses, and colleges may take that into account during the admissions process.

Are You Preparing for College?
Check out all of the useful books from The Princeton Review, including *The Best 389 Colleges, ACT Prep, SAT Prep,* and more!

OTHER RESOURCES

There are many resources available to help you improve your score on the AP Physics 1 Exam, not the least of which are your teachers. If you are taking an AP class, you may be able to get extra attention from your teacher, such as obtaining feedback on your essays. If you are not in an AP course, reach out to a teacher who teaches AP Physics 1 and ask if the teacher will review your essays or otherwise help you with content.

Train As You Will Test
The AP Physics 1 Exam is grueling, and you need to build up your endurance, so be sure to take practice tests throughout the school year.

Another wonderful resource is **AP Students,** the official site of the AP Exams. The scope of the information at this site is quite broad and includes:

- a course description, which includes details on what content is covered and sample questions
- sample questions from the AP Physics 1 Exam
- free-response question prompts and multiple-choice questions from previous years

The AP Students home page address is apstudents.collegeboard.org/

Finally, The Princeton Review offers tutoring for the AP Physics 1 Exam. Our expert instructors can help you refine your strategic approach and add to your content knowledge. For more information, call 1-800-2REVIEW.

DESIGNING YOUR STUDY PLAN

In Part I, you identified some areas for potential improvement. Let's now delve further into your performance on Test 1, with the goal of developing a study plan appropriate to your needs and time commitment.

Read the answers and explanations associated with the multiple-choice questions (starting at page 35). After you have done so, respond to the following questions:

Study Breaks Are Important
Don't burn yourself out before test day. Remember to take breaks every so often—go for a walk, listen to a favorite album, get some fresh air.

- Review the topic list on page 55. Next to each topic, indicate your rank of the topic as follows: "1" means "I need a lot of work on this," "2" means "I need to beef up my knowledge," and "3" means "I know this topic well."
- How many days/weeks/months away is your exam?
- What time of day is your best, most focused study time?
- How much time per day/week/month will you devote to preparing for your exam?
- When will you do this preparation? (Be as specific as possible: Mondays and Wednesdays from 3 to 4 P.M., for example.)
- Based on the answers above, will you focus on strategy (Part IV) or content (Part V) or both?
- What are your overall goals in using this book?

And Remember: Check for Updates!

In the past few years and the past few AP Exam administrations, the College Board announced a TON of updates and changes to assorted AP Exams and courses. Check your online Student Tools for late-breaking information. Also check the College Board's website for updated information: apcentral.collegeboard.org/courses/ap-physics-1?course=ap-physics-1

Looking to Guarantee a 4 or 5?
We now offer one-on-one tutoring for a guaranteed 5 or an online course for a guaranteed 4 on the AP Physics 1 Exam. For information on rates, availability, and to learn more about the guarantee, visit PrincetonReview.com/college/ap-test-prep

Part IV
Test-Taking Strategies for the AP Physics 1 Exam

PREVIEW

Review your Test 1 results and then respond to the following questions:

- How many multiple-choice questions did you miss even though you knew the answer?
- On how many multiple-choice questions did you guess blindly?
- How many multiple-choice questions did you miss after eliminating some answers and guessing based on the remaining answers?
- Did you find any of the free-response questions easier or harder than the others—and, if so, why?

Stay Up to Date!
For late-breaking information about test dates, exam formats, and any other changes pertaining to AP Physics 1, make sure to check the College Board's website at apstudents.collegeboard.org/courses/ap-physics-1-algebra-based/assessment

HOW TO USE THE CHAPTERS IN THIS PART

For the following Strategy chapters, think about what you are doing now before you read the chapters. As you read and engage in the directed practice, be sure to think critically about the ways you can modify your approach.

A QUICK NOTE ABOUT NAMING CONVENTIONS

Each classroom uses different conventions for naming things. In this book, we are using the AP conventions set by the College Board, but please note that your AP Physics 1 course teacher may use different naming conventions, so you should notice that.

For example, v_0 and v_i are usually the same thing but not always the same thing. Some teachers and books may treat these as the same, though. These mean velocity at time zero and initial velocity, respectively.

Take a look at this example:

> A car starts from rest and accelerates to 20 m/s in 5 seconds. The car holds this velocity for 10 seconds. The driver then speeds up to 28 m/s in 5 more seconds. What is the magnitude of the second acceleration?

v_0 = 0 miles per hour, but v_i in this instance will be 20 m/s because the problem is looking at the second acceleration. Be careful not to confuse v_0 and v_i. This may seem silly, but it's a common error so we wanted to address it right up front. Now, onto the good stuff!

Chapter 1
How to Approach Multiple-Choice Questions

CRACKING THE MULTIPLE-CHOICE SECTION

All the multiple-choice questions will have a similar format: each will be followed by four answer choices. At times, it may seem that there could be more than one possible correct answer. With the exception of the five multi-select questions, there is only one! Answers resulting from common mistakes are often included in the four answer choices to trap you.

Two Passes
The two-pass system will help you make the best use of your time and maximize your points by focusing first on the questions you know and can answer fairly quickly, and then revisiting the more time-consuming ones. Just be careful with your bubbling! (More on that below.)

The Two-Pass System

The AP Physics 1 Exam covers a broad range of topics. There's no way, even with our extensive review, that you will know everything about every topic in algebra-based physics. So, what should you do?

Adopt what we call the two-pass system. The two-pass system entails going through the test and answering the easy questions first. Save the more time-consuming questions for later. (Don't worry—you'll have time to do them later!) First, read the question and decide if it a "now" or "later" question. If you decide this is a "now" question, answer it in the test booklet. If it is a "later" question, come back to it. Once you have finished all the "now" questions on a double page, transfer the answers to your bubble sheet. Flip the page and repeat the process.

Once you've finished all the "now" questions, move on to the "later" questions. Start with the easier questions first. These are the ones that require calculations or that require you to eliminate the answer choices (in essence, the correct answer does not jump out at you immediately). Transfer your answers to your bubble sheet as soon as you answer these "later" questions.

Watch Out for Those Bubbles!

Because you're skipping problems, you need to keep careful track of the bubbles on your answer sheet. One way to accomplish this is by answering all the questions on a page and then transferring your choices to the answer sheet. If you prefer to enter them one by one, make sure you double-check the number beside the ovals before filling them in. We'd hate to see you lose points because you forgot to skip a bubble!

POE = BFF
Process of Elimination (POE) is your friend when it comes to multiple-choice questions. Instead of trying to pinpoint the right answer, focus instead on getting rid of the ones that are wrong.

Process of Elimination (POE)

On most tests, you need to know your material backward and forward in order to get the right answer. In other words, if you don't know the answer beforehand, you probably won't answer the question correctly. This is particularly true of fill-in-the-blank and essay questions. We're taught to think that the only way to get a question right is by knowing the answer. However, that's not the case on Section I of the AP Physics 1 Exam. You can get a perfect score on this portion of the test without knowing a single right answer—provided you know all the wrong answers!

What are we talking about? Process of Elimination is perhaps the most important technique to use on the multiple-choice section of the exam. Let's take a look at an example that probably looks quite familiar:

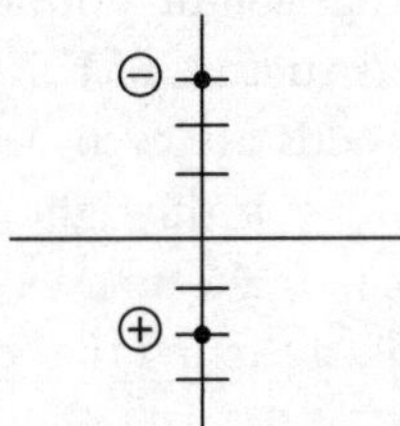

26. In the diagram above, a positive charge and negative charge are placed at $y = -2$ and $y = 3$, respectively. If the negative charge has a greater magnitude, then the only place of 0 net electric field would be

(A) along the positive x-axis
(B) along the negative x-axis
(C) along the positive y-axis
(D) along the negative y-axis

POE Walk-Through
Check out this example of POE in action. Remember this process when you are tackling drill questions and practice test questions later in this book.

Now, if this were a fill-in-the-blank-style question, you might be in a heap of trouble. But let's take a look at what you've got and how you can eliminate answer choices in order to reach the correct answer.

Remember that the electric field for any given particle will point toward a negative charge and away from a positive charge. Anywhere between the two charges here (regardless of x-coordinate) would have an electric field that pointed up. The negative charge would pull it up, and the positive one would push it up. This eliminates (A) and (B). Finally, the negative charge is greater magnitude. Therefore, the fields could only be in balance somewhere closer to the positive charge. This eliminates (C). So the correct answer must be (D).

We think we've illustrated our point: Process of Elimination is the best way to approach the multiple-choice questions. Even when you don't know the answer right off the bat, you'll surely know that two or three of the answer choices are not correct. What then?

Aggressive Guessing

As mentioned earlier, you are scored only on the number of questions you get right, so we know guessing can't hurt you. But can it help you? It sure can. Let's say you guess on four questions; odds are you'll get one right. So you've already increased your score by one point. Now, let's add POE into the equation. If you can eliminate as many as two answer choices from each question, your chances of getting them right increase, and so does your overall score. Remember, don't leave any bubbles blank on test day!

General Advice

Answering 50 multiple-choice questions in 90 minutes can be challenging. Make sure to pace yourself accordingly and remember that you do not need to answer every question correctly to do well. Exploit the multiple-choice structure of this section. There are three wrong answers and only one correct one (and your odds are even better with the multi-select questions!), so even if you don't know exactly which one is the right answer, you can eliminate some that you know for sure are wrong. Then you can make an educated guess from among the answers that are left and greatly increase your odds of getting that question correct.

Problems with graphs and diagrams are usually the fastest to solve, and problems with an explanation for each answer usually take the longest to work through. Do not spend too much time on any one problem or you may not get to easier problems further into the test.

The practice exams in this book are written to give you an idea of the format of the test, the difficulty of the questions, and to allow you to practice pacing yourself. Take them under the same conditions that you will encounter during the real exam.

REFLECT

Respond to the following questions:

- How long will you spend on multiple-choice questions?
- How will you change your approach to multiple-choice questions?
- What is your multiple-choice guessing strategy?
- Will you seek further help, outside of this book (such as a teacher, tutor, or AP Students), on how to approach the questions that you will see on the AP Physics 1 Exam?

Stay Tuned for Updates

In 2022, the College Board announced that some topics in AP Physics 1 and AP Physics 2 (fluids, waves) will realign in the next year or two. At the time of this printing, the realignment seems to have been pushed off to the 2023–2024 school year. To learn more about the specific information, visit the College Board's website for breaking news.

Chapter 2
How to Approach Free-Response Questions

FREE-RESPONSE SECTION

On the free-response section, be sure to show the graders what you're thinking. Write clearly—that is *very* important—and show your steps. If you make a mistake in one part and carry an incorrect result to a later part of the question, you can still earn valuable points if your method is correct. However, the graders cannot give you credit for work they can't follow or can't read. And, where appropriate, be sure to include units in your final answers.

Free-Response Terms Defined

Key Words in the Free-Response Section

Be sure to review these key terms (see the bold words!) so you know what you can expect from the free-response questions on this exam.

The College Board is very clear in what they will ask of you in the Free-Response section. As we have mentioned before, on the College Board website, each AP course and exam has a section and the College Board posts a ton of useful content there: The Course and Exam Description (a very lengthy PDF), practice questions, and specific information about the exam.

Here is a list of task verbs that are commonly used by the College Board in the free-response questions:

Calculate: Perform mathematical steps to arrive at a final answer, including algebraic expressions, properly substituted numbers, and correct labeling of units and significant figures. Also phrased as "What is?"

Compare: Provide a description or explanation of similarities and/or differences.

Derive: Perform a series of mathematical steps using equations or laws to arrive at a final answer.

Describe: Provide the relevant characteristics of a specified topic. Determine: Make a decision or arrive at a conclusion after reasoning, observation, or applying mathematical routines (calculations).

Evaluate: Roughly calculate numerical quantities, values (greater than, equal to, less than), or signs (negative, positive) of quantities based on experimental evidence or provided data. When making estimations, showing steps in calculations are not required.

Explain: Provide information about how or why a relationship, pattern, position, situation, or outcome occurs, using evidence and/or reasoning to support or qualify a claim. Explain "how" typically requires analyzing the relationship, process, pattern, position, situation, or outcome; whereas, explain "why" typically requires analysis of motivations or reasons for the relationship, process, pattern, position, situation, or outcome.

Justify: Provide evidence to support, qualify, or defend a claim, and/or provide reasoning to explain how that evidence supports or qualifies the claim.

Label: Provide labels indicating unit, scale, and/or components in a diagram, graph, model, or representation.

Plot: Draw data points in a graph using a given scale or indicating the scale and units, demonstrating consistency between different types of representations.

Sketch/Draw: Create a diagram, graph, representation, or model that illustrates or explains relationships or phenomena, demonstrating consistency between different types of representations. Labels may or may not be required.

State/Indicate/Circle: Indicate or provide information about a specified topic, without elaboration or explanation. Also phrased as "What...?" or "Would...?" interrogatory questions.

Verify: Confirm that the conditions of a scientific definition, law, theorem, or test are met in order to explain why it applies in a given situation. Also, use empirical data, observations, tests, or experiments to prove, confirm, and/or justify a hypothesis.

CRACKING THE FREE-RESPONSE SECTION

Section II is worth 50 percent of your grade on the AP Physics 1 Exam. This section is composed of five free-response questions. You're given a total of 90 minutes for this section. It is recommended that you spend the first 60 minutes on the three short-answer questions, and the next 30 minutes on the experimental design and qualitative/quantitative translation questions.

Clearly Explain and Justify Your Answers

Remember that your answers to the free-response questions are graded by readers and not by computers. Communication is a very important part of AP Physics 1. Compose your answers in precise sentences. Just getting the correct numerical answer is not enough. You should be able to explain your reasoning behind the technique that you selected and communicate your answer in the context of the problem. Even if the question does not explicitly say so, always explain and justify every step of your answer, including the final answer. Do not expect the graders to read between the lines. Explain everything as though somebody with no knowledge of physics is going to read it. Be sure to present your solution in a systematic manner using solid logic and appropriate language. And remember, although you won't earn points for neatness, the graders can't give you a grade if they can't read and understand your solution!

Use Only the Space You Need

Do not try to fill up the space provided for each question. The space given is usually more than enough. The people who design the tests realize that some students write in big letters and some students make mistakes and need extra space for corrections. So, if you have a complete solution, don't worry about the extra space. Writing more will not earn you extra credit. In fact, many students tend to go overboard and shoot themselves in the foot by making a mistake after they've already written the right answer.

Read the Whole Question!

Some questions might have several subparts. Try to answer them all, and don't give up on the question if one part is giving you trouble. For example, if the answer to part (b) depends on the answer to part (a), but you think you got the answer to part (a) wrong, you should still go ahead and do part (b) using your answer to part (a) as required. Chances are that the grader will not mark you wrong twice, unless it is obvious from your answer that you should have discovered your mistake.

Use Common Sense

Don't Forget Units!
In your free-response questions, you'll be doing calculations right there by hand. Once you get to your final answer, DO NOT FORGET to include units or dimensions in it.

Always use your common sense in answering questions. For example, on one free-response question that asked students to compute the weight of a newborn baby on the Moon, some students answered 70 pounds. It should have been immediately obvious that the answer was probably off by a decimal point. A 70-pound baby would be a giant! This is an important mistake that should be easy to fix. Some mistakes may not be so obvious from the answer. However, the grader will consider simple, easily recognizable errors to be very important.

Think Like a Grader

When answering questions, try to think about what kind of answer the grader is expecting. Look at past free-response questions and grading rubrics on the College Board website. These examples will give you some idea of how the answers should be phrased. The graders are told to keep in mind that there are two aspects to the scoring of free-response answers: showing a comprehensive knowledge of physics and communicating that knowledge. Again, responses should be written as clearly as possible in complete sentences. You don't need to show all the steps of a calculation, but you must explain how you got your answer and why you chose the technique you used.

Think Before You Write

Abraham Lincoln once said that if he had eight hours to chop down a tree, he would spend six of them sharpening his axe. Be sure to spend some time thinking about what the question is, what answers are being asked for, what answers might make sense, and what your intuition is before starting to write. These questions aren't meant to trick you, so all the information you need is given. If you think you don't have the right information, you may have misunderstood the question. In some calculations, it is easy to get confused, so think about whether your answers make sense in terms of what the question is asking. If you have some idea of what the answer should look like before starting to write, then you will avoid getting sidetracked and wasting time on dead-ends.

REFLECT

Respond to the following questions:

- How much time will you spend on the short free-response questions? What about the long free-response questions?
- What will you do before you begin writing your free-response answers?
- Will you seek further help, outside of this book (such as a teacher, tutor, or AP Students), on how to approach the questions that you will see on the AP Physics 1 Exam?

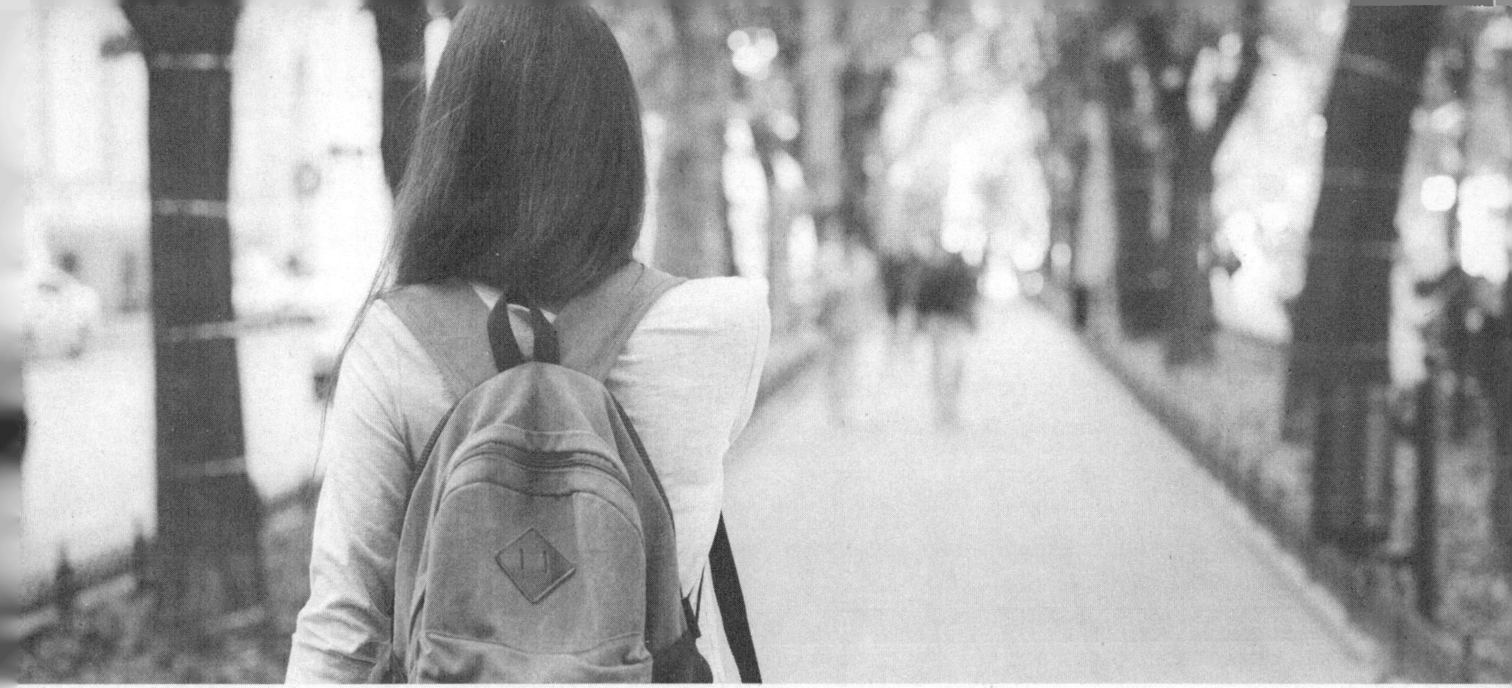

Part V Content Review for the AP Physics 1 Exam

Chapter 3
Vectors

Vectors will show up all over the place in our study of physics. Some physical quantities that are represented as vectors are displacement, velocity, acceleration, force, momentum, and electric and magnetic fields. Since vectors play such a recurring role, it's important to become comfortable working with them; the purpose of this chapter is to provide you with a command of the fundamental vector algebra we'll use in subsequent chapters. For now, we'll restrict our study to two-dimensional vectors (that is, ones that lie flat in a plane).

A Note About Vectors

In the College Board's AP Physics 1 Course and Exam Description (available as a PDF on their website), Vectors are not included in the 10 Unit structure (7 Units which will be tested and included on the test now). We cover them here, with their very own chapter, because they are foundational to physics and we think they are important to know and review. So let's dive in!

DEFINITION

A **vector** is a quantity that involves both magnitude and direction. A quantity that does not involve direction is a **scalar**. For example, the quantity *55 miles per hour* is a scalar, while the quantity *55 miles per hour to the north* is a vector. Other examples of scalars include mass, work, energy, power, temperature, and electric charge.

Vectors can be denoted in several ways, including:

$$\mathbf{A},\ \boldsymbol{A},\ \overline{A},\ \vec{\mathrm{A}}$$

In textbooks, you'll usually see one of the first two, but when it's handwritten, you'll see one of the last two.

Displacement (which is net distance traveled including direction) is an example of a vector:

$$\underbrace{\mathbf{A}}_{\text{displacement}} = \underbrace{\text{4 miles}}_{\text{magnitude}}\ \underbrace{\text{to the north}}_{\text{direction}} \qquad \mathbf{B} = \underbrace{\text{3 miles}}_{\text{magnitude}}\ \underbrace{\text{to the east}}_{\text{direction}}$$

Vectors obey the Commutative Law for Addition, which states:

Commutative Law for Addition

A + B = B + A

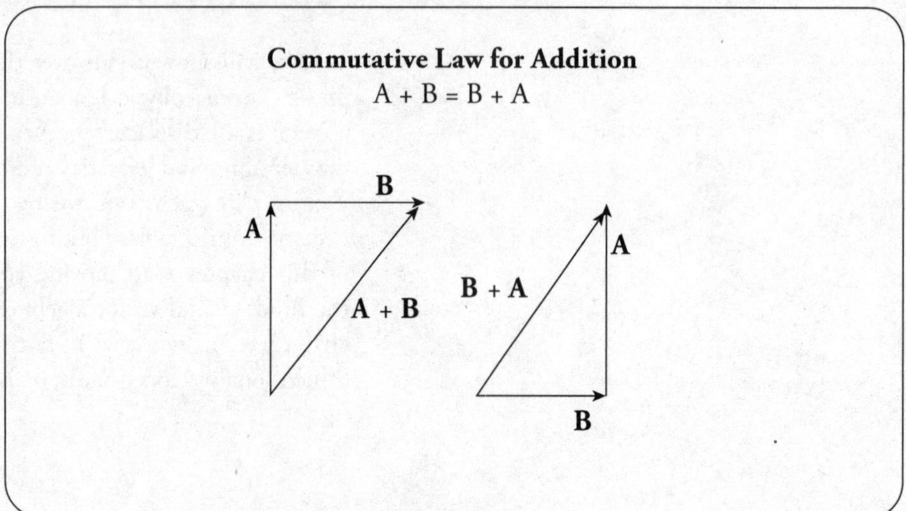

The vector sum **A** + **B** means the vector **A** followed by **B**. The vector sum of **B** + **A** means the vector **B** followed by **A**, and the result is an identical vector to **A** + **B**. Vectors are always added tail to end to find their sum, so **A** + **B** or **B** + **A**—both are examples of tail to end.

Two vectors are equal when they have the same magnitude and the same direction.

Example 1 Add the following two vectors:

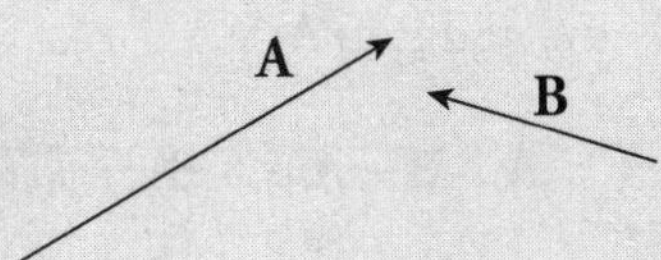

Solution. Place the tail of **B** at the tip of **A** and connect them:

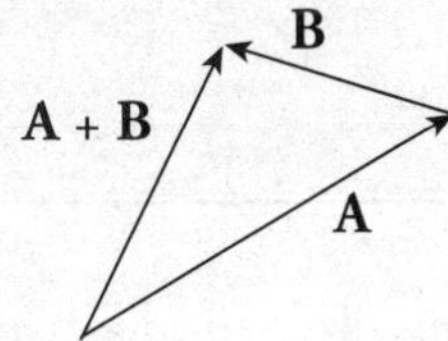

Also Worth Noting!

Scalar multiplication indicates a change in magnitude by the numerical multiple, and direction if there is a negative sign only (negative sign indicates opposite direction). An example of this is that of a car traveling east at a velocity of 100 km per hour. When multiplied by $k = -2$, the car's new velocity is 200 km per hour WEST.

SCALAR MULTIPLICATION

A vector can be multiplied by a scalar (that is, by a number), and the result is a vector. If the original vector is **A** and the scalar is *k*, then the scalar multiple *k***A** is as follows:

Scalar Multiplication

$$\text{magnitude of } k\mathbf{A} = |k| \times (\text{magnitude of } \mathbf{A})$$

$$\text{direction of } k\mathbf{A} = \begin{cases} \text{the same as } \mathbf{A} \text{ if } k \text{ is positive} \\ \text{the opposite of } \mathbf{A} \text{ if } k \text{ is negative} \end{cases}$$

Example 2 Sketch the scalar multiples 2**A**, $\frac{1}{2}$**A**, –**A**, and –3**A** of the vector **A**:

Solution.

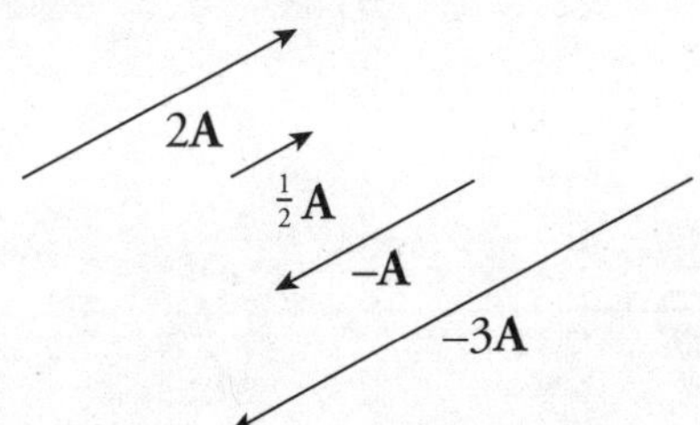

VECTOR SUBTRACTION

To subtract one vector from another, for example, to get **A** – **B**, simply form the vector –**B**, which is the scalar multiple (–1)**B**, and add it to **A**:

$$\mathbf{A} - \mathbf{B} = \mathbf{A} + (-\mathbf{B})$$

Tail To End!
By adding the negative of **B**, we are allowing the process to follow the tail to end convention that we discussed earlier.

Example 3 For the two vectors **A** and **B**, find the vector **A** – **B**.

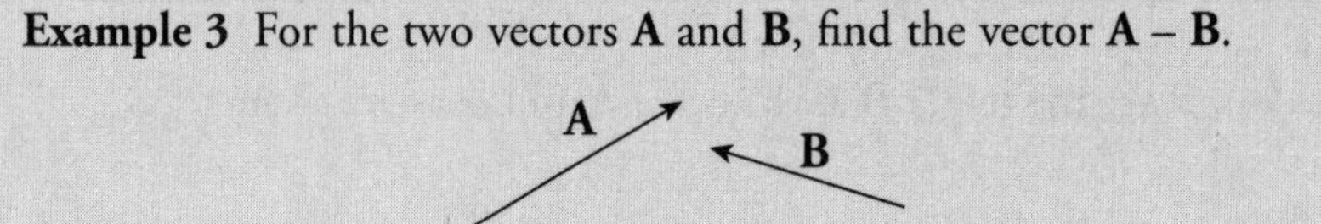

Solution. Flip **B** around (thereby forming –**B)** and add that vector to **A**:

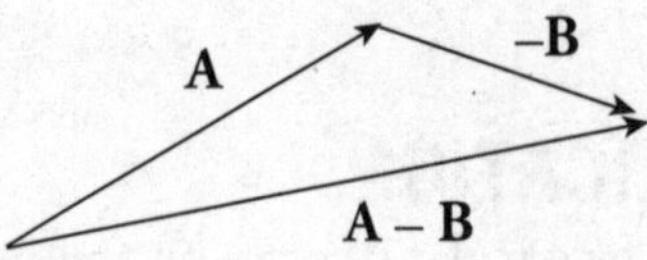

A Note About Direction
Make sure to pay attention to direction if you are not using a coordinate system. If you set a vector point to the right as positive, then you must set a vector pointing to the left as negative.

It is important to know that vector subtraction is **not** commutative; you must perform the subtraction in the order stated in the problem.

TWO-DIMENSIONAL VECTORS

Two-dimensional vectors are vectors that lie flat in a plane and can be written as the sum of a horizontal vector and a vertical vector. For example, in the following diagram, the vector **A** is equal to the horizontal vector **B** plus the vertical vector **C**:

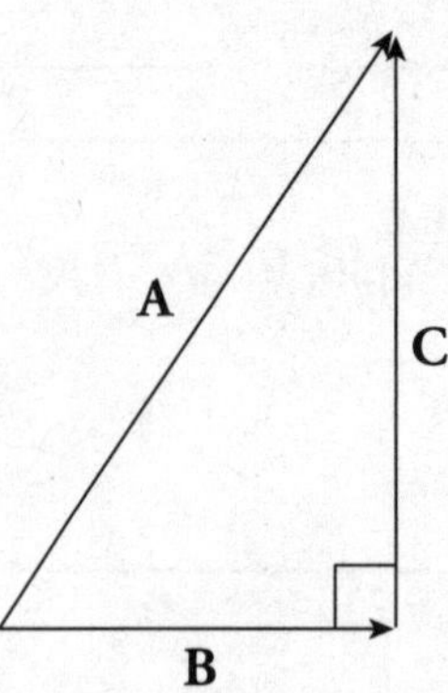

The horizontal vector is always considered a scalar multiple of what's called the **horizontal basis vector**, **i**, and the vertical vector is a scalar multiple of the **vertical basis vector**, **j**. Both of these special vectors have a magnitude of 1, and for this reason, they're called **unit vectors.** Unit vectors are often represented by placing a hat (caret) over the vector; for example, the unit **vectors i** and **j** are sometimes denoted **î** and **ĵ**.

Coordinate System
Think of *i* as your *x*-coordinate system and *j* as your *y*-coordinate system. *i* is just a unit vector that points in the positive *x* direction, and *j* is just a unit vector that points in the positive *y* direction.

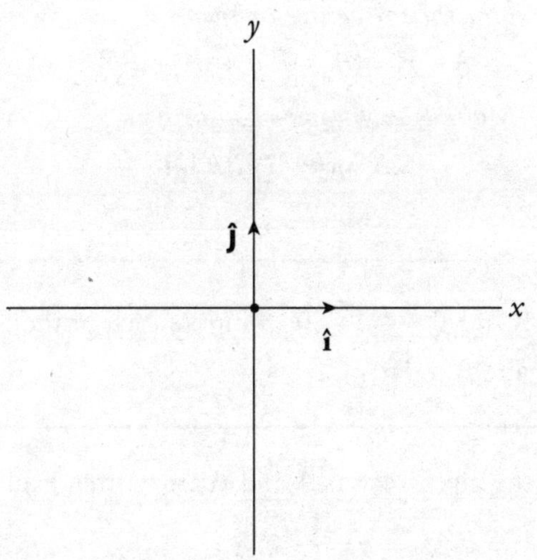

For instance, the vector **A** in the figure below is the sum of the horizontal vector **B** = 3**î** and the vertical vector **C** = 4**ĵ**.

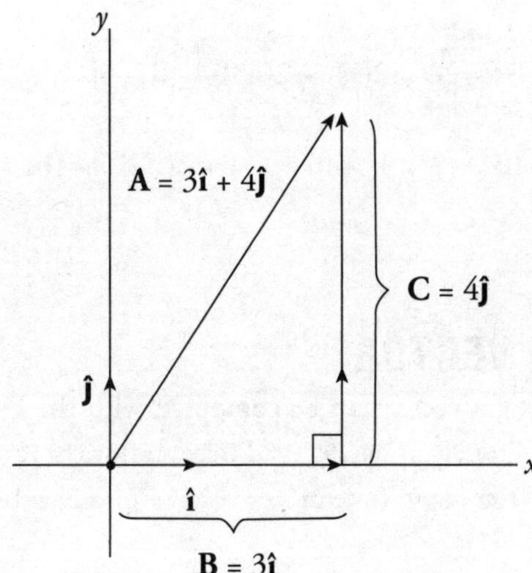

The vectors **B** and **C** are called the **vector components** of **A**, and the scalar multiples of **î** and **ĵ** which give **A**—in this case, 3 and 4—are called the **scalar components** of **A**. So vector **A** can be written as the sum $A_x\hat{\mathbf{i}} + A_y\hat{\mathbf{j}}$, where A_x and A_y are the scalar components of **A**. The component A_x is called the **horizontal** scalar component of **A**, and A_y is called the **vertical** scalar component of **A**. In general, any vector in a plane can be described in this manner.

VECTOR OPERATIONS USING COMPONENTS

The use of components makes the vector operations of addition, subtraction, and scalar multiplication pretty straightforward:

Vector addition: *Add the respective components.*

$$\mathbf{A} + \mathbf{B} = (A_x + B_x)\hat{\mathbf{i}} + (A_y + B_y)\hat{\mathbf{j}}$$

Vector subtraction: *Subtract the respective components.*

$$\mathbf{A} - \mathbf{B} = (A_x - B_x)\hat{\mathbf{i}} + (A_y - B_y)\hat{\mathbf{j}}$$

Scalar multiplication: *Multiply each component by k.*

$$k\mathbf{A} = (kA_x)\hat{\mathbf{i}} + (kA_y)\hat{\mathbf{j}}$$

Example 4 If $\mathbf{A} = 2\hat{\mathbf{i}} - 3\hat{\mathbf{j}}$ and $\mathbf{B} = -4\hat{\mathbf{i}} + 2\hat{\mathbf{j}}$, compute each of the following vectors: $\mathbf{A} + \mathbf{B}$, $\mathbf{A} - \mathbf{B}$, $2\mathbf{A}$, and $\mathbf{A} + 3\mathbf{B}$.

Solution. It's very helpful that the given vectors **A** and **B** are written explicitly in terms of the standard basis vectors $\hat{\mathbf{i}}$ and $\hat{\mathbf{j}}$:

$$\mathbf{A} + \mathbf{B} = (2 - 4)\hat{\mathbf{i}} + (-3 + 2)\hat{\mathbf{j}} = -2\hat{\mathbf{i}} - \hat{\mathbf{j}}$$

$$\mathbf{A} - \mathbf{B} = [2 - (-4)]\hat{\mathbf{i}} + (-3 - 2)\hat{\mathbf{j}} = 6\hat{\mathbf{i}} - 5\hat{\mathbf{j}}$$

$$2\mathbf{A} = 2(2)\hat{\mathbf{i}} + 2(-3)\hat{\mathbf{j}} = 4\hat{\mathbf{i}} - 6\hat{\mathbf{j}}$$

$$\mathbf{A} + 3\mathbf{B} = [2 + 3(-4)]\hat{\mathbf{i}} + [-3 + 3(2)]\hat{\mathbf{j}} = -10\hat{\mathbf{i}} + 3\hat{\mathbf{j}}$$

MAGNITUDE OF A VECTOR

The magnitude of a vector can be computed with the Pythagorean Theorem. The magnitude of vector **A** can be denoted in several ways: A or $|\mathbf{A}|$ or $\|\mathbf{A}\|$. In terms of its components, the magnitude of $\mathbf{A} = A_x\hat{\mathbf{i}} + A_y\hat{\mathbf{j}}$ is given by the equation

$$A = \sqrt{(A_x)^2 + (A_y)^2}$$

which is the formula for the length of the hypotenuse of a right triangle with sides of lengths A_x and A_y.

This is merely an interpretation of the Pythagorean Theorem. Make sure to brush up on geometry and trigonometry.

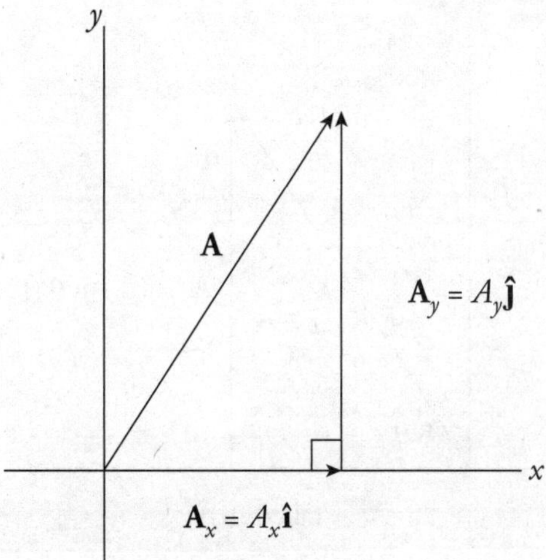

DIRECTION OF A VECTOR

The direction of a vector can be specified by the angle it makes with the positive *x*-axis. You can sketch the vector and use its components (and an inverse trig function) to determine the angle. For example, if θ denotes the angle that the vector $\mathbf{A} = 3\hat{\mathbf{i}} + 4\hat{\mathbf{j}}$ makes with the +*x*-axis, then $\tan\theta = 4/3$, so $\theta = \tan^{-1}(4/3) = 53.1°$.

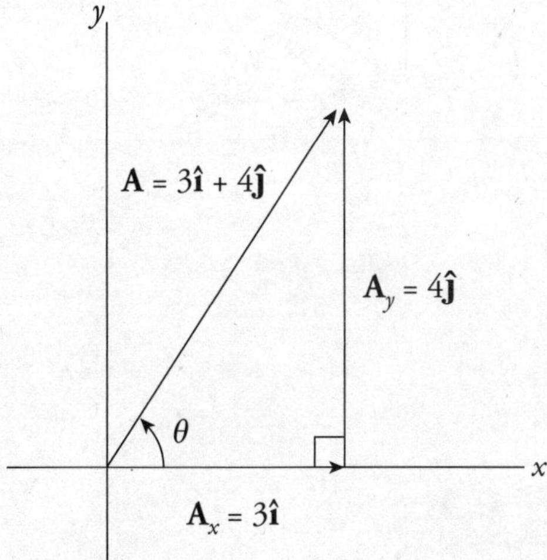

In general, the axis that θ is made to is known as the adjacent axis. The adjacent component is always going to get the cos θ. For example, if **A** makes the angle θ with the +*x*-axis, then its *x*- and *y*-components are $A\cos\theta$ and $A\sin\theta$, respectively (where A is the magnitude of **A**).

$$\mathbf{A} = \underbrace{(A\cos\theta)}_{A_x}\hat{\mathbf{i}} + \underbrace{(A\sin\theta)}_{A_y}\hat{\mathbf{j}}$$

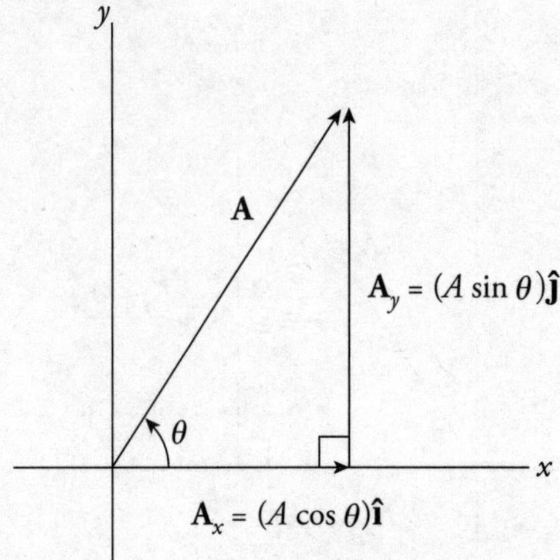

In general, any vector in the plane can be written in terms of two perpendicular component vectors. For example, vector **W** (shown below) is the sum of two component vectors whose magnitudes are $W \cos \theta$ and $W \sin \theta$:

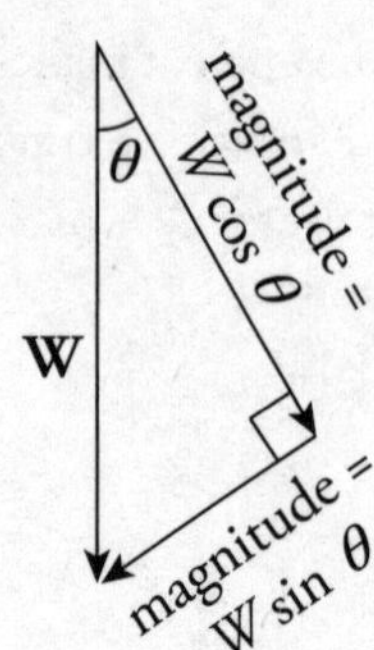

CHAPTER 3 KEY TERMS

vector
scalar
vector addition
vector subtraction
two-dimensional vector
horizontal basis vector
vertical basis vector
unit vector
vector components
scalar components

Chapter 3 Review Questions

Answers and explanations can be found in Chapter 11.

Section I: Multiple Choice

1. Two vectors, A and B, have the same magnitude, m, but vector A points north whereas vector B points east. What is the sum, A + B?

   (A) m, northeast

   (B) $m\sqrt{2}$, northeast

   (C) $m\sqrt{2}$, northwest

   (D) $2m$, northwest

2. If $\mathbf{F}_1 = -20\hat{\mathbf{j}}$, $\mathbf{F}_2 = -10\hat{\mathbf{i}}$, and $\mathbf{F}_3 = 5\hat{\mathbf{i}} + 10\hat{\mathbf{j}}$, what is the sum $\mathbf{F}_1 + \mathbf{F}_2 + \mathbf{F}_3$?

   (A) $-15\hat{\mathbf{i}} + 10\hat{\mathbf{j}}$

   (B) $-5\hat{\mathbf{i}} - 10\hat{\mathbf{j}}$

   (C) $5\hat{\mathbf{i}}$

   (D) $5\hat{\mathbf{i}} - 10\hat{\mathbf{j}}$

3. Both the x- and y-components of a vector are doubled. Which of the following describes what happens to the resulting vector?

   (A) Magnitude increases by a factor of $\sqrt{2}$
   (B) Magnitude increases by a factor of $\sqrt{2}$, and the direction changes.
   (C) Magnitude increases by a factor of 2.
   (D) Magnitude increases by a factor of 2, and the direction changes.

4. If vectors v_0 = 15 m/s north and v_f = 5 m/s south, what is $v_f - v_0$?

   (A) 10 m/s north

   (B) 10 m/s south

   (C) 20 m/s north

   (D) 20 m/s south

5. The magnitude of vector **A** is 10. Which of the following could be the components of **A**?

   (A) $A_x = 5, A_y = 5$
   (B) $A_x = 6, A_y = 8$
   (C) $A_x = 7, A_y = 9$
   (D) $A_x = 10, A_y = 10$

6. If the vector $\mathbf{A} = \hat{\mathbf{i}} - 2\hat{\mathbf{j}}$ and the vector $\mathbf{B} = 4\hat{\mathbf{i}} - 5\hat{\mathbf{j}}$, what angle does **A** + **B** form with the x-axis?

   (A) $\theta = \tan^{-1}\frac{3}{5}$

   (B) $\theta = \tan^{-1}\frac{7}{5}$

   (C) $\theta = \sin^{-1}\frac{3}{5}$

   (D) $\theta = \sin^{-1}\frac{5}{7}$

7. An object travels along the vector $d_1 = 4\text{m}\,\hat{\mathbf{i}} + 5\text{m}\,\hat{\mathbf{j}}$ and then along the vector $d_2 = 2\text{m}\,\hat{\mathbf{i}} - 3\text{m}\,\hat{\mathbf{j}}$. How far is the object from where it started?

   (A) 6.3 m
   (B) 8 m
   (C) 10 m
   (D) 14 m

8. If **A** = ↗ and **B** = ↗, which of the following best represents the direction of **A** – **B**?

   (A) →

   (B) ←

   (C) ↗

   (D) ↗

9. The x-component of vector **A** is 42, and the angle it makes with the positive x-direction is 50°. What is the y-component of vector **A**?

   (A) –65.3
   (B) –50.1
   (C) 50.1
   (D) 65.3

10. If two non-zero vectors are added together, and the resultant vector is zero, what must be true of the two vectors?

    (A) They have equal magnitude and are pointed in the same direction.
    (B) They have equal magnitude and are pointed in opposite directions.
    (C) They have different magnitudes and are pointed in opposite directions.
    (D) It is not possible for the sum of two non-zero vectors to be zero.

Section II: Free Response

1. Let the vectors, **A**, **B**, and **C** be defined by: $\mathbf{A} = 3\hat{\imath} + 6\hat{\jmath}$, $\mathbf{B} = -\hat{\imath} + 4\hat{\jmath}$, and $\mathbf{C} = 5\hat{\imath} - 2\hat{\jmath}$.

    (A) What is the magnitude of vector **A**?

    (B) Sketch the vector subtraction problem **B** – **C** and find the components of the resultant vector.

    (C) Find the components of **A** + 2**B**.

    (D) Express **A** – **B** – **C** as a magnitude and angle relative to the horizontal.

2. An ant walks 20 cm due north, 30 cm due east, and then 14 cm northeast.

    (A) Assuming that each portion of the ant's journey is a vector, sketch the ant's path.

    (B) How far has the ant traveled from its original position?

    (C) If the ant does not want to travel farther than 80 cm from its original position, how march farther north could it walk?

3. Consider an airplane taking off at an angle of 10° relative to the horizontal. After 2 s, the airplane has traveled a total of 140 m through the air.

    (A) Express the position, **p**, of the airplane compared with the point of liftoff as a vector using components and the $\hat{\imath}$ and $\hat{\jmath}$ basis vectors.

    (B) If the airplane had been taking off in a headwind, the position of the airplane would instead have been $\mathbf{p}' = \mathbf{p} + \mathbf{w}$, in which the impact of the wind, **w**, is the vector $\mathbf{w} = -30\hat{\imath}$. Draw a sketch to show the relationship between the vectors **p**′, **p**, and **w**.

    (C) What angle would the plane make with the horizontal as a result of the headwind after 2 s?

4. As a boat travels in a river, its velocity, **v**, is determined by the current, **c**, and its relative velocity compared to the water, **r**. For instance, at one point in the boat's journey, these vectors are related in the following way:

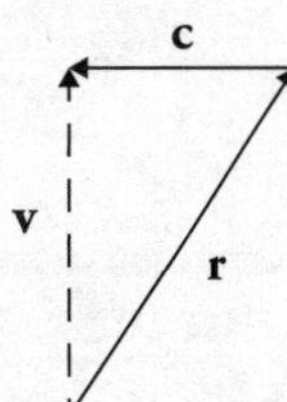

    (A) Write a vector equation that shows how **v** is related to **c** and **r**.

    (B) If the angle between **v** and **c** is 90°, as shown, write an equation for the magnitude of **v** as a function of **c** and **r**.

    (C) At a different point in the boat's journey, if the current is flowing due east at 5 m/s and the boat wants to travel due south at 10 m/s, what should the captain set its relative velocity to be (magnitude and direction)?

Summary

- Vectors are quantities that have both magnitude and direction. Many important physical quantities such as forces and velocities are vector quantities.
- Vectors can be represented graphically with an arrow, numerically with a magnitude and direction, or numerically with components.
- Vectors can be added (or subtracted) graphically by drawing the first vector and then starting the tail of the second vector at the end of the first (remembering to flip the direction of the second vector for subtraction).
- Vectors can be added (or subtracted) numerically by adding (or subtracting) individual components.
- Multiplying a vector by a scalar can change the length of the vector or flip the direction by 180° if the scalar is negative.
- If the magnitude, A, and angle relative to the horizontal, θ, are known, the x- and y-components can be calculated using:

$$A_x = A\cos\theta$$
$$A_y = A\sin\theta$$

- If the components, A_x and A_y are known, the magnitude, A, and angle relative to the horizontal, θ, can be calculated using:

$$A = \sqrt{(A_x)^2 + (A_y)^2}$$
$$\theta = \tan^{-1}\frac{A_y}{A_x}$$

Chapter 4
Kinematics

And yet it moves.

—Galileo Galilei

Galileo, the "father of modern physics," stated a distinction between the cause of motion and the description of motion. Kinematics is that modern description of motion that answers questions such as:

How far does this object travel?

How fast and in what direction does it move?

At what rate does its speed change?

Kinematics are the mathematical tools for describing motion in terms of displacement, velocity, and acceleration.

POSITION

An object's position is its location in a certain space. Since it is difficult to describe an object's location, in mathematics, we typically use a coordinate system to show where an object is located. Using a coordinate system with an origin also helps to determine positive and negative positions. Typically, we set the object in question as the origin and relate its surroundings to the object.

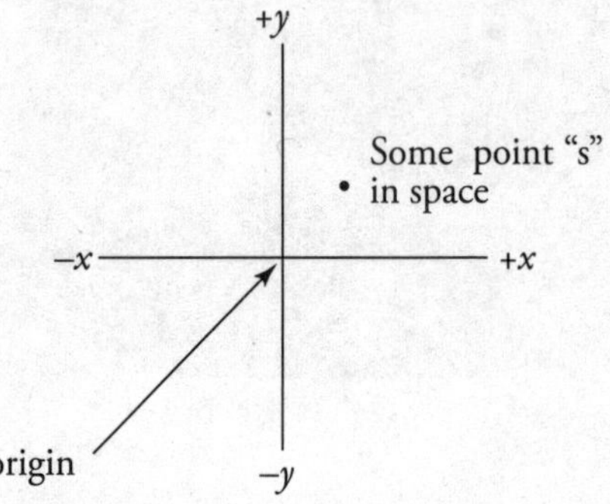

DISPLACEMENT

Pay Attention to What the Question Is Asking
Asking for displacement and total distance will give different answers to a problem.

Displacement is an object's change in position. It's the vector that points from the object's initial position to its final position, regardless of the path actually taken. Since displacement means *change in position*, it is generically denoted Δs, where Δ denotes *change in* and s means spatial location. The displacement is only the distance from an object's final position and initial position. The **total distance** takes into account the total path taken (meaning if an object went backward and then forward, those distances are included in the total distance). Since a distance is being measured, the SI unit for displacement is the meter $[\Delta s]$ = m.

Example 1 A rock is thrown straight upward from the edge of a 30 m cliff, rising 10 m, and then falling all the way down to the base of the cliff. Find the rock's displacement.

Solution. Displacement only refers to the object's initial position and final position, not the details of its journey. Since the rock started on the edge of the cliff and ended up on the ground 30 m below, its displacement is 30 m, downward. Its distance traveled is 50 m; 10 m on the way up and 40 m on the way down.

Example 2 An infant crawls 5 m east, then 3 m north, and then 1 m east. Find the magnitude of the infant's displacement.

Solution. Although the infant crawled a *total* distance of 5 + 3 + 1 = 9 m, this is not the displacement, which is merely the *net* distance traveled.

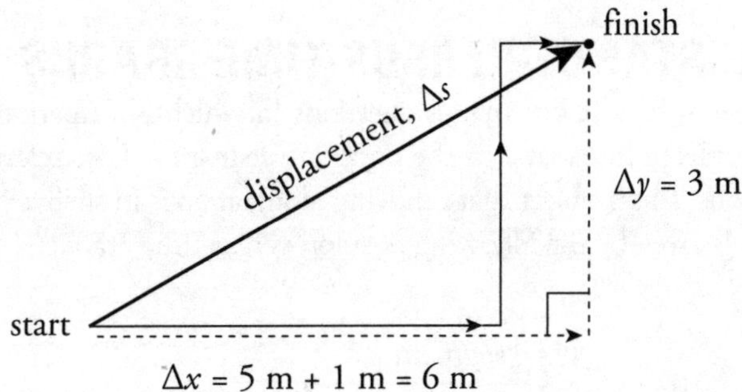

Math on the AP Physics 1 Exam

You must be proficient in basic trigonometry to do well on this test, so revisit that subject matter now if you need a refresher.

Using the Pythagorean Theorem, we can calculate that the magnitude of the displacement is

$$\Delta s = \sqrt{(\Delta x)^2 + (\Delta(y))^2} = \sqrt{(6\text{ m})^2 + (3\text{ m})^2} = \sqrt{45\text{ m}^2} = 6.7\text{ m}$$

Example 3 In a track-and-field event, an athlete runs exactly once around an oval track, a total distance of 500 m. Find the runner's displacement for the race.

Solution. If the runner returns to the same position from which she left, then her displacement is zero.

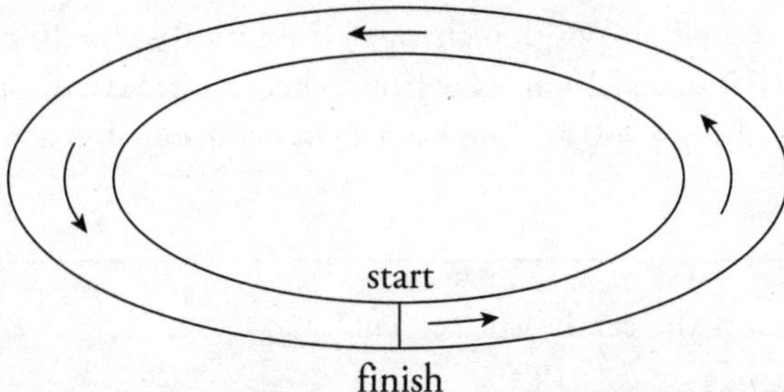

The *total* distance covered is 500 m, but the net distance—the displacement—is 0.

A Note About Notation

Δs is a more general term that works in any direction in space. The term x or "$\Delta x = x_f - x_i$" has a specific meaning that is defined in the x direction. However, to be consistent with AP notation and to avoid confusion between spatial location and speed, from this point on we will use x in our development of the concepts of speed, velocity, and acceleration. x is just a reference and can refer to any direction in terms of the object's motion (for example, straight down or straight up).

LOOKING AT DISTANCE-VERSUS-TIME GRAPHS

You should also be able to handle kinematics questions in which information is given graphically. One of the popular graphs in kinematics is the position-versus-time (also referred to as a *p*-versus-*t*) graph. For example, consider an object that's moving along an axis in such a way that its position x as a function of time t is given by the following position-versus-time graph:

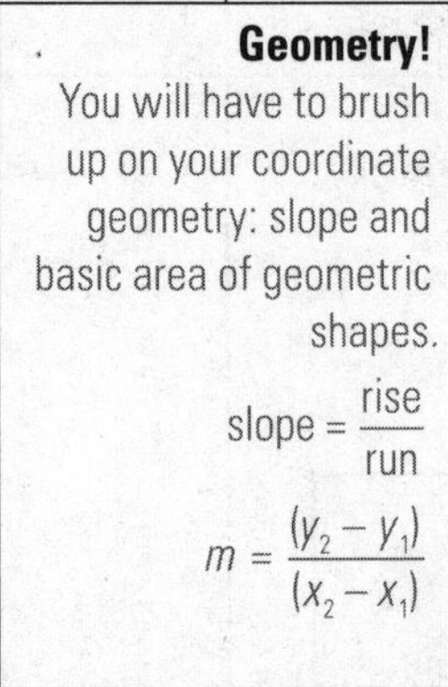

Geometry!
You will have to brush up on your coordinate geometry: slope and basic area of geometric shapes.

$$\text{slope} = \frac{\text{rise}}{\text{run}}$$

$$m = \frac{(y_2 - y_1)}{(x_2 - x_1)}$$

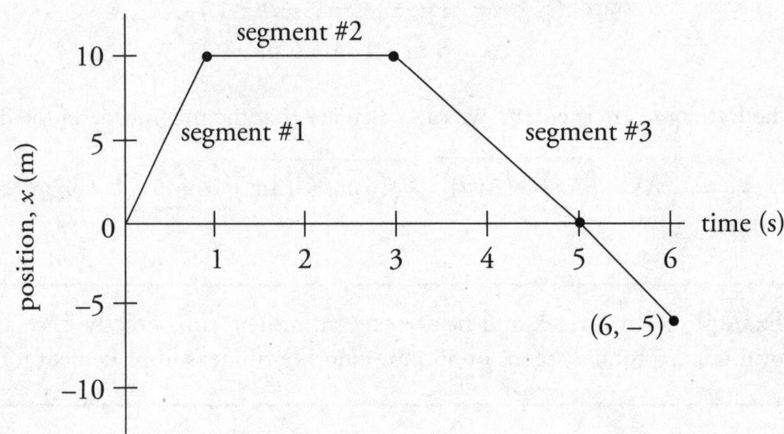

What does this graph tell us? It says that at time $t = 0$, the object was at position $x = 0$. Then, in the first second, its position changed from $x = 0$ to $x = 10$ m. Then, at time $t = 1$ s to 3 s, it stopped (that is, it stayed 10 m away from wherever it started). From $t = 3$ s to $t = 6$ s, it reversed direction, reaching $x = 0$ at time $t = 5$ s, and continued, reaching position $x = -5$ m at time $t = 6$ s.

Example 4 What is the total displacement?

Solution. Displacement is just the initial position subtracted from the final position. In this case, that's $-5 - 0 = -5$ m (or 5 meters to the left, or negative direction).

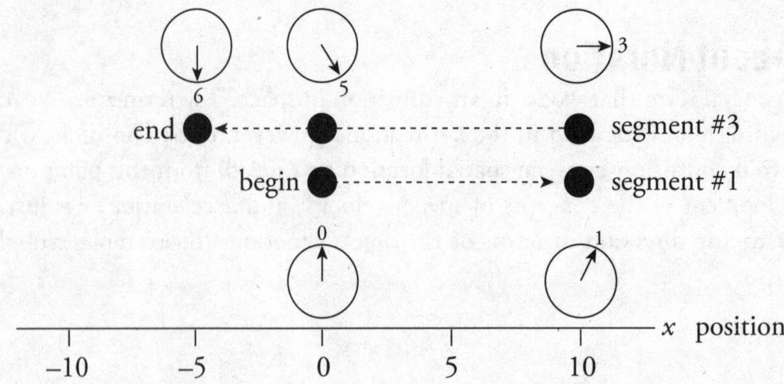

What is the total distance traveled?

10 m	+	15 m	=	25 meters
segment #1		segment #3		

SPEED AND VELOCITY

When we're in a moving car, the speedometer tells us how fast we're going; it gives us our speed. But what does it mean to have a speed of say, 10 m/s? It means that we're covering a distance of 10 meters every second. By definition, **average speed** is the ratio of the total distance traveled to the time required to cover that distance:

$$\text{average speed} = \frac{\text{total distance}}{\text{time}}$$

The car's speedometer doesn't care what direction the car is moving. You could be driving north, south, east, west, whatever; the speedometer would make no distinction. *55 miles per hour, north* and *55 miles per hour, east* register the same on the speedometer: 55 miles per hour. Remember that speed is scalar, so it does not take into account the direction.

However, we will also need to include *direction* in our descriptions of motion. We just learned about displacement, which takes both distance (net distance) and direction into account. The single concept that embodies both speed and direction is called **velocity**, and the definition of average velocity is

$$\text{average velocity} = \frac{\text{displacement}}{\text{time}}$$

$$\bar{\mathbf{v}} = \frac{\Delta \mathbf{x}}{\Delta t}$$

units = meter/seconds

Note that the bar over the **v** means *average*. Because $\Delta\mathbf{x}$ is a vector, $\bar{\mathbf{v}}$ is also a vector, and because Δt is a *positive* scalar, the direction of $\bar{\mathbf{v}}$ is the same as the direction of $\Delta\mathbf{x}$. The magnitude of the velocity vector is called the object's **speed**, and is expressed in units of meters per second (m/s).

Note the distinction between speed and velocity. *Velocity is speed plus direction.* An important note: the magnitude of the velocity is the speed. The magnitude of the average velocity is not called the average speed (as we will see in these next two examples).

Example 5 Assume that the runner in Example 3 completes the race in 1 minute and 18 seconds. Find her average speed and the magnitude of her average velocity.

Solution. *Average speed is total distance divided by elapsed time.* Since the length of the track is 500 m, the runner's average speed was (500 m)/(78 s) = 6.4 m/s. However, since her displacement was zero, her average velocity was zero also: $\bar{v} = \Delta x/\Delta t = (0 \text{ m})/(78 \text{ s}) = 0 \text{ m/s}$.

Example 6 Is it possible to move with constant speed but not constant velocity? Is it possible to move with constant velocity but not constant speed?

Solution. The answer to the first question is *yes*. For example, if you set your car's cruise control at 55 miles per hour but turn the steering wheel to follow a curved section of road, then the direction of your velocity changes (which means your velocity is not constant), even though your speed doesn't change.

The answer to the second question is *no*. Velocity means speed and direction; if the velocity is constant, then that means both speed and direction are constant. If speed were to change, then the velocity vector's magnitude would change.

ACCELERATION

When you step on the gas pedal in your car, the car's speed increases; step on the brake and the car's speed decreases. Turn the wheel, and the car's direction of motion changes. In all of these cases, the velocity changes. To describe this change in velocity, we need a new term: **acceleration**. In the same way that velocity measures the rate of change of an object's position, acceleration measures the rate of change of an object's velocity. An object's average acceleration is defined as follows:

$$\text{average acceleration} = \frac{\text{change in velocity}}{\text{time}}$$

$$\bar{a} = \frac{\Delta v}{\Delta t}$$

$$\text{units} = \text{meters/second}^2 = \text{m/s}^2$$

Remember: Velocity Is a Vector
Acceleration is a change in velocity, meaning a change in magnitude or direction.

Note that an object can accelerate even if its speed doesn't change. (Again, it's a matter of not allowing the everyday usage of the word *accelerate* to interfere with its technical, physics usage.) This is because acceleration depends on Δv, and the velocity vector v changes if (1) speed changes, or (2) direction changes, or (3) both speed and direction change. For instance, a car traveling around a circular racetrack is constantly accelerating even if the car's *speed* is constant, because the direction of the car's velocity vector is constantly changing.

Example 7 A car is traveling in a straight line along a highway at a constant speed of 80 miles per hour for 10 seconds. Find its acceleration.

Solution. Since the car is traveling at a constant velocity, its acceleration is zero. If there's no change in velocity, then there's no acceleration.

Example 8 A car is traveling along a straight highway at a speed of 20 m/s. The driver steps on the gas pedal and, 3 seconds later, the car's speed is 32 m/s. Find its average acceleration.

Solution. Assuming that the direction of the velocity doesn't change, it's simply a matter of dividing the change in velocity, $v_f - v_i$, 32 m/s – 20 m/s = 12 m/s, by the time interval during which the change occurred: $\bar{a} = \Delta v/\Delta t = (12 \text{ m/s})/(3 \text{ s}) = 4 \text{ m/s}^2$.

Example 9 Spotting a police car ahead, the driver of the car in Example 8 slows from 32 m/s to 20 m/s in 2 seconds. Find the car's average acceleration.

Solution. Dividing the change in velocity, 20 m/s – 32 m/s = –12 m/s, by the time interval during which the change occurred, 2 s, gives us $\bar{\mathbf{a}} = \Delta\mathbf{v}/\Delta t = (-12 \text{ m/s})/(2 \text{ s}) = -6 \text{ m/s}^2$. The negative sign here means that the direction of the acceleration is opposite the direction of the velocity, which describes slowing down.

Let's next consider an object moving along a straight axis in such a way that its velocity, *v*, as a function of time, *t*, is given by the following velocity-versus-time graph.

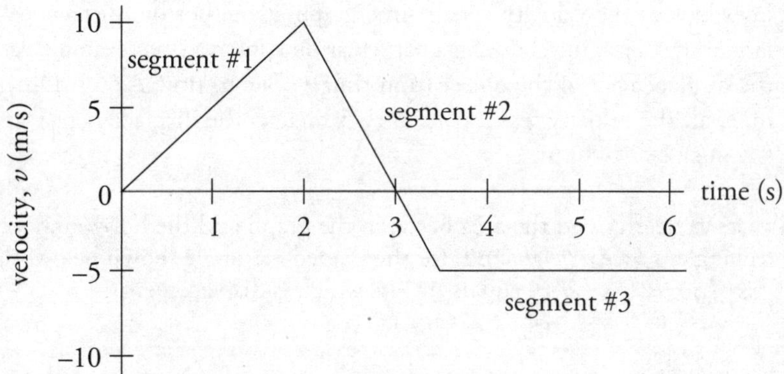

What does this graph tell us? It says that, at time $t = 0$, the object's velocity was $v = 0$. Over the first two seconds, its velocity increased steadily to 10 m/s. At time $t = 2$ s, the velocity then began to decrease (eventually becoming $v = 0$, at time $t = 3$ s). The velocity then became negative after $t = 3$ s, reaching $v = -5$ m/s at time $t = 3.5$ s. From $t = 3.5$ s on, the velocity remained a steady –5 m/s.

Segment #1: What can we ask about this motion? First, the fact that the velocity changed from $t = 0$ to $t = 2$ s tells us that the object accelerated. The acceleration during this time was

$$a = \frac{\Delta v}{\Delta t} = \frac{(10 - 0)\text{ m/s}}{(2 - 0)\text{ s}} = 5\text{ m/s}^2$$

Note, however, that the ratio that defines the acceleration, $\Delta v/\Delta t$, also defines the slope of the velocity-versus-time graph. Therefore,

> The slope of a velocity-versus-time graph gives the acceleration.

Segment #2: What was the acceleration from time $t = 2$ s to time $t = 3.5$ s? The slope of the line segment joining the point $(t, v) = (2\text{ s}, 10\text{ m/s})$ to the point $(t, v) = (3.5\text{ s}, -5\text{ m/s})$ is

$$a = \frac{\Delta v}{\Delta t} = \frac{(-5 - 10)\text{ m/s}}{(3.5 - 2)\text{ s}} = -10\text{ m/s}^2$$

Between time $t = 2$ s to time $t = 3.5$ s, the object experienced a negative acceleration. Between time $t = 2$ s to time $t = 3$ s, the object's velocity was slowed down till it reached 0 m/s. Between time $t = 3$ s to time $t = 3.5$ s, the object began increasing its velocity in the opposite direction.

When the slope of a velocity-versus-time graph is zero, the acceleration is zero.

Segment #3: After time $t = 3.5$ s, the slope of the graph is zero, meaning the object experienced zero acceleration. This, however, does not mean the object did not move since its velocity was constant and in the negative direction.

Let's look at the velocity versus time graph at the bottom of page 89 again: how far did the object travel during a particular time interval? For example, let's figure out the displacement of the object from time $t = 4$ s to time $t = 6$ s. During this time interval, the velocity was a constant –5 m/s, so the displacement was $\Delta x = v\Delta t = (-5\text{ m/s})(2\text{ s}) = -10$ m.

Geometrically, we've determined the area between the graph and the horizontal axis. After all, the area of a rectangle is *base* × *height* and, for the shaded rectangle shown below, the *base* is Δt, and the *height* is v. So, *base* × *height* equals $\Delta t \times v$, which is displacement.

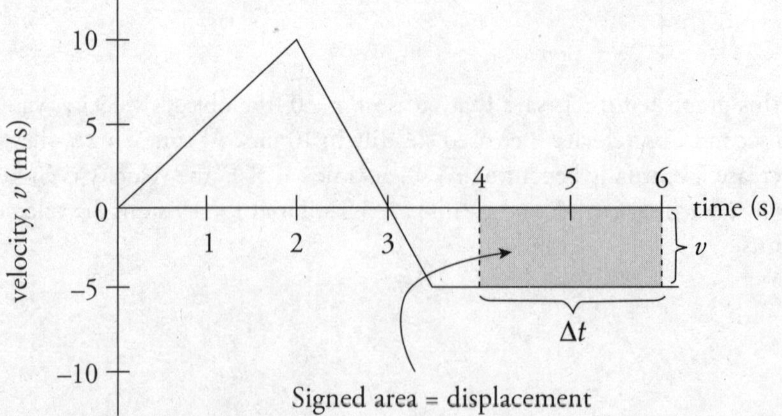

We say *signed area* because regions below the horizontal axis are negative quantities (since the object's velocity is negative, its displacement is negative). Thus, by counting areas above the horizontal axis as positive and areas below the horizontal axis as negative, we can make the following claim:

Given a velocity-versus-time graph, the area between the graph and the t-axis equals the object's displacement.

Not All Areas Are the Same!
Although the concept is strange, there are positive areas and negative areas. Think of area as a scalar quantity that can be negative and positive.

What is the object's displacement from time $t = 0$ to $t = 3$ s? Using the fact that displacement is the area bounded by the velocity graph, we figure out the area of the triangle shown below:

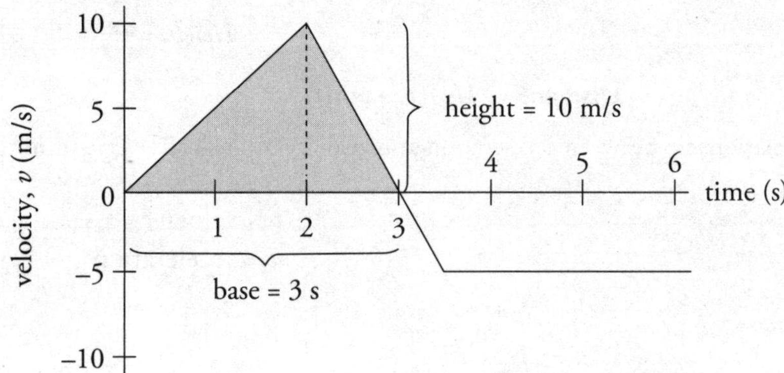

Since the area of a triangle is $\left(\frac{1}{2}\right)$ × base × height, we find that $\Delta x = \frac{1}{2}(3\text{ s})(10\text{ m/s}) = 15\text{ m}$.

Example 10 How far did the object travel from time 0 s to time t, given an initial velocity of v_o and a final velocity of v_f in the graph below?

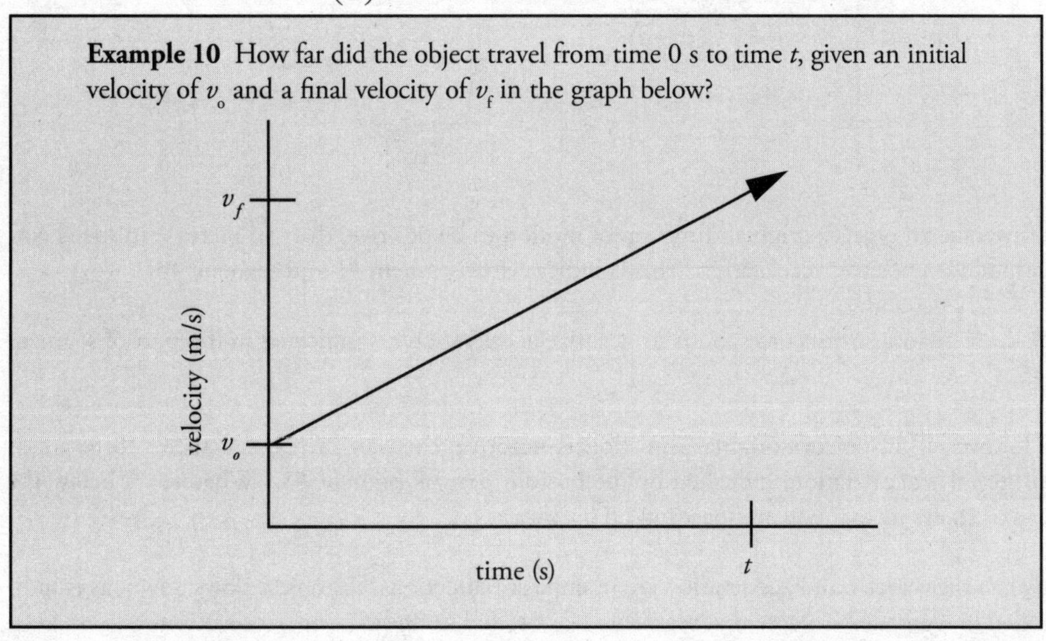

Solution.

Coming Soon
This is how the Big Five #1 equation is derived. More information about all that on the next page.

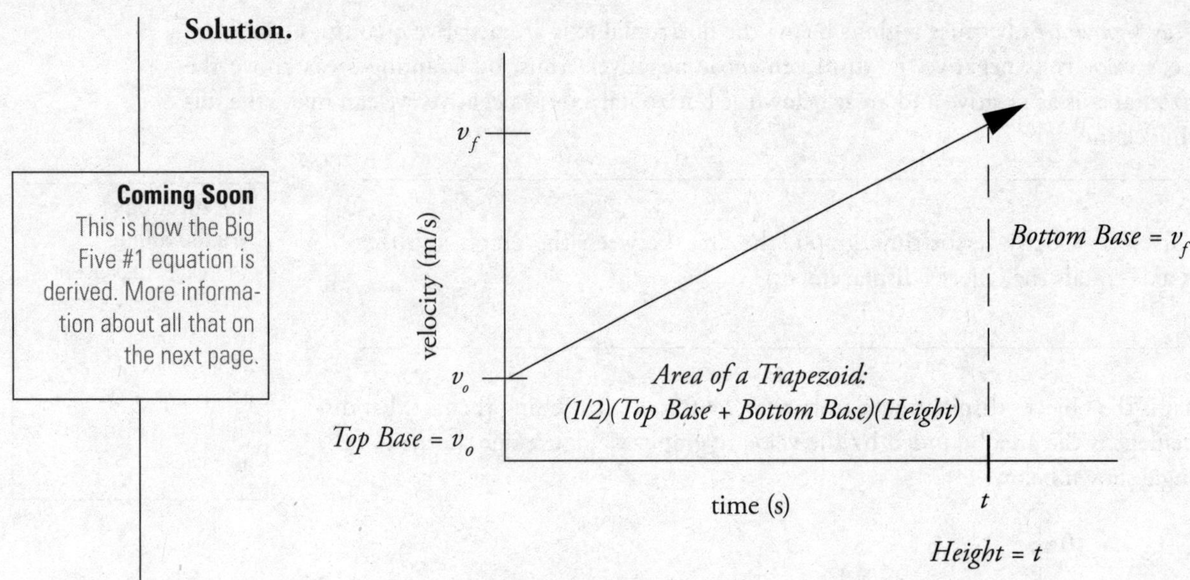

Distance = (1/2)(v_o + v_f)(t)

Let's consider the relationship between acceleration and velocity in a velocity-versus-time graph.

v = velocity
a = acceleration

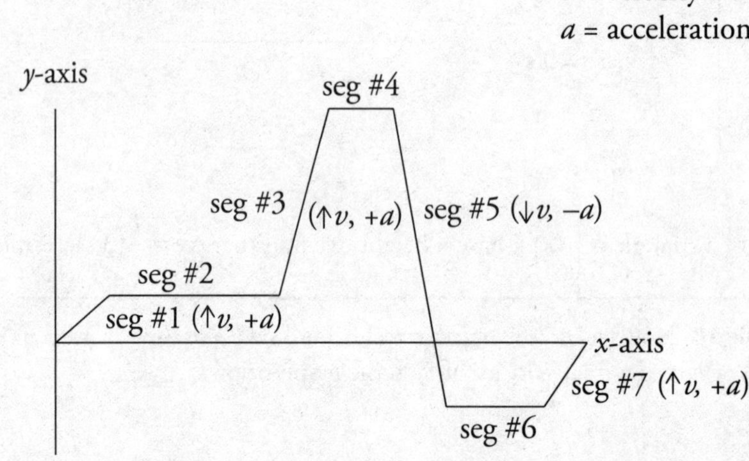

If we take an object's original direction of motion to be positive, then an increase in speed corresponds to positive acceleration. This is indicative of segment #1 and segment #3.

A decrease in velocity corresponds to negative acceleration (as indicated in top part of segment #5).

However, if an object's original direction is negative, then an increase in speed corresponds to negative acceleration, indicated by the bottom part of segment #5. Whatever is below the x-axis shows us that you are speeding up backward.

When the velocity and acceleration are in opposite directions, the object slows down, as is indicated in segment #7. Note that segments #2, #4, and #6 indicate no acceleration.

UNIFORMLY ACCELERATED MOTION AND THE BIG FIVE

The simplest type of motion to analyze is motion in which the acceleration is *constant,* also called **uniform accelerated motion.**

Another restriction that will make our analysis easier is to consider only motion that takes place along a straight line. In these cases, there are only two possible directions of motion. One is positive, and the opposite direction is negative. Most of the quantities we've been dealing with—displacement, velocity, and acceleration—are vectors, which means that they include both a magnitude and a direction. With straight-line motion, direction can be specified simply by attaching a + or – sign to the magnitude of the quantity. Therefore, although we will often abandon the use of bold letters to denote the vector quantities of displacement, velocity, and acceleration, the fact that these quantities include direction will still be indicated by a positive or negative sign.

In the real world, truly uniform acceleration hardly occurs. This can be attributed to multiple factors. However, for the purposes of the AP Physics 1 Exam, everything is treated in ideal situations unless otherwise noted.

Let's review the quantities we've seen so far. The fundamental quantities are position (x), velocity (v), and acceleration (a). Acceleration is a change in velocity, from an initial velocity (v_i or v_0) to a final velocity (v_f or simply v—with no subscript). And, finally, the motion takes place during some elapsed time interval, Δt. Also, if we agree to start our clocks at time $t_i = 0$, then $\Delta t = t_f - t_i = t - 0 = t$, so we can just write t instead of Δt in the first four equations. This simplification in notation makes these equations a little easier to write down. Therefore, we have five kinematics quantities: Δx, v_0, v, a, and Δt.

These five quantities are related by a group of five equations that we call the *Big Five.* They work in cases where acceleration is uniform, which are the cases we're considering.

When doing problems, writing down what is given and what you are solving for, can help determine which Big Five equation to use.

		Variable that's missing
Big Five #1:	$\Delta x = \frac{1}{2}(v_0 + v)t$	a
Big Five #2:	$v = v_0 + at$	x
Big Five #3:	$x = x_0 + v_0 t + \frac{1}{2}at^2$	v
Big Five #4:	$x = x_0 + vt - \frac{1}{2}at^2$	v_0
Big Five #5:	$v^2 = v_0^2 + 2a(x - x_0)$	t

Big Five #1 is the definition of velocity (this is the area under a velocity-versus-time graph, which will be covered a little later in this chapter). Big Five #2 is the definition of acceleration (this is the slope at any given moment of a velocity-versus-time graph).

Big Five #1 and #3 are simply the definitions of $\overline{v}$ and $\overline{a}$ written in different forms. The other Big Five equations can be derived from these two definitions.

Equations #1, #2, and #5 are the important three that can be used to solve any problem (equations #3 and #4 are derivations). However, it is advisable to memorize all five equations to speed up problem-solving during the test.

Example 11 An object with an initial velocity of 4 m/s moves along a straight axis under constant acceleration. Three seconds later, its velocity is 14 m/s. How far did it travel during this time?

Solution. We're given v_0, Δt, and v, and we're asked for x. So a is missing; it isn't given and it isn't asked for, so we use Big Five #1:

$$x = \overline{v}t = \frac{1}{2}(v_0 + v)t = \frac{1}{2}(4 \text{ m/s} + 14 \text{ m/s})(3 \text{ s}) = 27 \text{ m}$$

Structured Example Method to Solve Problems

Given:

v(initial) = 4 m/s
v(final) = 14 m/s
t = 3 s

Missing:

acceleration (a)

Unknown (What we are solving for):

distance (x)

Equation:

Big Five #1

It's okay to leave off the units in the middle of the calculation as long as you remember to include them in your final answer. **Leaving units off your final answer will cost you points on the AP Exam.**

Example 12 A car that's initially traveling at 10 m/s accelerates uniformly for 4 seconds at a rate of 2 m/s^2, in a straight line. How far does the car travel during this time?

Solution. We're given v_0, t, and a, and we're asked for x. So, v is missing; it isn't given, and it isn't asked for. Therefore, we use Big Five #3:

$$x = x_0 + v_0\Delta t + \frac{1}{2}a(\Delta t)^2 = (10 \text{ m/s})(4\text{s}) + \frac{1}{2}(2 \text{ m/s}^2)(4\text{ s})^2 = 56 \text{ m}$$

Example 13 A rock is dropped off a cliff that's 80 m high. If it strikes the ground with an impact velocity of 40 m/s, what acceleration did it experience during its descent?

Solution. If something is dropped, then that means it has no initial velocity: $v_0 = 0$. So, we're given v_0, Δx, and v, and we're asked for a. Since t is missing, we use Big Five #5:

$$v^2 = v_0^2 + 2a(x - x_0) \Rightarrow v^2 = 2a(x - x_0) \quad (\text{since } v_0 = 0)$$

$$a = \frac{v^2}{2(x - x_0)} = \frac{(40 \text{ m/s})^2}{2(80 \text{ m})} = 10 \text{ m/s}^2$$

Note that since a has the same sign as $(x - x_0)$, the acceleration vector points in the same direction as the displacement vector. This makes sense here, since the object moves downward, and the acceleration it experiences is due to gravity, which also points downward.

Initial Velocity?
Some problems do not give initial velocity in numerical terms. If you see the statement "starting from rest" or "dropped," you can assume the initial velocity is zero.

ADDITIONAL KINEMATIC GRAPHICAL ASPECTS

Not all graphs have nice straight lines as shown so far. Straight line segments represent constant slopes and therefore constant velocities, or accelerations, depending on the type of graph. What happens as an object changes its velocity in a position-versus-time graph? The "lines" become "curves." Let's look at a typical question that might be asked about such a curve.

Example 14

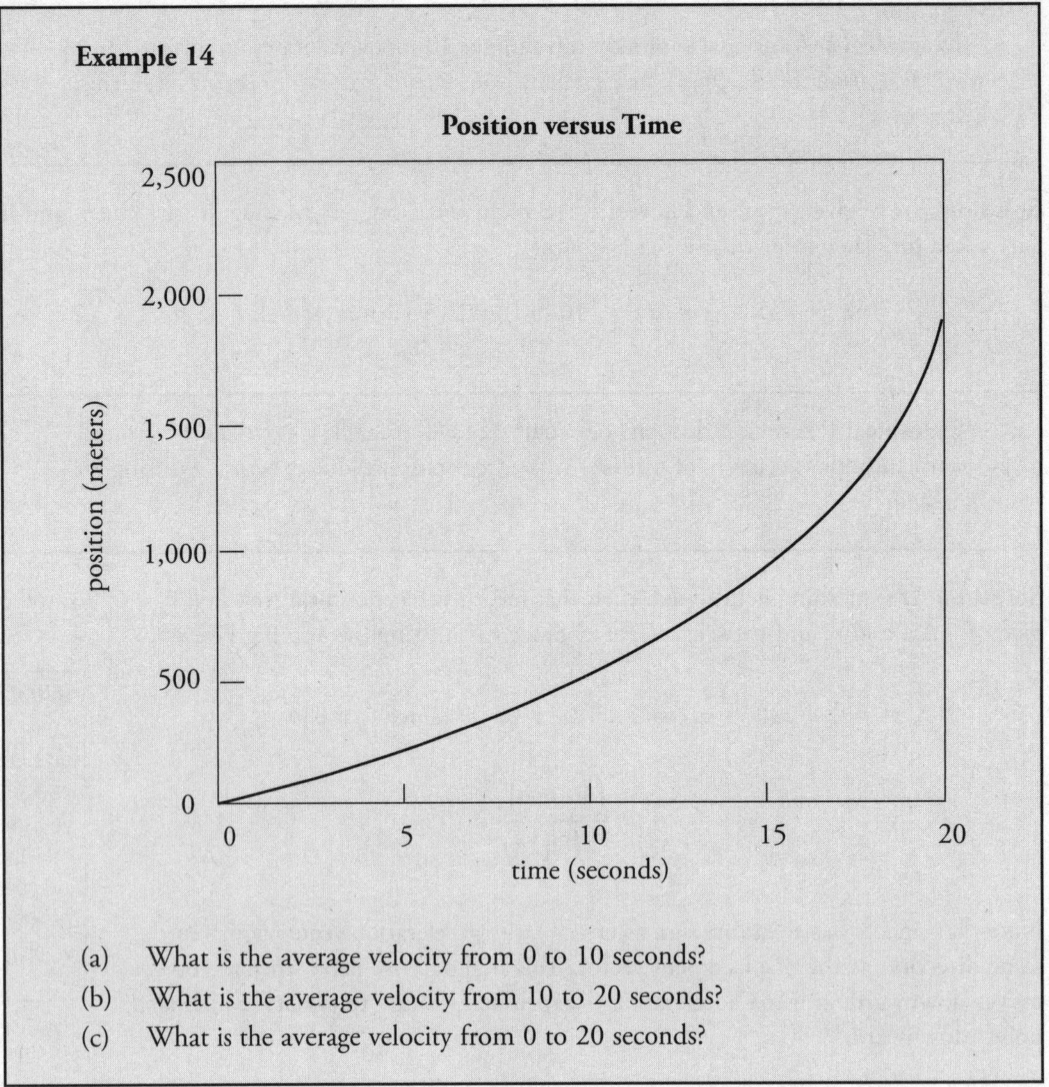

(a) What is the average velocity from 0 to 10 seconds?
(b) What is the average velocity from 10 to 20 seconds?
(c) What is the average velocity from 0 to 20 seconds?

Solution. This is familiar territory. To find the average velocity, use

(a) $v_{avg} = \frac{\Delta x}{\Delta t} \Rightarrow \frac{(500-0)}{(10-0)} \Rightarrow 50 \text{ m/s}$

(b) $v_{avg} = \frac{\Delta x}{\Delta t} \Rightarrow \frac{(2{,}000-500)}{(20-10)} \Rightarrow 150 \text{ m/s}$

(c) $v_{avg} = \frac{\Delta x}{\Delta t} \Rightarrow \frac{(2{,}000-0)}{(20-0)} \Rightarrow 100 \text{ m/s}$

The instantaneous velocity is the velocity at a given moment in time. When you drive in a car and look down at the speedometer, you see the magnitude of your instantaneous velocity at that time.

What Is a Tangent?
A line that touches a curved line at only one point.

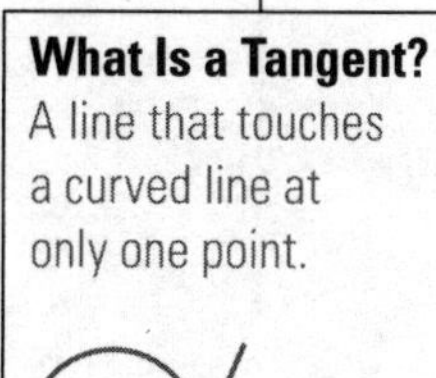

To find the instantaneous velocity at 10 seconds, we will need to approximate. The velocity from 9–10 seconds is close to the velocity at 10 seconds, but it is still a bit too slow. The velocity from 10–11 seconds is also close to the velocity at 10 seconds, but it is still a bit too fast. You can find the middle ground between these two ideas, or the slope of the line that connects the point before and the point after 10 seconds. This is very close to the instantaneous velocity. A true tangent line touches the curve at only one point, but this line is close enough for our purposes.

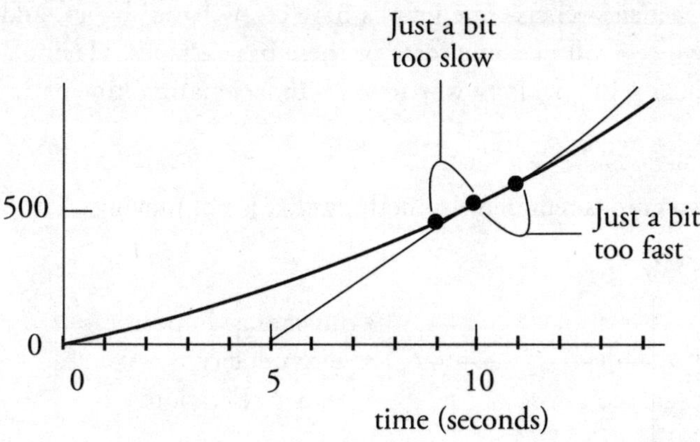

Example 15 Using the graph below, what is the instantaneous velocity at 10 seconds?

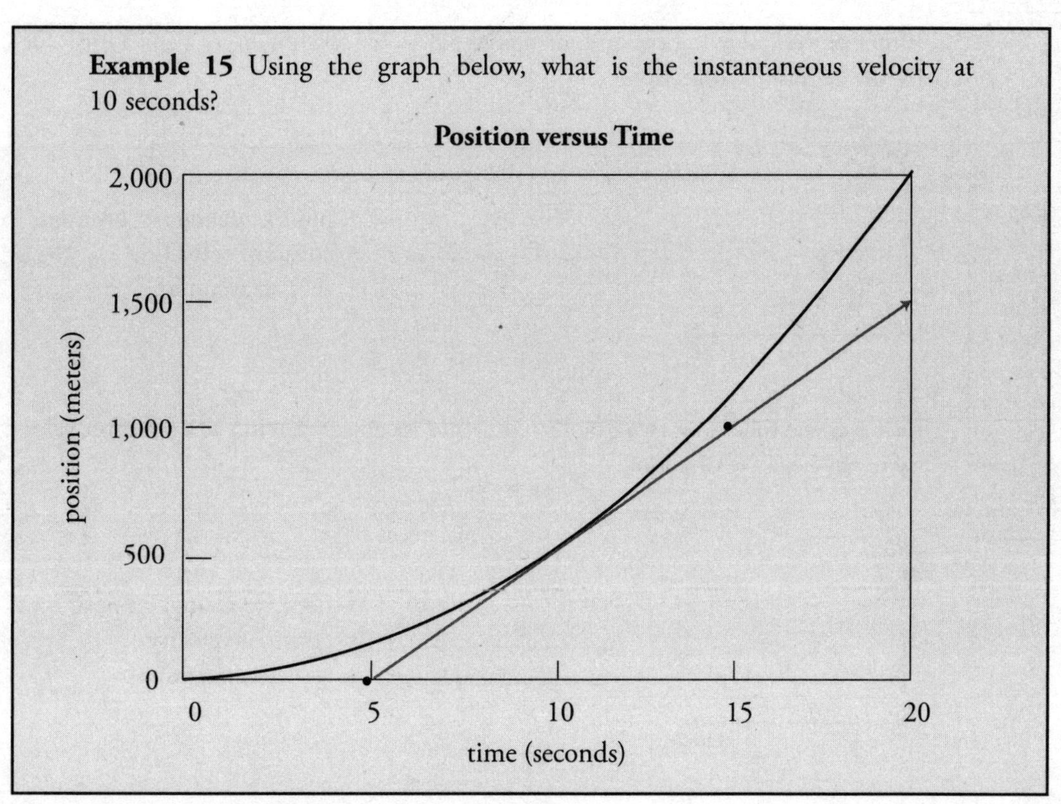

Solution. Draw a tangent line. Find the slope of the tangent by picking any two points on the tangent line. It usually helps if the points are kind of far apart, and it also helps if points can be found at "easy spots" such as (5, 0) and (15, 1,000) and not (6.27, 113) and (14.7, 983):

$$v_{ins} = v_{\tan} = \frac{\Delta x}{\Delta t} \Rightarrow \frac{(1{,}000-0)}{(15-5)} \Rightarrow 100 \text{ m/s}$$

QUALITATIVE GRAPHING

No Calculus
You may have seen the relationships of displacement, velocity, and acceleration described using the words "derivative" or "integral." That's calculus stuff that you'll learn one day (or may have learned already), but you don't need any of it for this test.

Beyond all the math, you are at a clear advantage when you start to recognize that position-versus-time and velocity-versus-time graphs have a few basic shapes, and that all the graphs you will see will be some form of these basic shapes. Having a feel for these building blocks will go a long way toward understanding kinematics graphs in physics.

Either of the following two graphs represents something that is not moving.

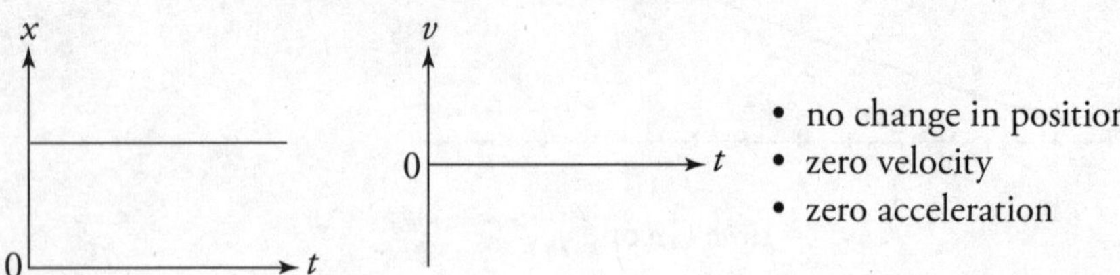

Either of the following two graphs represents an object moving at a constant velocity in the positive direction.

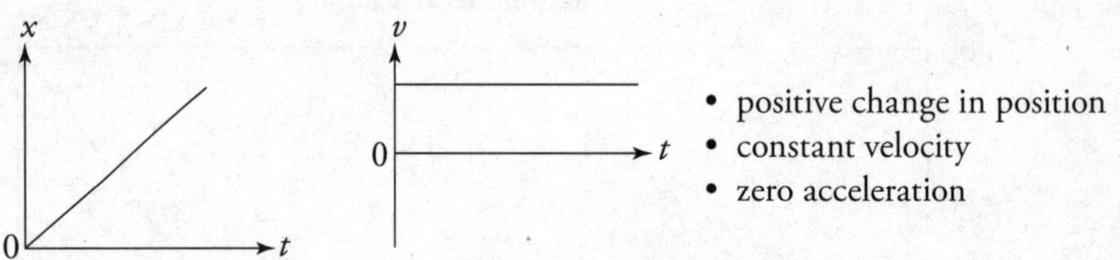

Know These Graphs
Familiarize yourself with these graphs so that you can quickly look at any graph and have a sense of what's going on as far as change in position, velocity, and acceleration immediately.

Either of the following two graphs represents an object moving at a constant velocity in the negative direction.

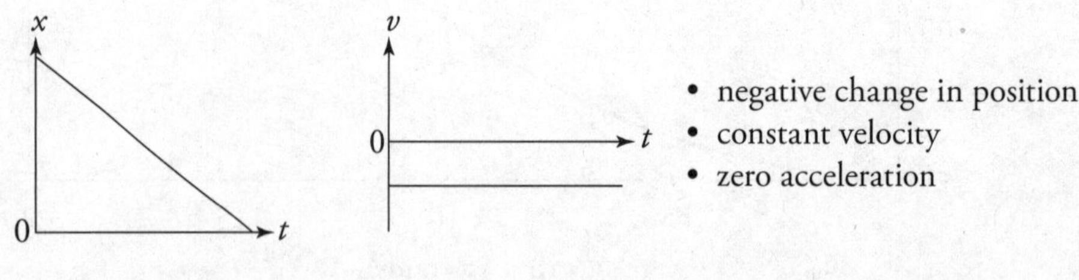

Either of the following two graphs represents an object speeding up in the positive direction.

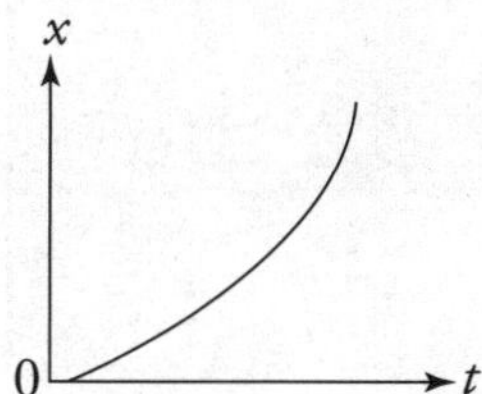

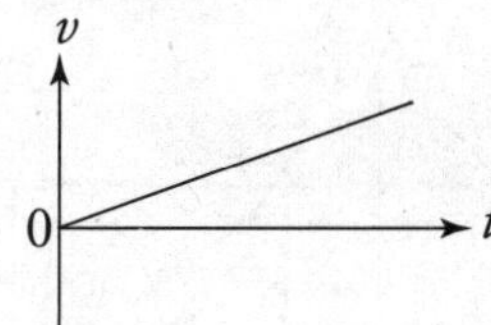

- positive change in position
- increasing velocity
- positive acceleration

Either of the following two graphs represents an object slowing down in the positive direction.

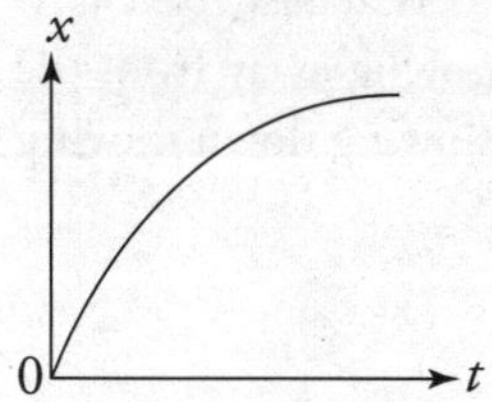

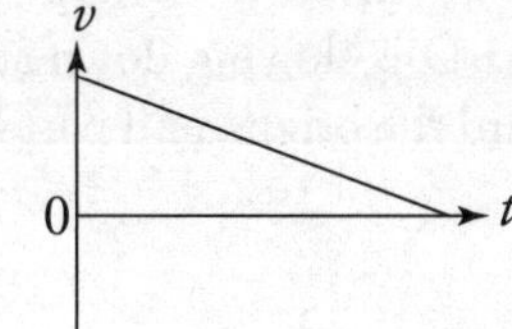

- positive change in position
- decreasing positive velocity
- negative acceleration

Either of the following two graphs represents an object slowing down in the negative direction.

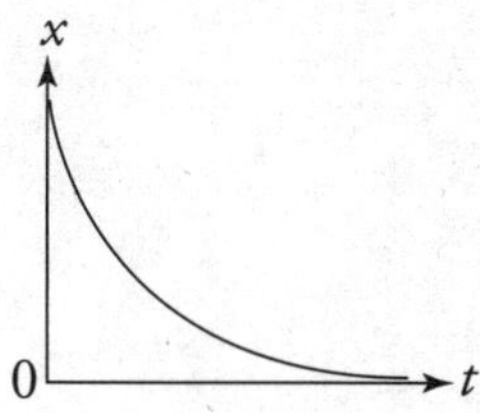

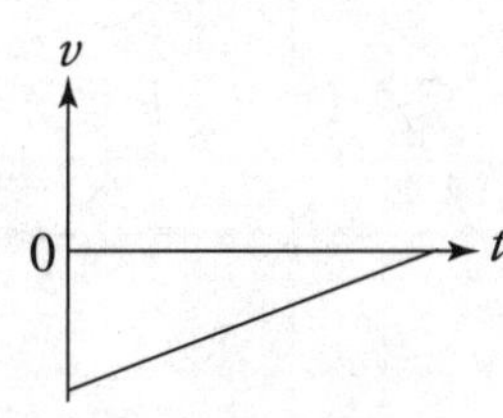

- negative change in position
- decreasing magnitude of negative velocity
- positive acceleration

Either of the following two graphs represents an object speeding up in the negative direction.

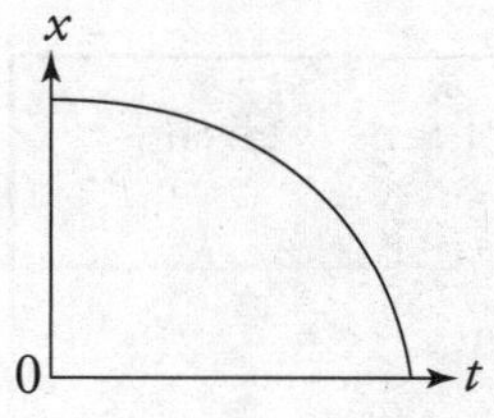

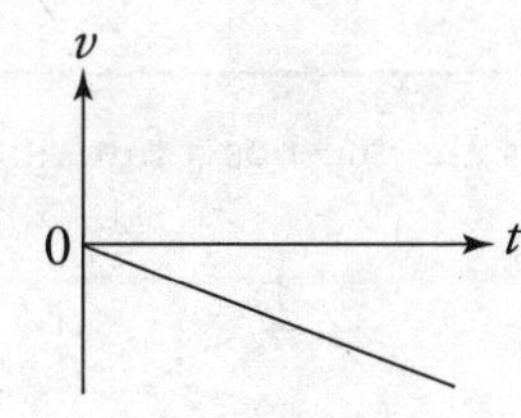

- negative change in position
- increasing magnitude of negative velocity
- negative acceleration

Example 16 Below is a position-versus-time graph. Describe in words the motion of the object and sketch the corresponding velocity-versus-time graph.

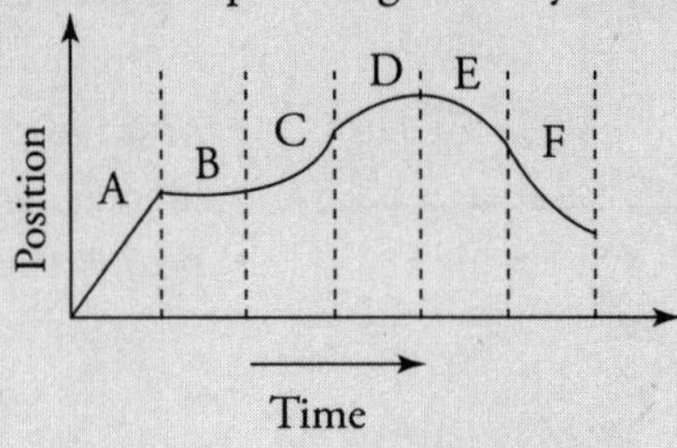

Solution. Part A is a constant speed moving away from the origin, part B is at rest, part C is speeding up moving away from the origin, part D is slowing down still moving away from the origin, part E is speeding up moving back toward the origin, and part F is slowing down moving back toward the origin.

Curves in a Position-Versus-Time Graph
Curving up represents a positive acceleration. Curving down represents a negative acceleration.

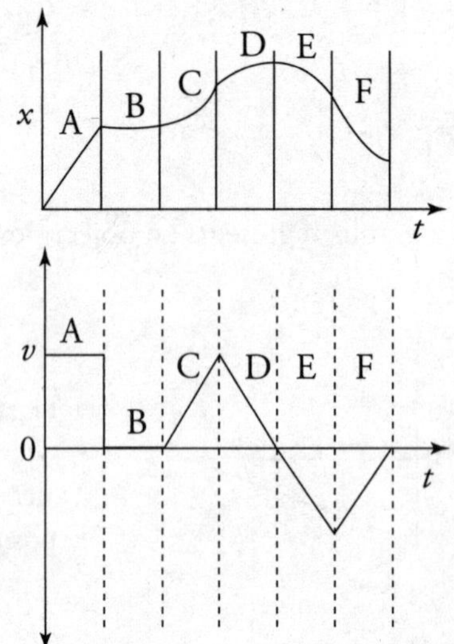

The area under an acceleration-versus-time (also referred to as an *a*-versus-*t*) graph gives the change in velocity.

One Slope to Rule Them All
The greater the slope at a point of a velocity-versus-time graph, the greater the acceleration is.

The greater the slope at a point of a position-versus-time graph, the greater the velocity is.

Example 17 The velocity of an object as a function of time is given by the following graph:

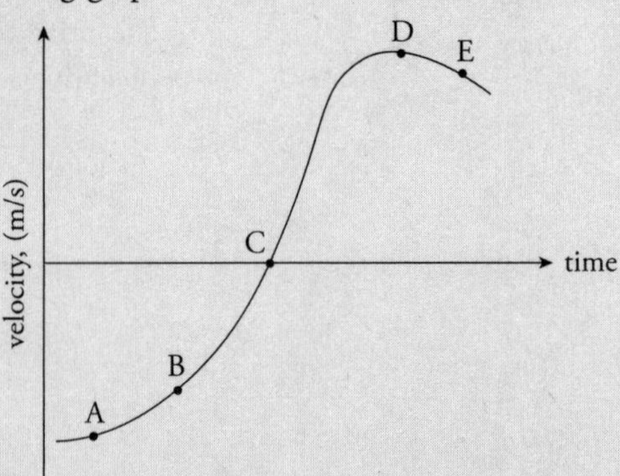

At which point (A, B, C, D, or E) is the magnitude of the acceleration the greatest?

Solution. The acceleration is the slope of the velocity-versus-time graph. Although this graph is not composed of straight lines, the concept of slope still applies; at each point, the slope of the curve is the slope of the tangent line to the curve. The slope is essentially zero at Points A and D (where the curve is flat), small and positive at B, and small and negative at E. The slope at Point C is large and positive, so this is where the object's acceleration is the greatest.

FREE FALL

The simplest real-life example of motion under pretty constant acceleration is the motion of objects in Earth's gravitational field, near the surface of the Earth and ignoring any effects due to the air (mainly air resistance). With these effects ignored, an object can fall *freely*. That is, it can fall experiencing only acceleration due to gravity. Near the surface of the Earth, the gravitational acceleration has a constant magnitude of about 9.8 m/s^2; this quantity is denoted g (for gravitational acceleration). On the AP Physics 1 Exam, you may use g = 10 m/s^2 as a simple approximation to g = 9.8 m/s^2. Even though you can use the exact value on your calculator, this approximation helps to save time, so in this book, we will always use g = 10 m/s^2. And, of course, the gravitational acceleration vector, **g**, points *downward.*

Since gravity points downward, for the sake of keeping things consistent throughout the rest of this chapter, we will set gravity as g = –10 m/s^2.

Example 18 A rock is dropped from a cliff 80 m above the ground. How long does it take to reach the ground?

Solution. We are given v_0 and asked for t. Unless you specifically see words to the contrary (such as, "you are on the Moon where the acceleration due to gravity is…"), assume you are also given a = –10 m/s^2. Because v is missing, and it isn't asked for, we can use Big Five equation #3.

$$y = y_0 + v_0 t + \frac{1}{2}at^2 \Rightarrow y = \frac{1}{2}at^2$$

If we set the origin y_0 = 0 at the base of the cliff and v_0 = 0, we get

$$t = \sqrt{\frac{2y}{a}}$$

$$t = \sqrt{\frac{2(-80\text{ m})}{(-10\text{ m/s}^2)}} = 4\text{ s}$$

Note: The negative in front of the 80 is inserted because the rock fell in the down direction.

A Note About Gravity
We don't always have to set gravity as negative. You are allowed to choose any coordinate system you wish as long as you match the positive or negative with gravity correctly.

Example 19 A baseball is thrown straight upward with an initial speed of 20 m/s. How high will it go?

Solution. We are given v_0, $a = -10\ \text{m/s}^2$ is implied, and we are asked for y. Now, neither t nor v is expressly given; however, we know the vertical velocity at the top is 0 (otherwise the baseball would still rise). Consequently, we use Big Five equation #5.

$$v^2 = v_0^2 + 2a(y - y_0) \Rightarrow -2ay = v_0^2$$

The time it takes for an object to be thrown straight up is the same as the time it takes for an object to fall down from the peak back into your hand.

We set $y_0 = 0$ and we know that $v = 0$, so that leaves us with

$$y = -\frac{v_0^2}{2a}$$

$$y = -\frac{(20\ \text{m/s})^2}{2(-10\ \text{m/s}^2)} = 20\ \text{m}$$

Example 20 One second after being thrown straight down, an object is falling with a speed of 20 m/s. How fast will it be falling 2 seconds later?

Solution. We're given v_0, a, and t and asked for v. Since x is missing, we use Big Five #2:

$$v = v_0 + at = (-20\ \text{m/s}) + (-10\ \text{m/s}^2)(2\ \text{s}) = -40\ \text{m/s}$$

The negative sign in front of the 40 simply indicates that the object is traveling in the down direction.

Example 21 If an object is thrown straight upward with an initial speed of 8 m/s and takes 3 seconds to strike the ground, from what height was the object thrown?

Solution. We're given a, v_0, and t and we need to find y_0. Because v is missing, we use Big Five #3:

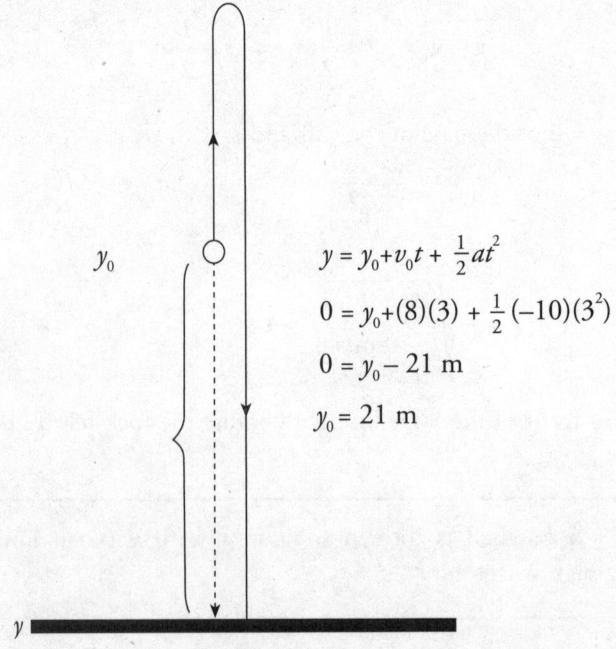

y_0
y
$y = y_0 + v_0t + \frac{1}{2}at^2$
$0 = y_0 + (8)(3) + \frac{1}{2}(-10)(3^2)$
$0 = y_0 - 21$ m
$y_0 = 21$ m

PROJECTILE MOTION

In general, an object that moves near the surface of the Earth will not follow a straight-line path (for example, a baseball hit by a bat, a golf ball struck by a club, or a tennis ball hit from the baseline). If we launch an object at an angle other than straight upward and consider only the effect of acceleration due to gravity, then the object will travel along a parabolic trajectory.

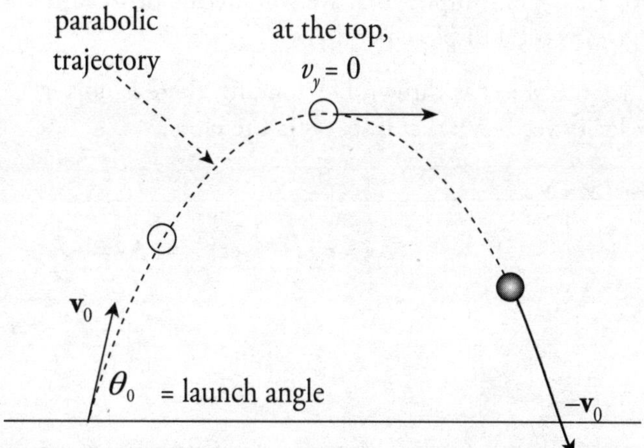

To simplify the analysis of parabolic motion, we analyze the horizontal and vertical motions separately, using the Big Five. This is the key to doing **projectile motion** problems.

A Note About Vectors
Vectors that are perpendicular to each other do not affect each other's magnitudes, only their directions.

Horizontal motion:	Vertical motion:
$\Delta x = v_{0x}t$	$y = y_0 + v_{0y}t - \frac{1}{2}gt^2$
$v_x = v_{0x}$ (constant!)	$v_y = v_{0y} - gt$
$a_x = 0$	$a_y = -g = -10 \text{ m/s}^2$

The quantity v_{0x}, which is the horizontal (or x) component of the initial velocity, is equal to $v_0 \cos \theta_0$, where θ_0 is the **launch angle**, the angle that the initial velocity vector, $\mathbf{v}_0$, makes with the horizontal. Similarly, the quantity v_{0y}, the vertical (or y) component of the initial velocity, is equal to $v_0 \sin \theta_0$.

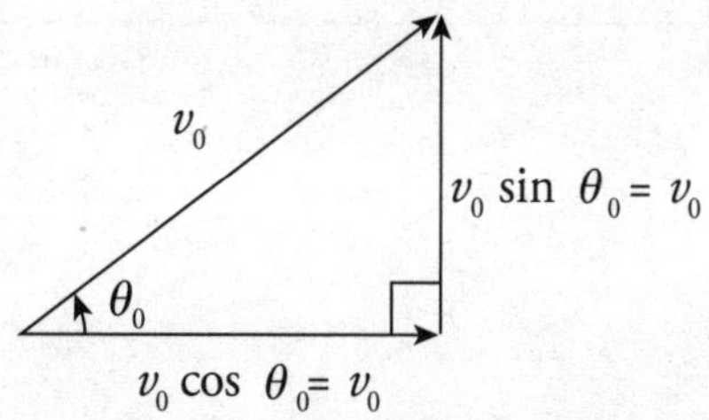

Note that time is the one value that is always the same for both the horizontal and vertical motion. If you're lost on a projectile motion question, solving for time is usually a good start.

> **Example 22** An object is thrown horizontally with an initial speed of 10 m/s. It hits the ground 4 seconds later. How far did it drop in 4 seconds?

Solution. The first step is to decide whether this is a *horizontal* question or a *vertical* question, since you must consider these motions separately. The question *How far did it drop?* is a *vertical* question, so the set of equations we will consider are those listed on the previous page under *vertical motion.* Next, *How far...?* implies that we will use the first of the vertical-motion equations, the one that gives vertical displacement, Δy.

> **REMEMBER THIS!**
> Horizontal velocity in standard parabolic motion is always constant.

Now, since the object is thrown horizontally, there is no vertical component to its initial velocity vector $\mathbf{v}_0$; that is, $v_{0y} = 0$. Therefore,

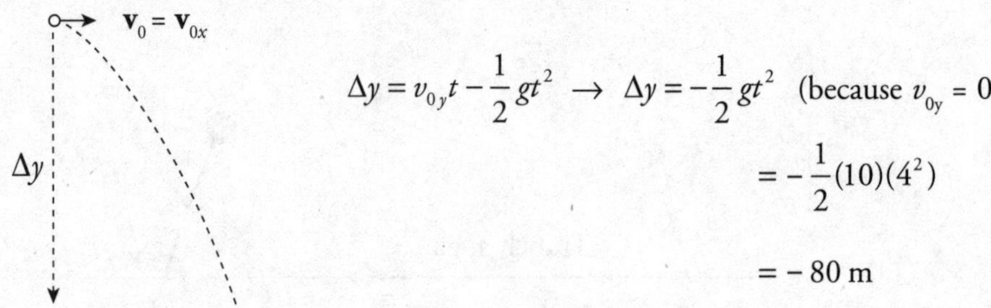

$$\Delta y = v_{0y}t - \frac{1}{2}gt^2 \rightarrow \Delta y = -\frac{1}{2}gt^2 \quad \text{(because } v_{0y} = 0\text{)}$$

$$= -\frac{1}{2}(10)(4^2)$$

$$= -80 \text{ m}$$

The fact that Δy is negative means that the displacement is *down.* Also, notice that the information given about v_{0x} is irrelevant to the question.

> **Example 23** From a height of 100 m, a ball is thrown horizontally with an initial speed of 15 m/s. How far does it travel horizontally in the first 2 seconds?

Solution. The question, *How far does it travel horizontally...?,* immediately tells us that we should use the first of the horizontal motion equations:

$$\Delta x = v_{0x}t = (15 \text{ m/s})(2 \text{ s}) = 30 \text{ m}$$

The information that the initial vertical position is 100 m above the ground is irrelevant (except for the fact that it's high enough that the ball doesn't strike the ground before the two seconds have elapsed).

> **Example 24** A projectile is traveling in a parabolic path for a total of 6 seconds. How does its horizontal velocity 1 s after launch compare to its horizontal velocity 4 s after launch?

Solution. The only acceleration experienced by the projectile is due to gravity, which is purely vertical, so that there is no horizontal acceleration. If there's no horizontal acceleration, then the horizontal velocity cannot change during flight, and the projectile's horizontal velocity 1 s after it's launched is the same as its horizontal velocity 3 s later.

For the vertical component of parabolic motion, you can treat it as throwing an object straight up and having it fall back down. Just remember that the object traveling to its peak is only a part of its time of travel. It still needs to fall.

Example 25 An object is projected upward with a 30° launch angle and an initial speed of 40 m/s. How long will it take for the object to reach the top of its trajectory? How high is this?

Solution. When the projectile reaches the top of its trajectory, its velocity vector is momentarily horizontal; that is, $v_y = 0$. Using the vertical-motion equation for v_y, we can set it equal to 0 and solve for t:

$$v_y \stackrel{\text{set}}{=} 0 \Rightarrow v_{0y} - gt = 0$$

$$t = \frac{v_{0y}}{g} = \frac{v_0 \sin \theta_0}{g} = \frac{(40 \text{ m/s}) \sin 30^\circ}{10 \text{ m/s}^2} = 2 \text{ s}$$

At this time, the projectile's vertical displacement is

$$\begin{aligned}\Delta y &= v_{0y}t - \frac{1}{2}(g)t^2 = (v_0 \sin \theta_0)t - \frac{1}{2}(g)t^2 \\ &= [(40 \text{ m/s}) \sin 30^\circ](2 \text{ s}) - \frac{1}{2}(10 \text{ m/s}^2)(2 \text{ s})^2 \\ &= 20 \text{ m}\end{aligned}$$

Example 26 An object is projected upward with a 30° launch angle from the ground and an initial speed of 60 m/s. For how many seconds will it be in the air? How far will it travel horizontally? Assume it returns to its original height.

Solution. The total time the object spends in the air is equal to twice the time required to reach the top of the trajectory (because the parabola is symmetrical). So, as we did in the previous example, we find the time required to reach the top by setting v_y equal to 0, and now double that amount of time:

$$v_y \stackrel{\text{set}}{=} 0 \Rightarrow v_{0y} - gt = 0$$

$$t = \frac{v_{0y}}{g} = \frac{v_0 \sin \theta_0}{g} = \frac{(60 \text{ m/s}) \sin 30^\circ}{10 \text{ m/s}^2} = 3 \text{ s}$$

Because the motion in this example is parabolic, $v_f = -v_0$. If you use that value for v_f in the equations here, you can solve for the entire flight time (making it unnecessary to multiply by 2 at the end). Both methods give the correct solution, so use whichever you find easier.

Therefore, the *total* flight time (that is, up and down) is $t_t = 2t = 2 \times (3 \text{ s}) = 6$ s.

Now, using the first horizontal motion equation, we can calculate the horizontal displacement after 6 seconds:

$$\Delta x = v_{0x}t_t = (v_0 \cos \theta_0)t_t = [(60 \text{ m/s})\cos 30°](6 \text{ s}) = 312 \text{ m}$$

Bonus Tips and Tricks...
Check us out on YouTube for additional test taking tips and must-know strategies at www.youtube.com/ThePrincetonReview

By the way, assuming it lands back at its original height, the full horizontal displacement of a projectile is called the projectile's **range**.

CHAPTER 4 KEY TERMS

displacement
total distance
average speed
velocity
speed
acceleration
uniform accelerated motion
projectile motion
launch angle
range

Chapter 4 Review Questions

Answers and explanations can be found in Chapter 11.

Section I: Multiple Choice

1. An object that's moving with constant speed travels once around a circular path. Which of the following statements are true concerning this motion? Select two answers.

    (A) The displacement is zero.
    (B) The average speed is zero.
    (C) The acceleration is zero.
    (D) The velocity is changing.

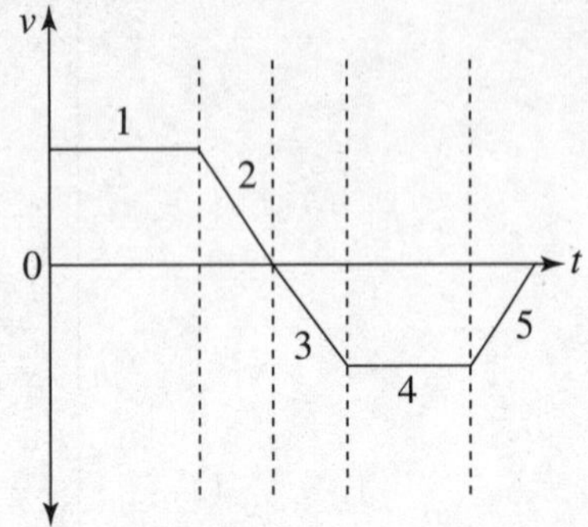

2. In section 5 of the velocity-versus-time graph above, the object is

    (A) speeding up moving in the positive direction
    (B) slowing down moving in the positive direction
    (C) speeding up moving in the negative direction
    (D) slowing down moving in the negative direction

3. Which of the following statements are true about uniformly accelerated motion? Select two answers.

    (A) If an object's acceleration is constant, then it must move in a straight line.
    (B) If an object's acceleration is zero, then its speed must remain constant.
    (C) If an object's speed remains constant, then its acceleration must be zero.
    (D) If an object's direction of motion is changing, then its acceleration is not zero.

4. A baseball is thrown straight upward. What is the ball's acceleration at its highest point?

    (A) $\frac{1}{2}g$, downward
    (B) g, downward
    (C) $\frac{1}{2}g$, upward
    (D) g, upward

5. How long would it take a car, starting from rest and accelerating uniformly in a straight line at 5 m/s^2, to cover a distance of 200 m?

    (A) 9.0 s
    (B) 10.5 s
    (C) 12.0 s
    (D) 15.5 s

6. A rock is dropped off a cliff and strikes the ground with an impact velocity of 30 m/s. How high was the cliff?

    (A) 20 m
    (B) 30 m
    (C) 45 m
    (D) 60 m

7. A stone is thrown horizontally with an initial speed of 10 m/s from a bridge. Assuming that air resistance is negligible, how long would it take the stone to strike the water 80 m below the bridge?

    (A) 1 s
    (B) 2 s
    (C) 4 s
    (D) 8 s

8. A soccer ball, at rest on the ground, is kicked with an initial velocity of 10 m/s at a launch angle of 30°. Calculate its total flight time, assuming that air resistance is negligible.
   (A) 0.5 s
   (B) 1 s
   (C) 2 s
   (D) 4 s

9. A stone is thrown horizontally with an initial speed of 30 m/s from a bridge. What is the stone's total speed when it enters the water 4 seconds later, assuming that air resistance is negligible?
   (A) 30 m/s
   (B) 40 m/s
   (C) 50 m/s
   (D) 60 m/s

10. Which one of the following statements is true concerning the motion of an ideal projectile launched at an angle of 45° to the horizontal?
    (A) The acceleration vector points opposite to the velocity vector on the way up and in the same direction as the velocity vector on the way down.
    (B) The speed at the top of the trajectory is zero.
    (C) The object's total speed remains constant during the entire flight.
    (D) The vertical speed decreases on the way up and increases on the way down.

11. A stone is thrown vertically upward with an initial speed of 5 m/s. What is the velocity of the stone 3 seconds later?
    (A) 25 m/s, upward
    (B) 25 m/s, downward
    (C) 35 m/s, upward
    (D) 35 m/s, downward

12. A car traveling at a speed of v_0 applies its brakes, skidding to a stop over a distance of x m. Assuming that the deceleration due to the brakes is constant, what would be the skidding distance of the same car if it were traveling with twice the initial speed?
    (A) $2x$ m
    (B) $3x$ m
    (C) $4x$ m
    (D) $8x$ m

Section II: Free Response

1. This question concerns the motion of a car on a straight track; the car's velocity as a function of time is plotted below.

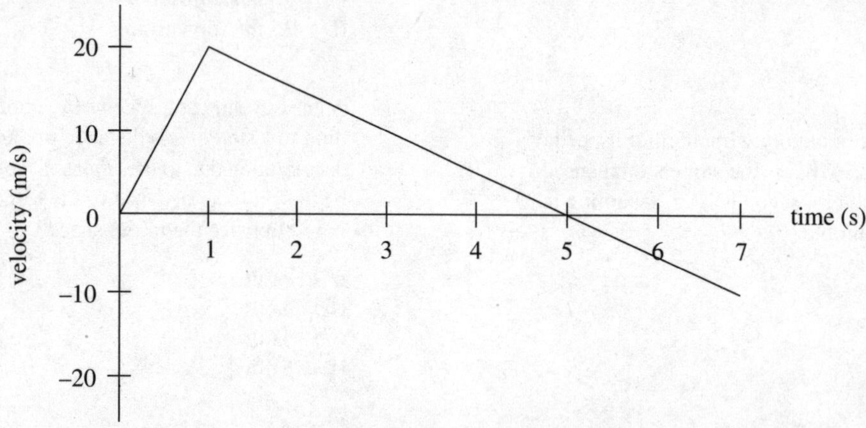

(A) Describe what happened to the car at time $t = 1$ s.

(B) How does the car's average velocity between time $t = 0$ and $t = 1$ s compare to its average velocity between times $t = 1$ s and $t = 5$ s?

(C) What is the displacement of the car from time $t = 0$ to time $t = 7$ s?

(D) Plot the car's acceleration during this interval as a function of time.

(E) Make a sketch of the object's position during this interval as a function of time. Assume that the car begins at $x = 0$.

2. Consider a projectile moving in a parabolic trajectory under constant gravitational acceleration. Its initial velocity has magnitude v_0, and its launch angle (with the horizontal) is θ_0.

(A) Calculate the maximum height, H, of the projectile.

(B) Calculate the (horizontal) range, R, of the projectile.

(C) For what value of θ_0 will the range be maximized?

(D) If $0 < h < H$, compute the time that elapses between passing through the horizontal line of height h in both directions (ascending and descending); that is, compute the time required for the projectile to pass through the two points shown in this figure:

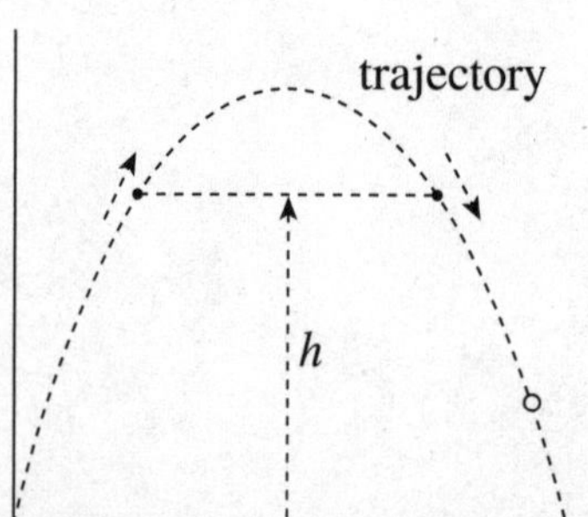

3. A cannonball is shot with an initial speed of 50 m/s at a launch angle of 40° toward a castle wall 220 m away. The height of the wall is 30 m. Assume that effects due to the air are negligible. (For this problem, use $g = 9.8\ \text{m/s}^2$.)

   (A) Show that the cannonball will strike the castle wall.

   (B) How long will it take for the cannonball to strike the wall?

   (C) At what height above the base of the wall will the cannonball strike?

4. A cannonball is fired with an initial speed of 40 m/s and a launch angle of 30° from a cliff that is 25 m tall.

   (A) What is the flight time of the cannonball?

   (B) What is the range of the cannonball?

   (C) In the two graphs below, plot the horizontal speed of the projectile versus time and the vertical speed of the projectile versus time (from the initial launch of the projectile to the instant it strikes the ground).

Horizontal Speed versus Time

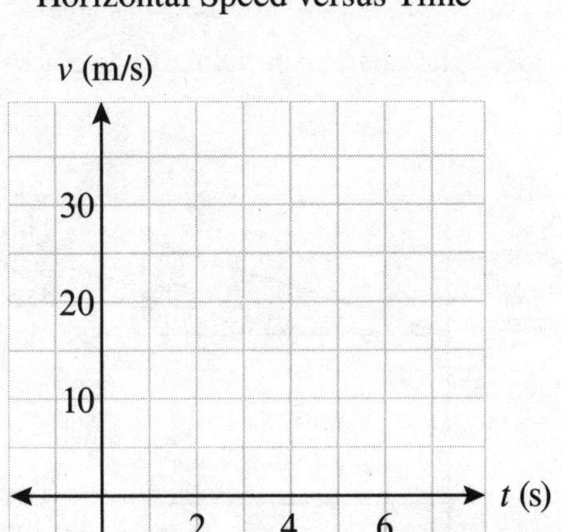

Vertical Speed versus Time

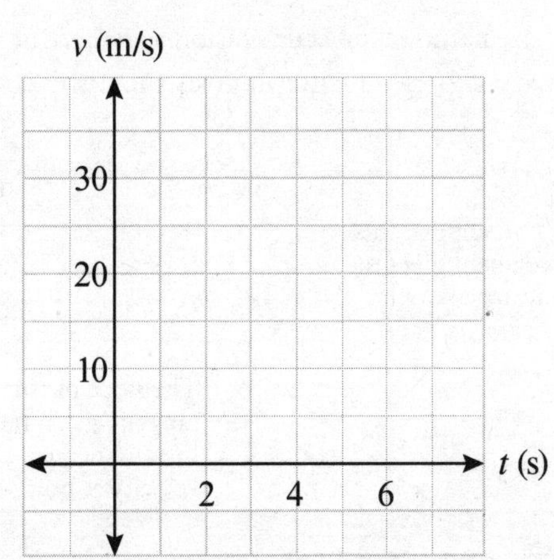

Summary

Graphs are very useful tools to help visualize the motion of an object. They can also help you solve problems once you learn how to translate from one graph to another. As you work with graphs, keep the following things in mind:

- Always look at a graph's axes first. This sounds obvious, but one of the most common mistakes students make is looking at a velocity-versus-time graph, thinking about it as if it were a position-versus-time graph.
- Don't ever assume one box is one unit. Look at the numbers on the axes.
- Lining up position-versus-time graphs *directly above* velocity-versus-time graphs and *directly above* acceleration-versus-time graphs is a must. This way you can match up key points from one graph to the next.
- The slope of an x versus t graph gives velocity. The slope of a v versus t graph gives acceleration.
- The area under an acceleration-versus-time graph gives the change in velocity. The area under a velocity-versus-time graph gives the displacement.

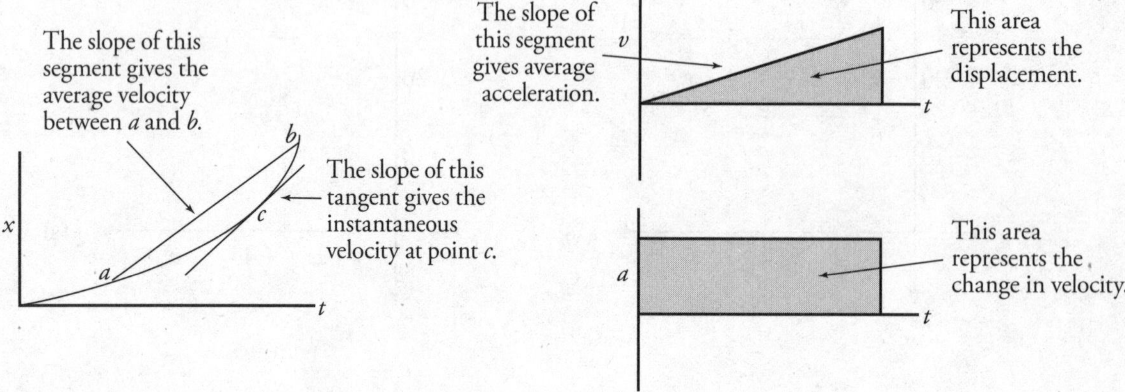

- The motion of an object in one dimension can be described using the Big Five equations. Look for what is given, determine what you're looking for, and use the equation that has those variables in them. Remember that sometimes there is hidden (assumed) information in the problem, such as $a = -10$ m/s^2.

Name of Equation	Equation	Missing Variable
Big Five #1:	$\Delta x = \frac{1}{2}(v_0 + v)t$	a
Big Five #2:	$v = v_0 + at$	x
Big Five #3:	$x = x_0 + v_0 t + \frac{1}{2}at^2$	v
Big Five #4:	$x = x_0 + vt - \frac{1}{2}at^2$	v_0
Big Five #5:	$v^2 = v_0^2 + 2a(x - x_0)$	t

- For projectiles, it is important to separate the horizontal and vertical components.

Horizontal Motion:	Vertical Motion:
$x = v_x t$	$y = y_0 + v_0 t + \frac{1}{2}gt^2$
$v_x = v_{0x}$ = constant	$v_y = v_{0y} + gt$
$a_x = 0$	$a = g = -10$ m/s^2

- At any given moment, the relationship between v, v_x, and v_y is given by

$v_x = v\cos\theta$ $\qquad$ $v_y = v\sin\theta$

$v^2 = v_x^2 + v_y^2$ and $\theta = \tan^{-1}\left(\frac{v_y}{v_x}\right)$

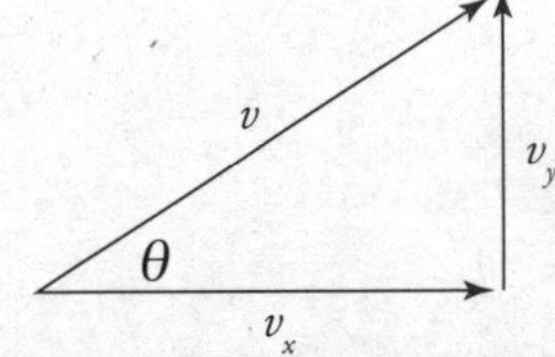

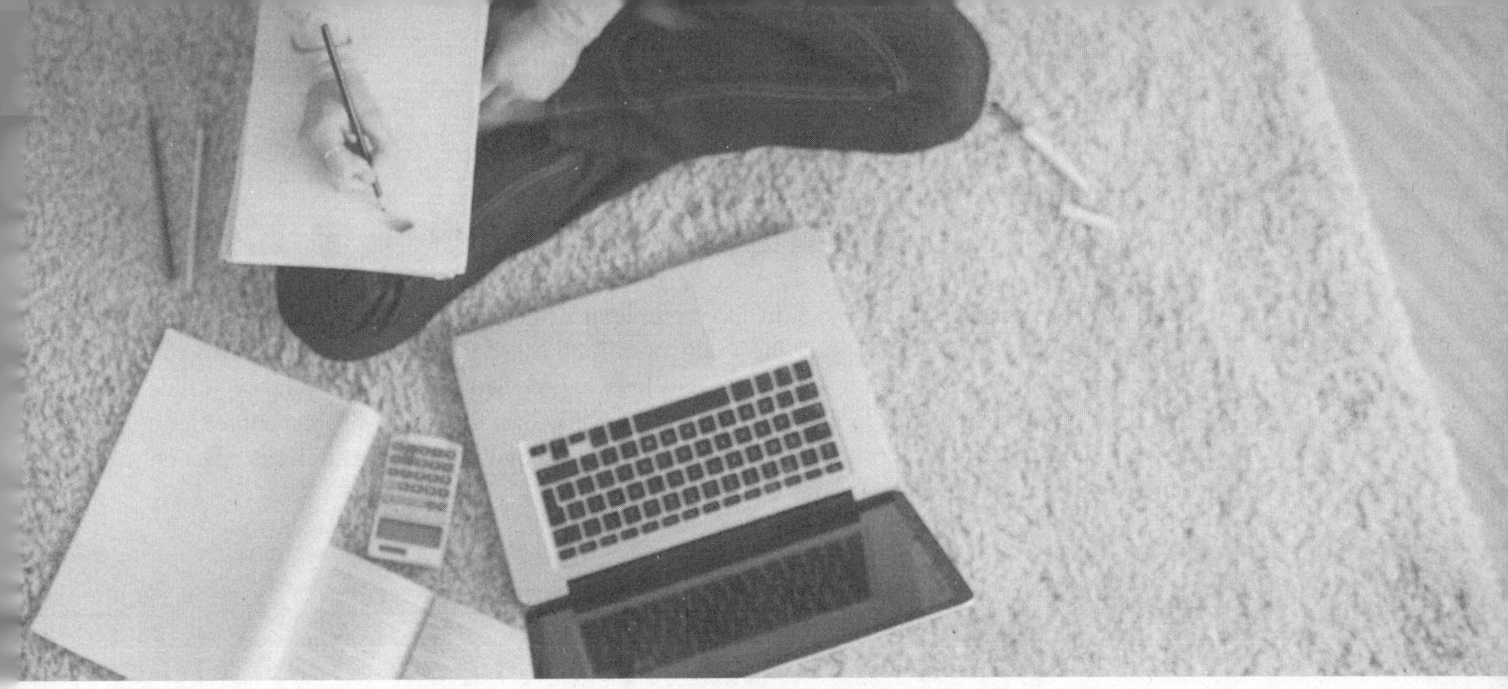

Chapter 5
Dynamics

"Nature is pleased with simplicity. And nature is no dummy."

—Sir Isaac Newton

The Englishman Sir Isaac Newton published a book in 1687 called *Philosphiae Naturalis Principia Mathematica* (The Mathematical Principles of Natural Philosophy)—referred to nowadays as simply *The Principia*—which began the modern study of physics as a scientific discipline. Three of the laws that Newton stated in *The Principia* form the basis for dynamics and are known simply as Newton's Laws of Motion.

In kinematics, we discovered the nature of how objects move, but there still existed the question of why objects move the way they do. For this fundamental task of understanding the cause of motion, we must turn our attention to dynamics.

INTRODUCTION TO FORCES

An interaction between two bodies—a push or a pull—is force. If an apple falls from a tree, it falls to the ground. If you pull on a door handle, it opens a door. If you push on a crate, you move it. In all these cases, some force is required for these actions to happen. In the first case of an apple falling from a tree, the Earth is exerting a downward pull called gravitational force. When you stand on the floor, the floor provides an upward force called the normal force. When you slide the crate across the floor, the floor exerts a frictional force against the crate. Whether a force is an action-at-a-distance force, like gravity, or a contact force, like pulling a door handle, its effect on objects is described by Newton's laws.

NEWTON'S FIRST LAW

Newton's First Law says that *an object will continue in its state of motion unless compelled to change by a force impressed upon it.* That is, unless an unbalanced force acts on an object, the object's velocity will not change: if the object is at rest, then it will stay at rest; if it is moving, then it will continue to move at a constant speed in a straight line.

This property of objects, their natural resistance to changes in their state of motion, is called **inertia**. In fact, Newton's First Law is often referred to as the **Law of Inertia**.

Example 1 What net force is required to maintain a 5,000 kg object moving at a constant velocity of magnitude 7,500 m/s?

If an object is experiencing zero acceleration (its velocity is not increasing or decreasing), then the object is experiencing zero net force.

Solution. Newton's First Law says that any object will continue in its state of motion unless an unbalanced force acts on it. Therefore, no net force is required to maintain a 5,000 kg object moving at a constant velocity of magnitude 7,500 m/s.

You might be asking, "If no net force is needed to keep a car moving at a constant speed, why does the driver need to press down on the gas pedal in order to maintain a constant speed?" There is a big difference between *force* and *net force*. As the car moves forward, there is a frictional force (more on that shortly) opposite the direction of motion that would be slowing the car down. The gas supplies energy to the engine to spin the tires to exert a forward force on the car to counteract friction and make the net force zero, which maintains a constant speed.

Here's another way to look at it: constant velocity means $\mathbf{a} = 0$, so the equation $F_{net} = ma$ immediately gives $F_{net} = 0$.

NEWTON'S SECOND LAW

Newton's Second Law predicts what will happen when an unbalanced force *does* act on an object: the object's velocity will change; the object will accelerate. More precisely, it says that its acceleration, **a**, will be directly proportional to the strength of the total—or *net*—force ($\mathbf{F}_{net}$) and inversely proportional to the object's mass, *m*:

$$\mathbf{a} = \mathbf{F/m} \text{ or}$$

$$\mathbf{F}_{net} = ma \text{ or } \Sigma F = ma$$

$$\text{Units} = \text{Newtons} = \text{kg} \cdot \text{m/s}^2$$

This is the most important equation in mechanics!

Two identical boxes, one empty and one full, have different masses. The box that's full has the greater mass, because it contains more stuff; more stuff, more mass. Mass (*m*) is measured in *kilograms*, abbreviated as kg. (Note: An object whose mass is 1 kg weighs about 2.2 pounds.) If a given force produces some change in velocity for a 2 kg object, then a 1 kg object would experience twice that change. Note that mass is a proxy for the extent of inertia inherent in an object, and thus inertia is a reflection of an object's mass.

Forces are represented by vectors; they have magnitude and direction. If several different forces act on an object simultaneously, then the net force, $\mathbf{F}_{net}$, is the vector sum of all these forces. (The phrase *resultant force* is also used to mean *net force.*)

Since $\mathbf{F}_{net} = ma$, and *m* is a *positive* scalar, the direction of **a** always matches the direction of $\mathbf{F}_{net}$.

Finally, since $F = ma$, the units for F equal the units of m times the units of a:

$$[F] = [m][a]$$

$$= \text{kg·m/s}^2$$

A force of 1 kg·m/s² is renamed 1 **newton** (abbreviated as N). A medium-size apple weighs about 1 N.

The difference between weight and mass is often difficult to grasp at first, but we'll address that issue in a few pages.

Example 2 How much force is required to cause an object of mass 2 kg to have an acceleration of 4 m/s²?

Solution. According to Newton's Second Law, $F_{net} = ma = (2 \text{ kg})(4 \text{ m/s}^2) = 8 \text{ N}$.

NEWTON'S THIRD LAW

This is the law that's commonly remembered as *to every action, there is an equal, but opposite, reaction.* More precisely, if Object 1 exerts a force on Object 2, then Object 2 exerts a force back on Object 1, equal in strength but opposite in direction. These two forces, $\mathbf{F}_{1\text{-on-}2}$ and $\mathbf{F}_{2\text{-on-}1}$, are called an **action/reaction pair.**

Example 3 Object 1 exerts a 300 N force to the right on object 2. Both objects are stationary. What is the force that object 2 exerts on object 1?

Solution. Newton's Third Law states that the force from object 2 on object 1 is equal in magnitude but opposite in direction to the force from object 1 on object 2. Therefore, object 2 exerts a 300 N force to the left on object 1.

The motion of the objects is irrelevant. Newton's Third Law is concerned with individual forces acting on objects, unlike the first and second laws that deal with net forces.

Example 4 An object feels two forces: one of strength 8 N pulling to the left and one of strength 20 N pulling to the right. If the object's mass is 4 kg, what is its acceleration?

Solution. Forces are represented by vectors and can be added and subtracted. Therefore, an 8 N force to the left added to a 20 N force to the right yields a net force of 20 – 8 = 12 N to the right. Then Newton's Second Law gives $\mathbf{a} = \mathbf{F}_{net}/m$ = (12 N to the right)/(4 kg) = 3 m/s^2 to the right.

NEWTON'S LAWS: A SUMMARY

These three laws will appear time and time again in later concepts. Let's sum up those three laws in easier terms.

Newton's First Law

- Moving stuff keeps moving; resting stuff keeps resting
- Law of Inertia: objects naturally resist changes in their velocities
- Measure of inertia is mass

Newton's Second Law

$$\mathbf{F}_{net} = m\boldsymbol{a}$$

- Force (**F**) acting on an object is equal to the mass (*m*) of an object times its acceleration (***a***)
- Forces are vectors
- Units: Newton (N) = kg·m/s^2

Newton's Third Law

For action/reaction pairs:

$$F_{(1\text{ on }2)} = -F_{(2\text{ on }1)}$$

the forces are equal but in opposite directions

WEIGHT

Mass and weight are not the same thing—there is a clear distinction between them in physics—but they are often used interchangeably in everyday life. The **weight** of an object is the gravitational force exerted on it by the Earth (or by whatever planet it happens to be on). **Mass,** by contrast, is a measure of the quantity of matter that comprises an object. An object's mass does not change with location. Weight changes depending on location. For example, you weigh less on the Moon than you do on Earth.

Amass Your Mass Facts (Groan)
Also, remember that mass is a proxy for inertia.

Since weight is a force, we can use $\mathbf{F} = ma$ to compute it. What acceleration would the gravitational force impose on an object? The gravitational acceleration, of course! Therefore, setting $\mathbf{a} = \mathbf{g}$, the equation $\mathbf{F} = ma$ becomes

$$\mathbf{F}_w = mg \text{ or } F_g = mg$$

This is the equation for the weight of an object of mass *m*. (Weight is often symbolized as F_g, rather than $\mathbf{F}_w$.) Notice that mass and weight are proportional but not identical. Furthermore, mass is measured in kilograms, while weight is measured in newtons.

Example 5 What is the mass of an object that weighs 500 N?

Solution. Since weight is m multiplied by g, mass is F_w (weight) divided by g. Therefore,

$$m = F_w/g = (500\text{ N})/(10\text{ m/s}^2) = 50\text{ kg}$$

Example 6 A person weighs 150 pounds. Given that a pound is a unit of weight equal to 4.45 N, what is this person's mass?

Solution. This person's weight in newtons is (150 lb)(4.45 N/lb) = 667.5 N, so his mass is

$$m = F_w/g = (667.5\text{ N})/(10\text{ m/s}^2) = 66.75\text{ kg}$$

Example 7 A book whose mass is 2 kg rests on a table. Find the magnitude of the force exerted by the table on the book.

Solution. The book experiences two forces: the downward pull of the Earth's gravity and the upward, supporting force exerted by the table. Since the book is at rest on the table, its acceleration is zero, so the net force on the book must be zero. Therefore, the magnitude of the support force must equal the magnitude of the book's weight, which is $F_w = mg = (2\text{ kg})(10\text{ m/s}^2) = 20$ N.

NORMAL FORCE

When an object is in contact with a surface, the surface exerts a contact force on the object. The component of the contact force that's *perpendicular* to the surface is called the **normal force** on the object. (In physics, the word *normal* means *perpendicular.*) The normal force is what prevents objects from falling through tabletops or you from falling through the floor. The normal force is denoted by $\mathbf{F}_N$, or simply by **N**. (If you use the latter notation, be careful not to confuse it with N, the abbreviation for the newton.)

There is no formula to directly calculate normal force. Instead, you have to use Newton's laws and your knowledge of the other forces at work to find a value for it in a given situation.

Example 8 A book whose mass is 2 kg rests on a table. Find the magnitude of the normal force exerted by the table on the book.

Solution. The book experiences two forces: the downward pull of Earth's gravity and the upward, supporting force exerted by the table. Since the book is at rest on the table, its acceleration is zero, so the net force on the book must be zero. Therefore, the magnitude of the support force must equal the magnitude of the book's weight, which is $F_w = mg = (2)(10) = 20$ N. This means the normal force must be 20 N as well: $\mathbf{F}_N = 20$ N. (Note that this is a repeat of Example 6, except now we have a name for the "upward, supporting force exerted by the table"; it's called the normal force.)

AN OVERALL STRATEGY

The previous examples are the lowest level of understanding Newton's laws. They are pretty straightforward thinking sometimes referred to as "plug and chug." Most of physics is not that simple. Frequently there is more than one force acting on an object, and many times angles are involved. Following the below strategy can greatly increase your chance of success for all but the most trivial of Newton's Second Law problems.

I. You must be able to visualize what's going on. Make a sketch if it helps, but definitely make a free-body diagram by doing the following:

A. Draw a dot to represent the object. Draw arrows going away from the dot to represent any (all) forces acting on the object.

i. Anything touching the object exerts a force.

a. If the thing touching the object is a rope, ropes can only pull. Draw the force accordingly.

b. If the thing touching the object is a table, ramp, floor, or some other flat surface, a surface can exert two forces.

1. The surface exerts a force perpendicular to itself toward the object. This force is always present if two things are in contact, and it is called the normal force.

2. If there is kinetic friction present, then the surface exerts a force on the object that is parallel to the surface and opposite to the direction of motion.

ii. Some things can exert a force without touching an object. For example, the Earth pulls down on everything via the mystery of gravity. Electricity and magnetism also exert their influences without actually touching. Unless you hear otherwise, gravity points down!

iii. If you know one force is bigger than another, you should draw that arrow longer than the smaller force's arrow.

iv. Don't draw a velocity and mistake it for a force. No self-respecting velocity vector hangs out in a free-body diagram! Oh, and there is no such thing as the force of inertia.

II. Clearly define an appropriate coordinate system. Be sure to break up each force that does not lie on an axis into its *x*- and *y*-components.

III. Write out Newton's Second Law in the form of $\sum F_x = ma_x$ and/or $\sum F_y = ma_y$, using the forces identified in the free-body diagram to fill in the appropriate forces.

IV. Do the math.

College Board Says...
These items sound a lot like Big Idea #3 from the College Board's AP Physics 1 Course and Exam Description!

As you go through the following examples, notice how this strategy is used.

Example 9 Draw a free-body diagram for each of the following situations:

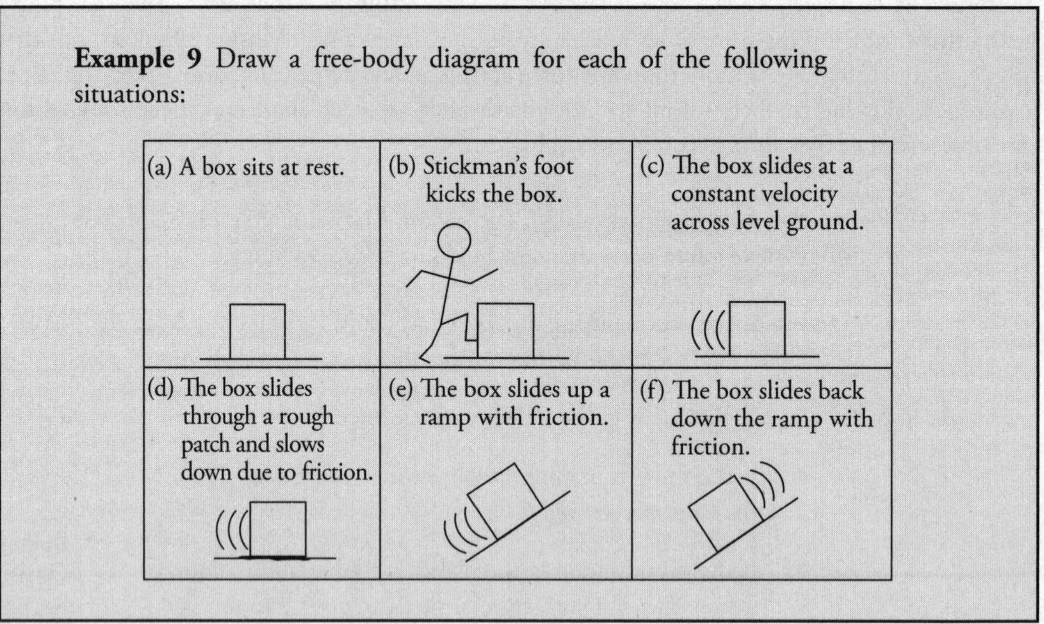

(a) A box sits at rest.
(b) Stickman's foot kicks the box.
(c) The box slides at a constant velocity across level ground.
(d) The box slides through a rough patch and slows down due to friction.
(e) The box slides up a ramp with friction.
(f) The box slides back down the ramp with friction.

Solution. Notice that gravity always points down (even on ramps). Also the normal force is perpendicular to the surface (even on ramps). Do not always put the normal opposite the direction of gravity, because the normal is relative to the surface, which may be tilted. Finally, friction is always parallel to the surface (or perpendicular to the normal) and tends to point in the opposite direction from motion.

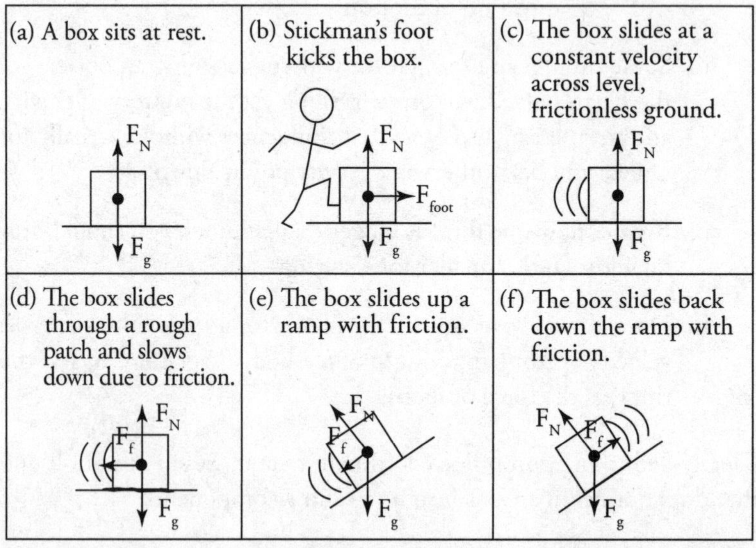

Example 10 A can of paint with a mass of 6 kg hangs from a rope. If the can is to be pulled up to a rooftop with an acceleration of 1 m/s^2, what must the tension in the rope be?

Solution. First draw a picture. Represent the object of interest (the can of paint) as a heavy dot, and draw the forces that act on the object as arrows connected to the dot. This is called a **free-body (or force) diagram.**

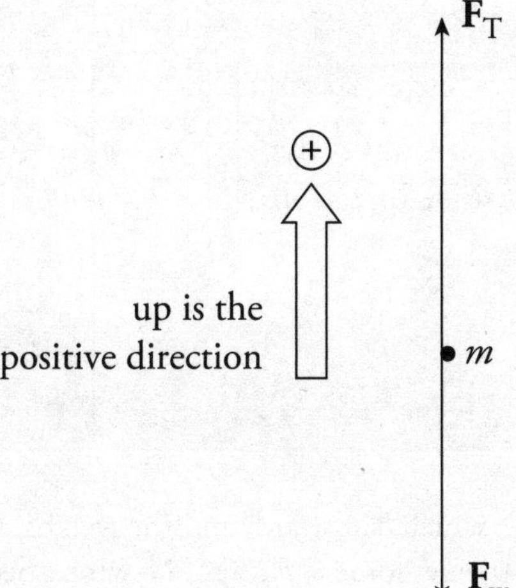

We have the tension force in the rope, $\mathbf{F}_T$ (also symbolized merely by **T**), which is upward, and the weight, $\mathbf{F}_w$, which is downward. Calling *up* the positive direction, the net force is $F_T - F_w$. The second law, $F_{net} = ma$, becomes $F_T - F_w = ma$, so

$$F_T = F_w + ma = mg + ma = m(g + a) = 6(10 + 1) = 66 \text{ N}$$

Tension's direction should be drawn opposite to the force creating the tension. If weight pulls down on the rope, then tension is up.

Example 11 A can of paint with a mass of 6 kg hangs from a rope. If the can is pulled up to a rooftop with a constant velocity of 1 m/s, what must the tension in the rope be?

Solution. The phrase "constant velocity" automatically means $a = 0$ and, therefore, $F_{net} = 0$. In the diagram above, $\mathbf{F}_T$ would need to have the same magnitude as $\mathbf{F}_w$ in order to keep the can moving at a constant velocity. Thus, in this case, $F_T = F_w = mg = (6)(10) = 60$ N.

Example 12 How much tension must a rope have to lift a 50 N object with an acceleration of 10 m/s²?

Solution. First draw a free-body diagram:

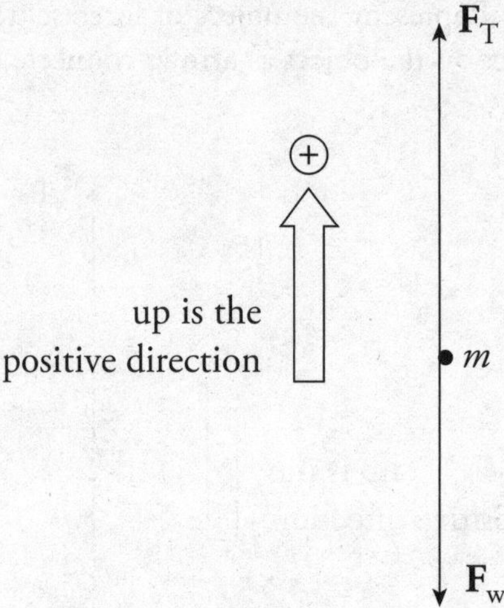

We have the tension force, $\mathbf{F}_T$, which is upward, and the weight, $\mathbf{F}_w$, which is downward. Calling *up* the positive direction, the net force is $F_T - F_w$. Newton's Second Law, $F_{net} = ma$, becomes $F_T - F_w = ma$, so $F_T = F_w + ma$. Remembering that $m = F_w/g$, we find that

$$F_T = F_w + ma = F_w + \frac{F_w}{g}a = 50 \text{ N} + \frac{50 \text{ N}}{10 \text{ m/s}^2}\left(10 \text{ m/s}^2\right) = 100 \text{ N}$$

FRICTION

When an object is in contact with a surface, the surface exerts a contact force on the object. The component of the contact force that's *parallel* to the surface is called the called the **friction** (or **friction force**) on the object. Friction, like the normal force, arises from electrical interactions between atoms of which the object is composed and those of which the surface is composed.

We'll look at two main categories of friction: (1) **static friction** and (2) **kinetic (sliding) friction**. Static friction results from the weak electrostatic bonds formed between the surfaces when the object is at rest. These bonds need to be broken before the object will slide. As a result, the static friction force is higher than the kinetic friction force. When the object begins to slide, the weak electronegative bonds cannot form fast enough, so the kinetic friction is lower. It is easier to keep an object moving than it is to start an object moving because the kinetic μ is lower than the static μ. Static friction occurs when there is no relative motion between the object and the surface (no sliding); kinetic friction occurs when there *is* relative motion (when there's sliding).

The strength of the friction force depends, in general, on two things: the nature of the surfaces and the strength of the normal force. The nature of the surfaces is represented by the **coefficient of friction,** which is denoted by μ (*mu*) and has no units. The greater this number is, the stronger the friction force will be. For example, the coefficient of friction between rubber-soled shoes and a wooden floor is 0.7, but between rubber-soled shoes and ice, it's only 0.1. Also, since kinetic friction is generally weaker than static friction (it's easier to keep an object sliding once it's sliding than it is to start the object sliding in the first place), there are two coefficients of friction: one for static friction (μ_s) and one for kinetic friction (μ_k). For a given pair of surfaces, it's virtually always true that $\mu_k < \mu_s$. The strengths of these two types of friction forces are given by the following equations:

$$F_{\text{static friction, max}} = \mu_s F_N$$

$$F_{\text{kinetic friction}} = \mu_k F_N$$

Real World Physics
If you've ever tried to slide something heavy across the floor (like moving a couch), then you should already be familiar with static friction. You have to push pretty hard to get it moving, but then it slides a bit more easily once you get it started.

Note that the equation for the strength of the static friction force is for the *maximum* value only. This is because static friction can vary, precisely counteracting weaker forces that attempt to move an object. For example, suppose an object experiences a normal force of F_N = 100 N and the coefficient of static friction between it and the surface it's on is 0.5. Then, the *maximum* force that static friction can exert is (0.5)(100 N) = 50 N. However, if you push on the object with a force of, say, 20 N, then the static friction force will be 20 N (in the opposite direction), *not* 50 N; the object won't move. The net force on a stationary object must be zero. Static friction can take on all values, up to a certain maximum, and you must overcome the maximum static friction force to get the object to slide. The direction of $\mathbf{F}_{\text{kinetic friction}} = \mathbf{F}_{\text{f (kinetic)}}$ is opposite to that of motion (sliding), and the direction of $\mathbf{F}_{\text{static friction}} = \mathbf{F}_{\text{f (static)}}$ is usually, but not always, opposite to that of the intended motion.

Example 13 A crate of mass 20 kg is sliding across a wooden floor. The coefficient of kinetic friction between the crate and the floor is 0.3.

(a) Determine the strength of the friction force acting on the crate.

(b) If the crate is being pulled by a force of 90 N (parallel to the floor), find the acceleration of the crate.

Good to Know
Example 13 illustrates how static friction can vary.

Solution. First draw a free-body diagram:

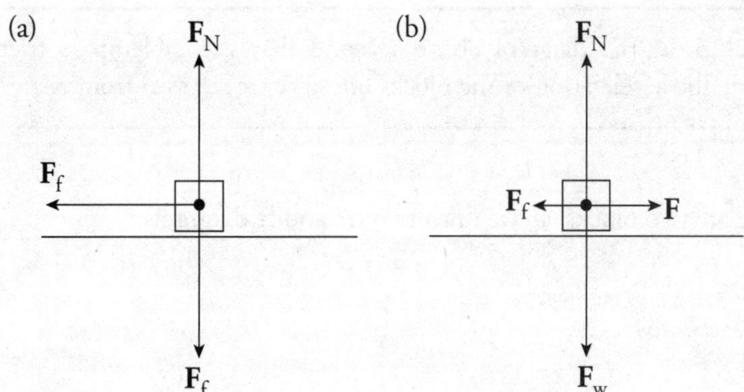

(a) The normal force on the object balances the object's weight, so $F_N = mg$ = (20 kg)(10 m/s²) = 200 N. Therefore, $F_{(\text{kinetic})} = \mu_k F_N$ = (0.3)(200 N) = 60 N.

(b) The net horizontal force that acts on the crate is $F - F_f$ = 90 N – 60 N = 30 N, so the acceleration of the crate is $a = F_{\text{net}}/m$ = (30 N)/(20 kg) = 1.5 m/s².

Example 14 A crate of mass 100 kg rests on the floor. The coefficient of static friction is 0.4. If a force of 250 N (parallel to the floor) is applied to the crate, what's the magnitude of the force of static friction on the crate?

Solution. The normal force on the object balances its weight, so $F_N = mg = (100 \text{ kg})(10 \text{ m/s}^2) = 1{,}000 \text{ N}$. Therefore, $F_{\text{static friction, max}} = F_{\text{f (static), max}} = \mu_s F_N = (0.4)(1{,}000 \text{ N}) = 400 \text{ N}$. This is the *maximum* force that static friction can exert, but in this case it's not the actual value of the static friction force. Since the applied force on the crate is only 250 N, which is less than the $F_{\text{f (static), max}}$, the force of static friction will be less also: $F_{\text{f (static)}} = 250 \text{ N}$, and the crate will not slide.

PULLEYS

Pulleys are devices that change the direction of the tension force in the cords that slide over them. Pulley systems multiply the force by however many strings are pulling on the object.

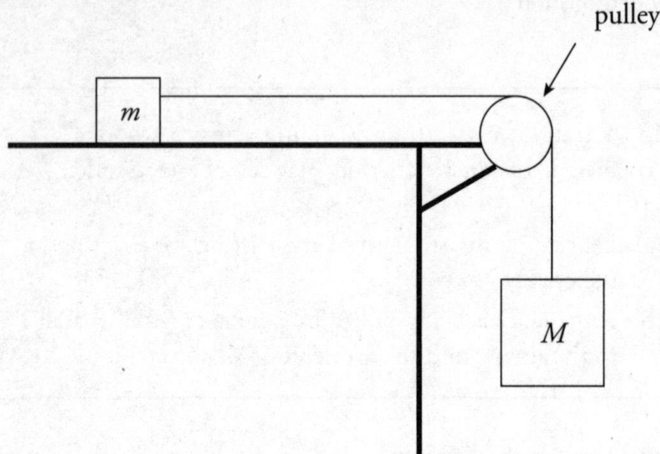

Example 15 In the diagram above, assume that the tabletop is frictionless. Determine the acceleration of the blocks once they're released from rest.

Solution. There are two blocks, so we draw two free-body diagrams:

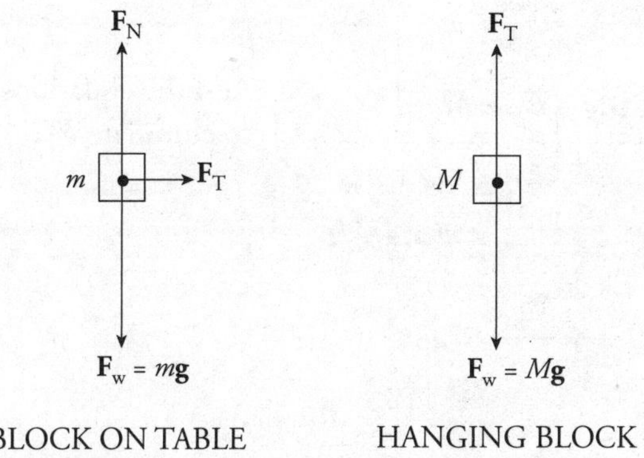

To get the acceleration of each one, we use Newton's Second Law, $\mathbf{F}_{net} = m\mathbf{a}$.

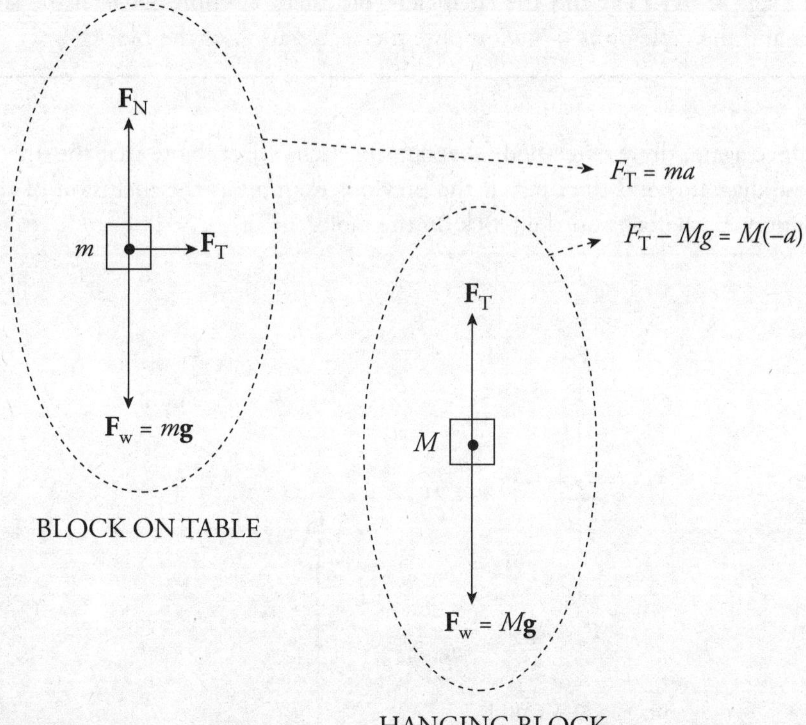

Note that there are two unknowns, F_T and a, but we can eliminate F_T by combining the two equations, and then we can solve for a.

Newton's Second Law is defined as $\mathbf{F} = m\mathbf{a}$, whereas the force required to accelerate an object is equal to the mass of the object multiplied by the magnitude of the desired acceleration. If the net force in a system is not zero, then the object in the system is accelerating. Newton's Second Law can be used to calculate the acceleration of objects with unbalanced forces. Once the net force is determined, substitute this force and mass into the $\mathbf{F} = m\mathbf{a}$ ($\mathbf{F}_{net} = m\mathbf{a}$) equation to determine the acceleration of the object.

$$\begin{aligned} F_T &= ma \\ Mg - F_T &= Ma \\ \hline Mg &= ma + Ma \\ &= a(m + M) \end{aligned}$$

Add the equations to eliminate F_T.

$$\frac{Mg}{m + M} = a$$

> **Example 16** Using the same diagram as in the previous example, assume that m = 2 kg, M = 10 kg, and the coefficient of kinetic friction between the small block and the tabletop is 0.5. Compute the acceleration of the blocks.

Solution. Once again, draw a free-body diagram for each object. Note that the only difference between these diagrams and the ones in the previous example is the inclusion of the force of (kinetic) friction, $\mathbf{F}_f$, that acts on the block on the table.

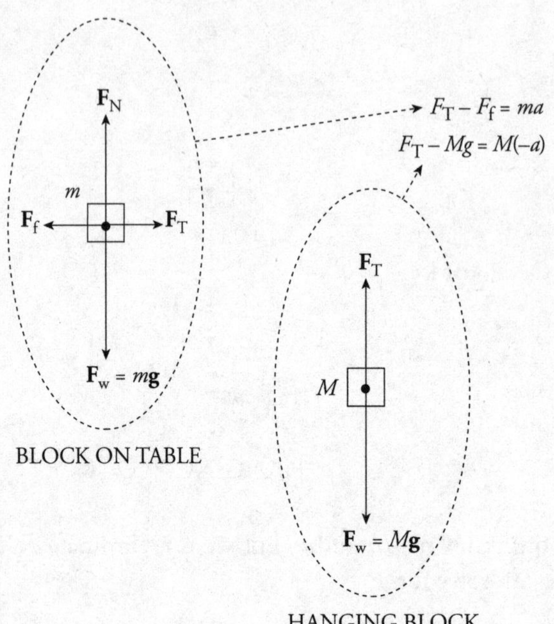

As before, we have two equations that contain two unknowns (a and F_T):

$$F_T - F_f = ma \quad (1)$$
$$F_T - Mg = M(-a) \quad (2)$$

Subtract the equations (thereby eliminating F_T) and solve for a. Note that, by definition, $F_f = \mu F_N$, and from the free-body diagram for m, we see that $F_N = mg$, so $F_f = \mu mg$:

$$Mg - F_f = ma + Ma$$

$$Mg - \mu mg = a(m + M)$$

$$\frac{M - \mu m}{m + M}g = a$$

Substituting in the numerical values given for m, M, and μ, we find that $a = \frac{3}{4}g$ (or 7.5 m/s^2).

Example 17 In the previous example, calculate the tension in the cord.

Solution. Since the value of a has been determined, we can use either of the two original equations to calculate F_T. Using Equation (2), $F_T - Mg = M(-a)$ (because it's simpler), we find

$$F_T = Mg - Ma = Mg - M \bullet \frac{3}{4}g = \frac{1}{4}Mg = \frac{1}{4}(10)(10) = 25\text{ N}$$

As you can see, we would have found the same answer if Equation (1) had been used:

$$\begin{aligned} F_T - F_f = ma \Rightarrow \ F_T = F_f + ma = \mu mg + ma = \mu mg + m \cdot \frac{3}{4}g &= mg\left(\mu + \frac{3}{4}\right) \\ &= (2)(10)(0.5 + 0.75) \\ &= 25\text{ N} \end{aligned}$$

INCLINED PLANES

An **inclined plane** is basically a ramp. If you look at the forces acting on a block that sits on a ramp using a standard coordinate system, it initially looks straightforward. However, part of the normal force acts in the x direction, part acts in the y direction, and the block has acceleration in both the x and y directions. If friction is present, it also has components in both the x and y directions. The math has the potential to be quite cumbersome.

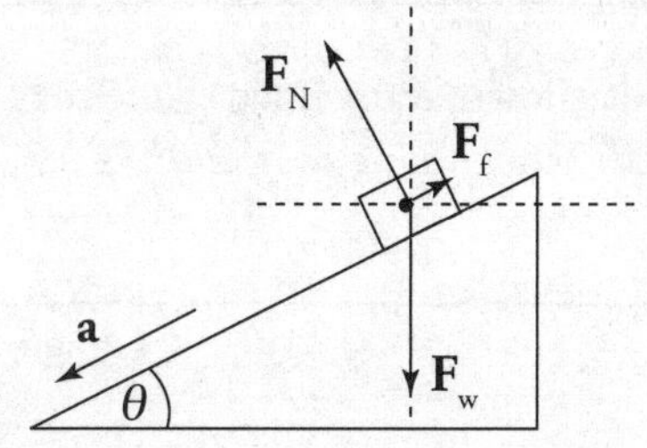

This is a non-rotated coordinate system—notice $\mathbf{F}_f$, $\mathbf{F}_N$, and **a** will each have to be broken into *x*- and *y*-components.

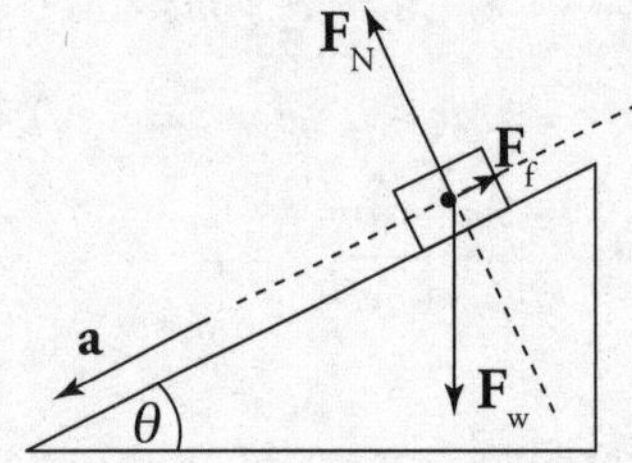

This is a rotated coordinate system—$\mathbf{F}_N$ acts only perpendicular to the ramp, and $\mathbf{F}_f$ and acceleration acts only parallel to the ramp, so only $\mathbf{F}_w$ must be broken into components. If an object of mass m is on the ramp, then the force of gravity on the object, $\mathbf{F}_w = m\mathbf{g}$, has two components: one that's parallel to the ramp ($mg \sin \theta$) and one that's normal to the ramp ($mg \cos \theta$), where θ is the incline angle. The force driving the block down the inclined plane is the component of the block's weight that's parallel to the ramp: $mg \sin \theta$.

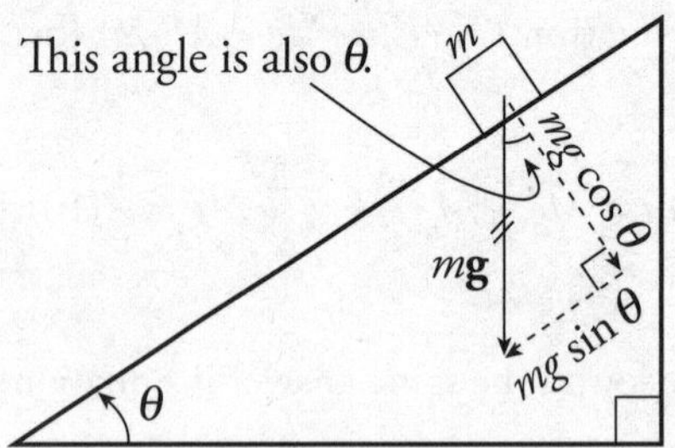

> **Example 18** A block slides down a frictionless, inclined plane that makes a 30° angle with the horizontal. Find the acceleration of this block.

Solution. Let m denote the mass of the block, so the force that pulls the block down the incline is $mg \sin \theta$, and the block's acceleration down the plane is

$$a = \frac{F}{m} = \frac{mg \sin \theta}{m} = g \sin \theta = g \sin 30^\circ = \frac{1}{2} g = 5 \text{ m/s}^2$$

> **Example 19** A block slides down an inclined plane that makes a 30° angle with the horizontal. If the coefficient of kinetic friction is 0.3, find the acceleration of the block.

Solution. First draw a free-body diagram. Notice that, in the diagram shown below, the weight of the block, $F_w = mg$, has been written in terms of its scalar components: $F_w \sin\theta$ parallel to the ramp and $F_w \cos\theta$ normal to the ramp:

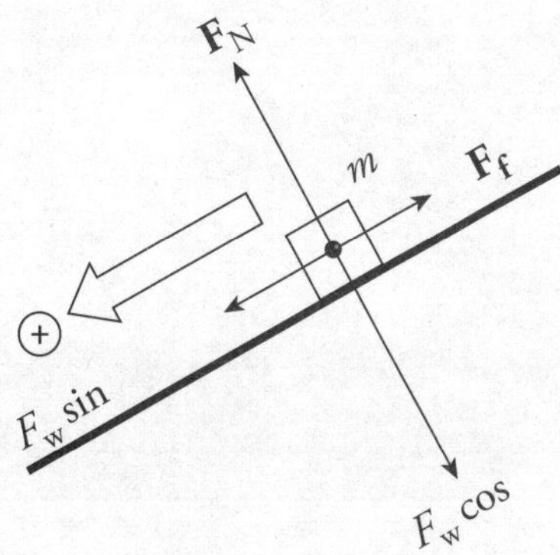

The force of friction, $\mathbf{F}_f$, that acts up the ramp (opposite to the direction in which the block slides) has magnitude $F_f = \mu F_N$. But the diagram shows that $F_N = F_w \cos\theta$, so $F_f = \mu(mg\cos\theta)$. Therefore, the net force down the ramp is

$$F_w \sin\theta - F_f = mg\sin\theta - \mu mg\cos\theta = mg(\sin\theta - \mu\cos\theta)$$

Then, setting F_{net} equal to ma, we solve for a:

$$\begin{aligned} a = \frac{F_{net}}{m} &= \frac{mg(\sin\theta - \mu\cos\theta)}{m} \\ &= g(\sin\theta - \mu\cos\theta) \\ &= (10\text{ m/s}^2)(\sin 30° - 0.3\cos 30°) \\ &= 2.4\text{ m/s}^2 \end{aligned}$$

Got a Question?
For answers to test-prep questions for all your tests and additional test-taking tips, subscribe to our YouTube channel at www.youtube.com/ThePrincetonReview

CHAPTER 5 KEY TERMS

Newton's First Law
inertia
Law of Inertia
Newton's Second Law
newton (N)
action/reaction pair
Newton's Third Law
weight
mass
normal force
free-body (or force) diagram
friction
static friction
kinetic (sliding) friction
coefficient of friction
pulleys
inclined plane

Chapter 5 Review Questions

Answers and explanations can be found in Chapter 11.

Section I: Multiple Choice

1. A person standing on a horizontal floor feels two forces: the downward pull of gravity and the upward supporting force from the floor. These two forces

   (A) have equal magnitudes and form an action/reaction pair
   (B) have equal magnitudes but do not form an action/reaction pair
   (C) have unequal magnitudes and form an action/reaction pair
   (D) have unequal magnitudes and do not form an action/reaction pair

2. A person who weighs 800 N steps onto a scale that is on the floor of an elevator car. If the elevator accelerates upward at a rate of 5 m/s^2, what will the scale read?

   (A) 400 N
   (B) 800 N
   (C) 1,000 N
   (D) 1,200 N

3. A frictionless inclined plane of length 20 m has a maximum vertical height of 5 m. If an object of mass 2 kg is placed on the plane, which of the following best approximates the net force it feels?

   (A) 5 N
   (B) 10 N
   (C) 15 N
   (D) 20 N

4. A 20 N block is being pushed across a horizontal table by an 18 N force. If the coefficient of kinetic friction between the block and the table is 0.4, find the acceleration of the block.

   (A) 0.5 m/s^2
   (B) 1 m/s^2
   (C) 5 m/s^2
   (D) 7.5 m/s^2

5. The coefficient of static friction between a box and a ramp is 0.5. The ramp's incline angle is 30º. If the box is placed at rest on the ramp, the box will do which of the following?

   (A) Accelerate down the ramp
   (B) Accelerate briefly down the ramp but then slow down and stop
   (C) Move with constant velocity down the ramp
   (D) Not move

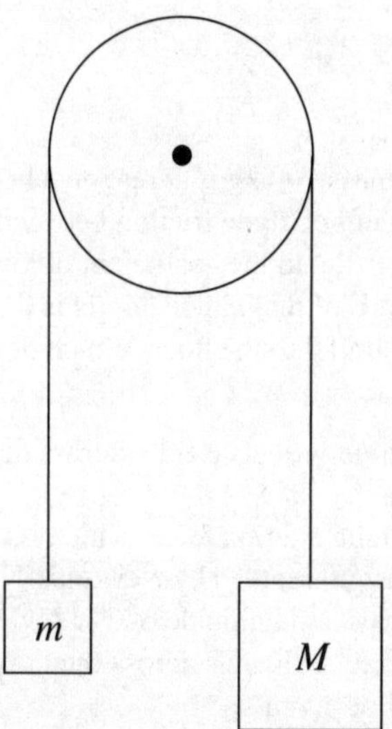

6. Assuming a frictionless, massless pulley, determine the acceleration of the blocks once they are released from rest.

   (A) $\frac{m}{M+m}g$
   (B) $\frac{M}{m}g$
   (C) $\frac{M+m}{M-m}g$
   (D) $\frac{M-m}{M+m}g$

7. If all of the forces acting on an object balance so that the net force is zero, then

(A) the object must be at rest
(B) the object's speed will decrease
(C) the object's direction of motion can change, but not its speed
(D) none of the above will occur

8. A block of mass m is at rest on a frictionless, horizontal table placed in a laboratory on the surface of the Earth. An identical block is at rest on a frictionless, horizontal table placed on the surface of the Moon. Let **F** be the net force necessary to give the Earth-bound block an acceleration of **a** across the table. Given that g_{Moon} is one-sixth of g_{Earth}, the force necessary to give the Moon-bound block the same acceleration **a** across the table is

(A) **F**/6
(B) **F**/3
(C) **F**
(D) 6**F**

9. A crate of mass 100 kg is at rest on a horizontal floor. The coefficient of static friction between the crate and the floor is 0.4, and the coefficient of kinetic friction is 0.3. A force **F** of magnitude 344 N is then applied to the crate, parallel to the floor. Which of the following is true?

(A) The crate will accelerate across the floor at 0.5 m/s^2.
(B) The static friction force, which is the reaction force to **F** as guaranteed by Newton's Third Law, will also have a magnitude of 344 N.
(C) The crate will slide across the floor at a constant speed of 0.5 m/s.
(D) The crate will not move.

10. Two crates are stacked on top of each other on a horizontal floor; Crate #1 is on the bottom, and Crate #2 is on the top. Both crates have the same mass. Compared to the strength of the force $\mathbf{F}_1$ necessary to push only Crate #1 at a constant speed across the floor, the strength of the force $\mathbf{F}_2$ necessary to push the stack at the same constant speed across the floor is greater than $\mathbf{F}_1$ because

(A) the normal force on Crate #1 is greater
(B) the coefficient of kinetic friction between Crate #1 and the floor is greater
(C) the coefficient of static friction between Crate #1 and the floor is greater
(D) the weight of Crate #1 is greater

Section II: Free Response

1. This question concerns the motion of a crate being pulled across a horizontal floor by a rope. In the diagram below, the mass of the crate is m, the coefficient of kinetic friction between the crate and the floor is μ, and the tension in the rope is $\mathbf{F}_T$.

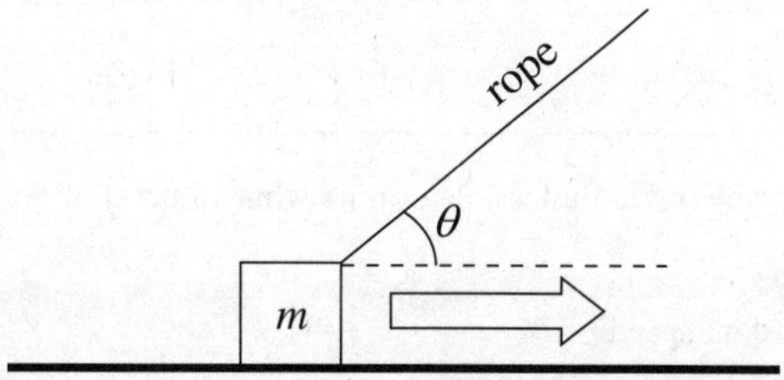

(A) Draw and label all of the forces acting on the crate.

(B) Compute the normal force acting on the crate in terms of m, F_T, θ, and g.

(C) Compute the acceleration of the crate in terms of m, F_T, θ, μ, and g.

2. In the diagram below, a massless string connects two blocks—of masses m_1 and m_2, respectively—on a flat, frictionless tabletop. A force **F** pulls on Block #2, as shown:

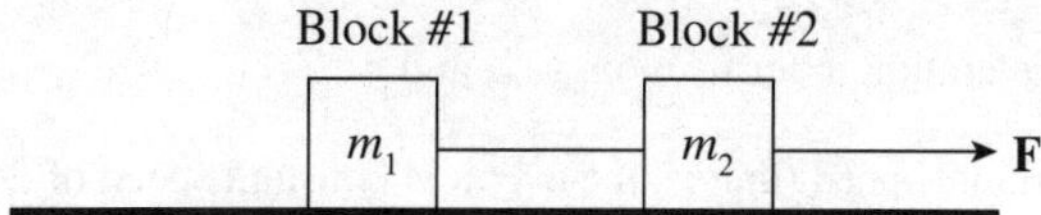

(A) Draw and label all of the forces acting on Block #1.

(B) Draw and label all of the forces acting on Block #2.

(C) What is the acceleration of Block #1? Please state your answer in terms of F, m_1, and m_2.

(D) What is the tension in the string connecting the two blocks? Please state your answer in terms of F, m_1, and m_2.

(E) If the string connecting the blocks were not massless, but instead had a mass of m, determine

i. the acceleration of Block #1, in terms of F, m, m_1, and m_2.

ii. the difference between the strength of the force that the connecting string exerts on Block #2 and the strength of the force that the connecting string exerts on Block #1. Please state your answer in terms of F, m, m_1, and m_2.

3. In the figure shown, assume that the pulley is frictionless and massless.

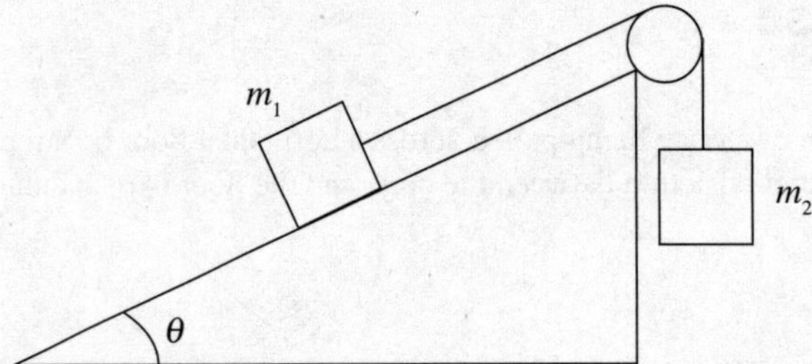

(A) If the surface of the inclined plane is frictionless, determine what value(s) of θ will cause the box of mass m_1 to

i. accelerate up the ramp
ii. slide up the ramp at constant speed

(B) If the coefficient of kinetic friction between the surface of the inclined plane and the box of mass m_1 is μ_k, derive (but do not solve) an equation satisfied by the value of θ, which will cause the box of mass m_1 to slide up the ramp at constant speed.

4. A skydiver is falling with speed v_0 through the air. At that moment (time $t = 0$), she opens her parachute and experiences the force of air resistance whose strength is given by the equation $F = kv$, in which k is a proportionality constant and v is her descent speed. The total mass of the skydiver and equipment is m. Assume that g is constant throughout her descent.

(A) Draw and label all the forces acting on the skydiver after her parachute opens.

(B) Determine the skydiver's acceleration in terms of m, v, k, and g.

(C) Determine the skydiver's terminal speed (that is, the eventual constant speed of descent).

(D) Sketch a graph of v as a function of time, being sure to label important values on the vertical axis.

Summary

- Forces are not needed to maintain motion. Forces cause objects to change their motion, whether that means speeding up, slowing down, or changing direction. This idea is expressed by Newton's Second Law: $F_{net} = ma$.
- Weight is commonly referred to as the force of gravity F_w or $F_g = mg$, where g is –10 m/s^2.
- Friction is defined by $F_f = \mu N$ and comes in two types—static and kinetic where $\mu_{kinetic} < \mu_{static}$.
- The normal force (N or sometimes F_N) is frequently given by N or $F_N = mg \cos \theta$, where θ is the angle between the horizontal axis and the surface on which the object rests.
- To solve almost any problem involving Newton's Second Law, use the following strategy:

  I. You must be able to visualize what's going on. Make a sketch if it helps. Make sure to make a free-body diagram (or FBD).

  II. Clearly define an appropriate coordinate system.

  III. Write out Newton's Second Law in the form of $\sum F_x = ma_x$ and/or $\sum F_y = ma_y$, using the forces identified in the FBD to fill in the appropriate forces.

  IV. Do the math.

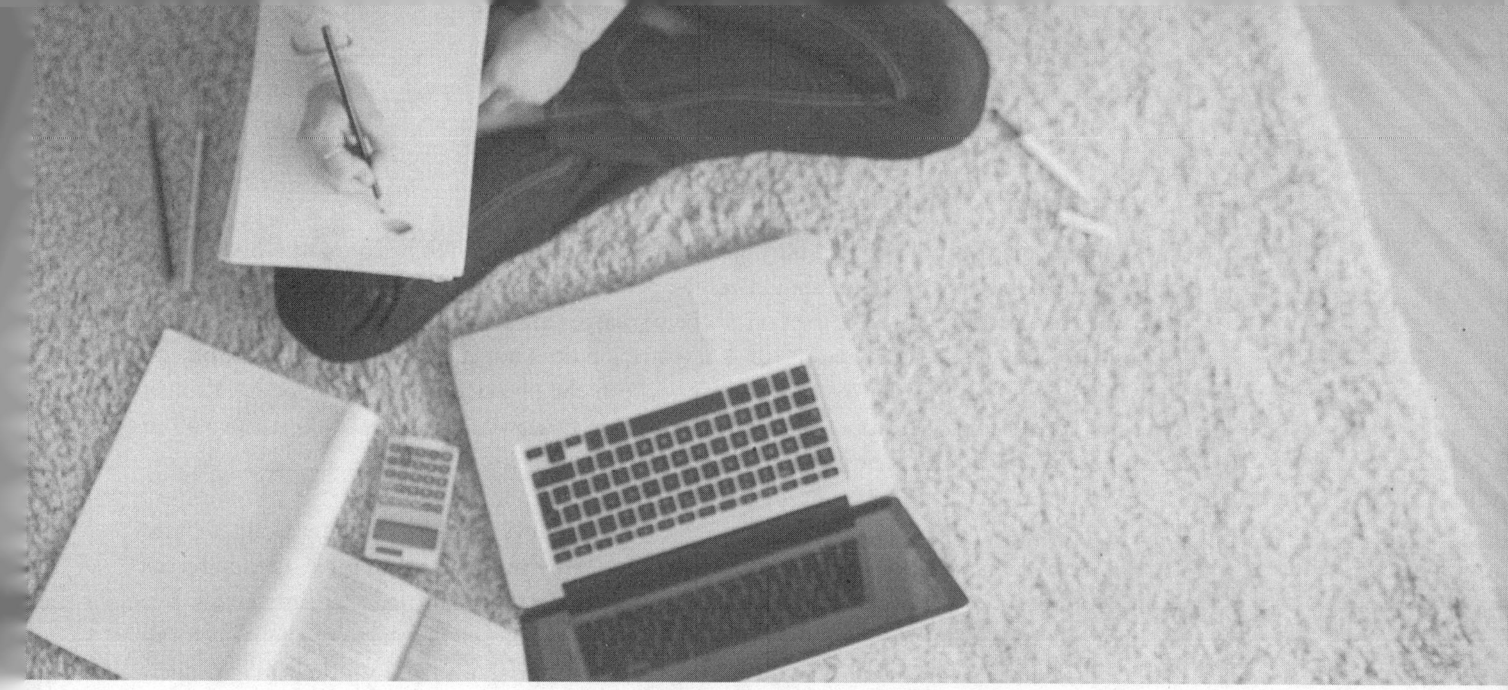

Chapter 6
Circular Motion and Gravitations

"I can calculate the motion of heavenly bodies but not the madness of people."

—Sir Isaac Newton

Since man has looked up to the sky, we have always tried to find a reason as to why celestial objects move the way they do. Kepler discovered that planets make elliptical orbits. Later, Newton realized the objects not moving in straight lines must experience an outside force. What force was making these orbits elliptical? Our first concepts explained linear motion. Then we added a second dimension and our motion became parabolic. Now we will explore when objects begin to undergo circular motion. Then we will have a better understanding of the orbit of our Moon around the Earth and the Earth around the Sun.

UNIFORM CIRCULAR MOTION

Let's simplify matters and consider the object's speed around its path to be constant. This is called **uniform circular motion**. You should remember that although the speed may be constant, the velocity is not, because the direction of the velocity is always changing. Since the velocity is changing, there must be acceleration. This acceleration does not change the speed of the object; it changes only the direction of the velocity to keep the object on its circular path. Also, in order to produce an acceleration, there must be a force; otherwise, the object would move off in a straight line (Newton's First Law).

Take a look at the figures below. The figure on the left shows an object moving along a circular trajectory, along with its velocity vectors at two nearby points. The vector $\mathbf{v}_1$ is the object's velocity at time $t = t_1$, and $\mathbf{v}_2$ is the object's velocity vector a short time later (at time $t = t_2$). The velocity vector is always tangential to the object's path (whatever the shape of the trajectory). Notice that since we are assuming constant speed, the lengths of $\mathbf{v}_1$ and $\mathbf{v}_2$ (their magnitudes) are the same.

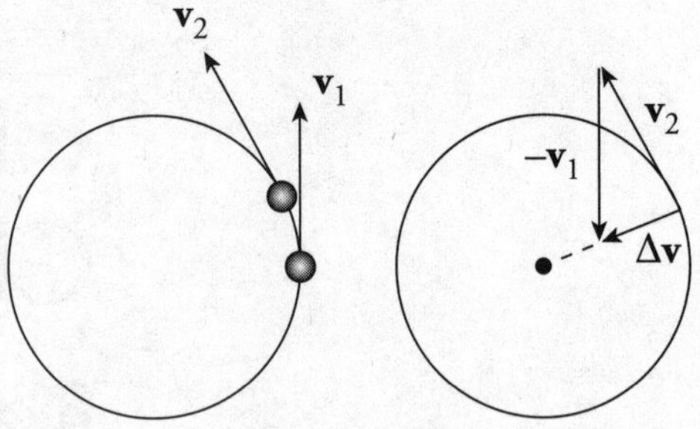

Non-Uniform Circular Objects
Planets and most objects do not undergo uniform circular motion. They usually follow elliptical orbits with varying speeds. For the purposes of the AP Physics 1 Exam, this will not be tested.

Since $\Delta\mathbf{v} = \mathbf{v}_2 - \mathbf{v}_1$ points toward the center of the circle (see the figure on the right), so does the acceleration, since $\mathbf{a} = \Delta\mathbf{v}/\Delta t$. Because the acceleration vector points toward the center of the circle, it's called **centripetal acceleration**, or $\mathbf{a}_c$. The centripetal acceleration is what turns the velocity vector to keep the object traveling in a circle. The magnitude of the centripetal acceleration depends on the object's speed, v, and the radius, r, of the circular path according to the equation

$$a_c = \frac{v^2}{r}$$

Example 1 An object of mass 5 kg moves at a constant speed of 6 m/s in a circular path of radius 2 m. Find the object's acceleration and the net force responsible for its motion.

Solution. By definition, an object moving at constant speed in a circular path is undergoing uniform circular motion. Therefore, it experiences a centripetal acceleration of magnitude v^2/r, always directed toward the center of the circle:

$$a_c = \frac{v^2}{r} = \frac{(6\text{ m/s})^2}{2\text{ m}} = 18\text{ m/s}^2$$

The force that produces the centripetal acceleration is given by Newton's Second Law, coupled with the equation for centripetal acceleration:

$$F_c = ma_c = m\frac{v^2}{r}$$

Such a Slacker
Centripetal force is the only force you need to know for the AP Physics 1 Exam that can NEVER do work. This is because, by definition, it is always perpendicular to the direction of the object's motion.

This equation gives the magnitude of the force. As for the direction, recall that because **F** = *m***a**, the directions of **F** and **a** are always the same. Since centripetal acceleration points toward the center of the circular path, so does the force that produces it. Therefore, it's called **centripetal force.** The centripetal force acting on this object has a magnitude of $F_c = ma_c$ = (5 kg)(18 m/s²) = 90 N.

Centripetal Force and Centrifugal Force
Centripetal force points into the center of the circle, and centrifugal points away from the circle. Centrifugal force is referred to as a fictitious force since it is not a real force. Centripetal force is the net force from the physical forces acting on the object. Neither one should ever be drawn on a force diagram.

Example 2 A 10.0 kg mass is attached to a string that has a breaking strength of 200 N. If the mass is whirled on a frictionless surface in a horizontal circle of radius 80 cm, what maximum speed can it have?

Solution. The first thing to do in problems like this is to identify what forces produce the centripetal acceleration. Notice that this is a horizontal circle. We can limit our examination to the horizontal (*x*) direction. Because gravity and the normal force exert forces in the *y* direction, they can be ignored. If we were given a problem with a vertical circle, we would have to include the effects of gravity, which will be demonstrated in Example 4. In this example, the tension in the string produces the centripetal force:

$$\mathbf{F}_T \text{ provides } \mathbf{F}_c \Rightarrow F_T = \frac{mv^2}{r} \Rightarrow v = \sqrt{\frac{rF_T}{m}} \Rightarrow v_{max} = \sqrt{\frac{rF_{T,\,max}}{m}}$$

$$= \sqrt{\frac{(0.80 \text{ m})(200 \text{ N})}{10 \text{ kg}}}$$

$$= 4 \text{ m/s}$$

Notice the unit change from 80 cm to 0.80 m. As a general rule, stick to kilograms, meters, and seconds because the newton is composed of these units.

Example 3 An athlete who weighs 800 N is running around a curve at a speed of 5.0 m/s in an arc whose radius of curvature, *r*, is 5.0 m. Find the centripetal force acting on him. What provides the centripetal force? What could happen to him if *r* were smaller?

Solution. Using the equation for the strength of the centripetal force, we find that

$$F_c = m\frac{v^2}{r} = \frac{F_w}{g} \cdot \frac{v^2}{r} = \frac{800 \text{ N}}{10 \text{ N/kg}} \cdot \frac{(5.0 \text{ m/s})^2}{5.0 \text{ m}} = 400 \text{ N}$$

In this case, static friction provides the centripetal force. If the radius of curvature of the arc were smaller, then the centripetal force required to keep him running in a circle would increase. If the centripetal force increased enough, it might exceed what the force of static friction could provide, at which point he would slip.

Example 4 A roller-coaster car enters the circular-loop portion of the ride. At the very top of the circle (where the people in the car are upside down), the speed of the car is 15 m/s, and the acceleration points straight down. If the diameter of the loop is 40 m and the total mass of the car (plus passengers) is 1,200 kg, find the magnitude of the normal force exerted by the track on the car at this point.

Solution. There are two forces acting on the car at its topmost point: the normal force exerted by the track and the gravitational force, both of which point downward. At the top of the loop, the gravitational force and the normal force point downward. This is because the normal force acts perpendicular to the surface of the track, and the gravitational force is always directed downward.

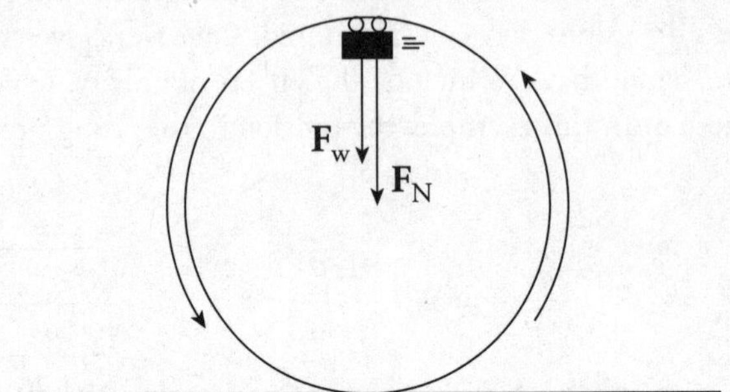

The combination of these two forces, $\mathbf{F}_N + \mathbf{F}_w$, provides the centripetal force:

$$\begin{aligned} F_N + F_w = \frac{mv^2}{r} \Rightarrow F_N &= \frac{mv^2}{r} - F_w \\ &= \frac{mv^2}{r} - mg \\ &= m\left(\frac{v^2}{r} - g\right) \\ &= (1{,}200 \text{ kg})\left[\frac{(15 \text{ m/s})^2}{\frac{1}{2}(40 \text{ m})} - 10 \text{ m/s}^2\right] \\ &= 1{,}500 \text{ N} \end{aligned}$$

Example 5 In the previous example, if the net force on the car at its topmost point is straight down, why doesn't the car fall straight down?

Solution. Remember that force tells an object how to accelerate. If the car had zero velocity at this point, then it would certainly fall straight down, but the car has a nonzero velocity (to the left) at this point. The fact that the acceleration is downward means that, at the next moment, v will point down to the left at a slight angle, ensuring that the car remains on a circular path, in contact with the track.

Example 6 How would the normal force change in Example 4 if the car was at the bottom of the circle?

Solution. There are still two forces acting on the car: the gravitational force still points downward, but the normal force pushes 90 degrees to the surface (upward). These forces now oppose one another. The combination of these two forces still provides the centripetal force. Because the centripetal acceleration points inward, we will make anything that points toward the center of the circle positive and anything that points away from the circle negative. Therefore, our equation becomes

$$F_N - F_w = \frac{mv^2}{r} \Rightarrow F_N = \frac{mv^2}{r} + F_w$$

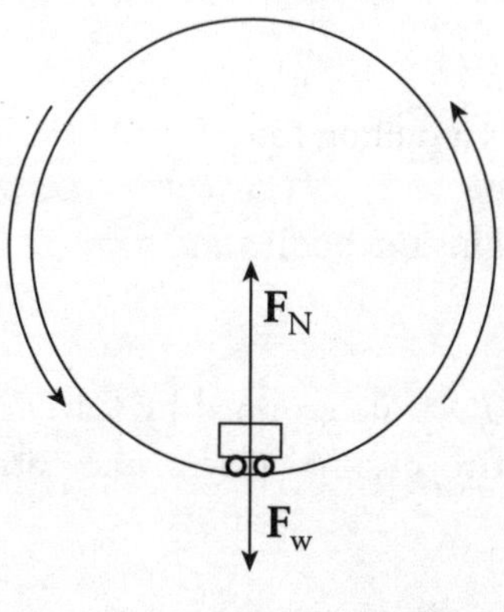

$$= \frac{mv^2}{r} + mg$$

$$= m\left(\frac{v^2}{r} + g\right)$$

$$= 1{,}200 \text{ kg}\left[\frac{(15 \text{ m/s})^2}{\frac{1}{2}(40 \text{ m})} + 10 \text{ m/s}^2\right]$$

$$= 25{,}500 \text{ N}$$

$\mathbf{F_N}$

$\mathbf{F_w}$

Notice the big difference between this answer and the answer from Example 4. This is why you would feel very little force between you and the seat at the top of the loop, but you would feel a big slam at the bottom of the loop.

NEWTON'S LAW OF GRAVITATION

A Mass by Any Other Name
Technically, the masses in Newton's Law of Gravitation refer to gravitational mass, whereas the one in Newton's Second Law is an inertial mass. Fortunately, they've been experimentally verified to be the same, so that we don't have to keep track of the difference!

Newton eventually formulated a law of gravitation: any two objects in the universe exert an attractive force on each other—called the **gravitational force**—whose strength is proportional to the product of the objects' masses and inversely proportional to the square of the distance between them as measured from center to center. If we let G be the **universal gravitational constant**, then the strength of the gravitational force is given by the following equation:

$$F_G = \frac{G m_1 m_2}{r^2}$$

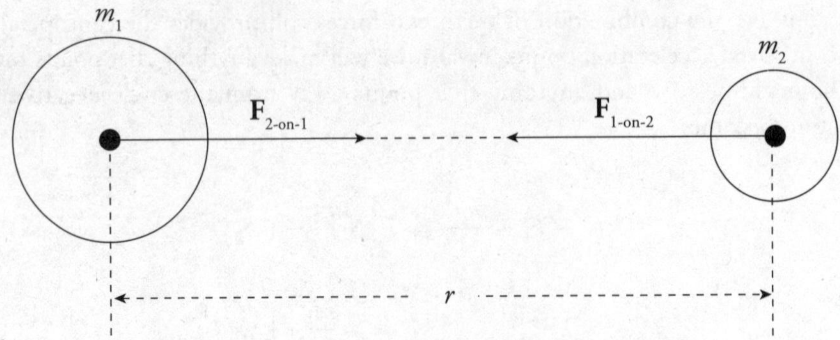

Gravity is always a pulling force.

The forces $\mathbf{F}_{1\text{-on-}2}$ and $\mathbf{F}_{2\text{-on-}1}$ act along the line that joins the bodies and form an action/reaction pair.

The first reasonably accurate numerical value for G was determined by Cavendish more than 100 years after Newton's law was published. To two decimal places, the currently accepted value of G is

$$G = 6.67 \times 10^{-11}\ \text{N} \cdot \text{m}^2/\text{kg}^2$$

Close Enough
Some might wonder why 10 m/s^2 is used for gravity everywhere on Earth. What if we're on top of a mountain? The difference between sea level and the peak of Mount Everest is less than half a percent. Super accurate measurement nitpicking is for chemistry class, not physics.

Why is g equal to 9.8 m/s^2? Generally, the acceleration due to gravity on an object of m at a distance r from a planet of mass M is

$$\vec{g} = \frac{\vec{F}_g}{m} = \frac{G\frac{Mm}{r^2}}{m} = \frac{GM}{r^2}$$

The force of gravity is attractive, so the direction of acceleration is toward the center of the planet. The acceleration due to gravity $\vec{g}$ is also known as a gravitational field. At any point in space, multiplying $\vec{g}$ by the mass of an object at that point gives the gravitational force acting on that object. Since the Earth is a sphere, r is approximately equal to the radius of the Earth at all points on its surface. Using the mass of the Earth and the radius of the Earth gives a magnitude of 9.8 m/s^2.

Example 7 Given that the radius of the Earth is 6.37×10^6 m, determine the mass of the Earth.

Solution. Consider a small object of mass m near the surface of the Earth (mass M). Its weight is mg, but its weight is just the gravitational force it feels due to the Earth, which is GMm/R^2. Therefore,

$$mg = G\frac{Mm}{R^2} \quad \Rightarrow \quad M = \frac{gR^2}{G}$$

Since we know that g = 10 m/s^2 and $G = 6.67 \times 10^{-11}$ N·m^2/kg^2, we can substitute to find

$$M = \frac{gR^2}{G} = \frac{(10\text{ m/s}^2)(6.37 \times 10^6\text{ m})^2}{6.67 \times 10^{-11}\text{ N}\cdot\text{m}^2/\text{kg}^2} = 6.1 \times 10^{24}\text{ kg}$$

Example 8 We can derive the expression $g = GM/R^2$ by equating mg and GMm/R^2 (as we did in the previous example), and this gives the magnitude of the *absolute gravitational acceleration*, a quantity that's sometimes denoted g_0. The notation g is acceleration, but with the spinning of the Earth taken into account. Show that if an object is at the equator, its *measured weight* (that is, the weight that a scale would measure), mg, is less than its *true weight*, mg_0, and compute the weight difference for a person of mass m = 60 kg.

Solution. Imagine looking down at the Earth from above the North Pole.

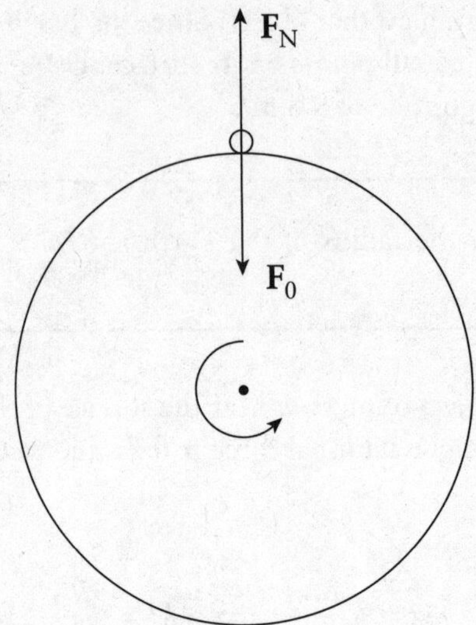

Relationships Are Important
Not just in daily life but especially in physics. Recognizing the relationship between variables in formulas means at times you will not need to do much calculation.

The net force toward the center of the Earth is $\mathbf{F}_0 - \mathbf{F}_N$, which provides the centripetal force on the object. Therefore,

$$F_0 - F_N = \frac{mv^2}{R}$$

Since $v = 2\pi R/T$, where T is the Earth's rotation period, we have

$$F_0 - F_N = \frac{m}{R}\left(\frac{2\pi R}{T}\right)^2 = \frac{4\pi^2 mR}{T^2}$$

or, since $F_0 = mg_0$ and $F_N = mg$,

$$mg_0 - mg = \frac{4\pi^2 mR}{T^2}$$

Since the quantity $\frac{4\pi^2 mR}{T^2}$ is positive, mg must be less than mg_0. The difference between mg_0 and mg, for a person of mass $m = 60$ kg, is only

$$\frac{4\pi^2 mR}{T^2} = \frac{4\pi^2(60 \text{ kg})(6.37 \times 10^6 \text{ m})}{\left(24 \text{ hr} \times \frac{60 \text{ min}}{\text{hr}} \times \frac{60 \text{ s}}{\text{min}}\right)^2} = 2.0 \text{ N}$$

and the difference between g_0 and g is

$$g_0 - g = \frac{mg_0 - mg}{m} = \frac{4\pi^2 R}{T^2} = \frac{4\pi^2(6.37 \times 10^6 \text{ m})}{\left(24 \text{ hr} \times \frac{60 \text{ min}}{\text{hr}} \times \frac{60 \text{ s}}{\text{min}}\right)^2} = 0.034 \text{ m/s}^2$$

Note that this difference is so small (< 0.3%) that it can usually be ignored.

Example 9 Communications satellites are often parked in geosynchronous orbits above Earth's surface. These satellites have orbit periods that are equal to Earth's rotation period, so they remain above the same position on Earth's surface. Determine the altitude that a satellite must have to be in a geosynchronous orbit above a fixed point on Earth's equator. (The mass of the Earth is 5.98×10^{24} kg.)

Solution. Let m be the mass of the satellite, M the mass of Earth, and R the distance from the center of Earth to the position of the satellite. The gravitational pull of Earth provides the centripetal force on the satellite, so

$$G\frac{Mm}{R^2} = \frac{mv^2}{R} \quad \Rightarrow \quad G\frac{M}{R} = v^2$$

The orbit speed of the satellite is $2\pi R/T$, so

$$G\frac{M}{R} = \left(\frac{2\pi R}{T}\right)^2$$

which implies that

$$G\frac{M}{R} = \frac{4\pi^2 R^2}{T^2} \quad \Rightarrow \quad 4\pi^2 R^3 = GMT^2 \quad \Rightarrow \quad R = \sqrt[3]{\frac{GMT^2}{4\pi^2}}$$

Now the key feature of a geosynchronous orbit is that its period matches Earth's rotation period, $T = 24$ hr. Substituting the numerical values of G, M, and T into this expression, we find that

$$R = \sqrt[3]{\frac{GMT^2}{4\pi^2}} = \sqrt[3]{\frac{(6.67\times 10^{-11})(5.98\times 10^{24})(24\cdot 60\cdot 60)^2}{4\pi^2}}$$
$$= 4.23\times 10^7 \text{ m}$$

Therefore, if r_E is the radius of Earth, then the satellite's altitude above Earth's surface must be

$$h = R - r_E = (4.23 \times 10^7 \text{ m}) - (6.37 \times 10^6 \text{ m}) = 3.59 \times 10^7 \text{ m}$$

Example 10 The Moon orbits Earth in a (nearly) circular path at (nearly) constant speed. If M is the mass of Earth, m is the mass of the Moon, and r is the Moon's orbit radius, find an expression for the Moon's orbit speed.

Solution: We begin by answering the question, What produces the centripetal force? The answer is the gravitational pull by the Earth. We now simply translate our answer into an equation, like this:

gravitational pull	produces	the centripetal force
F_{grav}	=	$m(v^2/r)$

Since we know $F_{grav} = GMm/r^2$, we get

$$F_{grav} = F_c$$

$$G(Mm/r^2) = m(v^2/r)$$

$$G(M/r) = v^2$$

$$v = (GM/r)^{(1/2)}$$

Notice that the mass of the Moon, m, cancels out. So, any object orbiting at the same distance from the Earth as the Moon must move at the same speed as the Moon.

CHAPTER 6 KEY TERMS

uniform circular motion
centripetal acceleration
centripetal force
Newton's Law of Gravitation
gravitational force
universal gravitational constant

Chapter 6 Review Questions

Answers and explanations can be found in Chapter 11.

Section I: Multiple Choice

1. An object moves at constant speed in a circular path. Which of the following statements are true? Select two answers.

(A) The velocity is changing.
(B) The velocity is constant.
(C) The magnitude of acceleration is constant.
(D) The magnitude of acceleration is changing.

Questions 2–3 refer to the situation below.

A 60 cm rope is tied to the handle of a bucket which is then whirled in a vertical circle. The mass of the bucket is 3 kg.

2. At the lowest point in its path, the tension in the rope is 50 N. What is the speed of the bucket?

(A) 1 m/s
(B) 2 m/s
(C) 3 m/s
(D) 4 m/s

3. What is the critical speed below which the rope would become slack when the bucket reaches the highest point in the circle?

(A) 0.6 m/s
(B) 1.8 m/s
(C) 2.4 m/s
(D) 4.8 m/s

4. An object moves at a constant speed in a circular path of radius r at a rate of 1 revolution per second. What is its acceleration?

(A) 0
(B) $2\pi^2 r$
(C) $2\pi^2 r^2$
(D) $4\pi^2 r$

5. If the distance between two point particles is doubled, then the gravitational force between them

(A) decreases by a factor of 4
(B) decreases by a factor of 2
(C) increases by a factor of 2
(D) increases by a factor of 4

6. At the surface of Earth, an object of mass m has weight w. If this object is transported to an altitude that is twice the radius of Earth, then at the new location,

(A) its mass is m and its weight is $w/2$
(B) its mass is $m/2$ and its weight is $w/4$
(C) its mass is m and its weight is $w/4$
(D) its mass is m and its weight is $w/9$

7. A moon of mass m orbits a planet of mass $100m$. Let the strength of the gravitational force exerted by the planet on the moon be denoted by F_1, and let the strength of the gravitational force exerted by the moon on the planet be F_2. Which of the following is true?

(A) $F_1 = 100F_2$
(B) $F_1 = 10F_2$
(C) $F_1 = F_2$
(D) $F_2 = 10F_1$

8. The dwarf planet Pluto has 1/500 the mass and 1/15 the radius of Earth. What is the value of g (in m/s^2) on the surface of Pluto?

(A) $\frac{50}{225}$
(B) $\frac{50}{15}$
(C) $\frac{15}{50}$
(D) $\frac{225}{50}$

9. A moon of Jupiter has a nearly circular orbit of radius R and an orbit period of T. Which of the following expressions gives the mass of Jupiter?

(A) $\dfrac{4\pi^2 R}{T^2}$

(B) $\dfrac{2\pi R^3}{(GT^2)}$

(C) $\dfrac{4\pi R^2}{(GT^2)}$

(D) $\dfrac{4\pi^2 R^3}{(GT^2)}$

10. Two large bodies, Body A of mass m and Body B of mass $4m$, are separated by a distance R. At what distance from Body A, along the line joining the bodies, would the gravitational force on an object be equal to zero? (Ignore the presence of any other bodies.)

(A) $\dfrac{R}{16}$

(B) $\dfrac{R}{8}$

(C) $\dfrac{R}{4}$

(D) $\dfrac{R}{3}$

11. You are looking at a top view of a planet orbiting the Sun in a clockwise direction. Which of the following would describe the velocity, acceleration, and force acting on the planet due to the Sun's pull at point P?

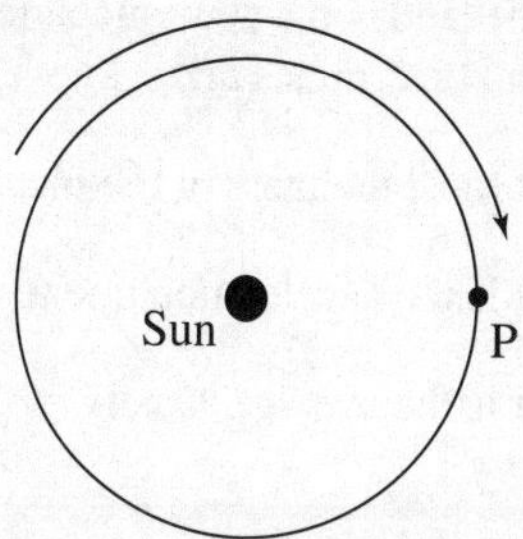

(A) $v\downarrow$ $a\uparrow$ $F\uparrow$
(B) $v\downarrow$ $a\leftarrow$ $F\leftarrow$
(C) $v\downarrow$ $a\rightarrow$ $F\rightarrow$
(D) $v\uparrow$ $a\leftarrow$ $F\leftarrow$

12. Which of the following statements are true for a satellite in outer space orbiting the Earth in uniform circular motion? Select two answers.

(A) There are no forces acting on the satellite.
(B) The force of gravity is the only force acting on the satellite.
(C) The force of gravity is balanced by outward force of the object.
(D) The mass of the satellite has no effect on the orbital speed.

Section II: Free Response

1. A robot probe lands on a new, uncharted planet. It has determined the diameter of the planet to be 8×10^6 m. It weighs a standard 1 kg mass and determines that 1 kg weighs only 5 newtons on this new planet.

   (A) What must the mass of the planet be?

   (B) What is the acceleration due to gravity on this planet? Express your answer in both m/s^2 and g's (where 1 g = 10 m/s^2).

   (C) What is the average density of this planet?

2. The Earth has a mass of 6×10^{24} kg and orbits the Sun in 3.15×10^7 seconds at a constant radius of 1.5×10^{11} m.

   (A) What is the Earth's centripetal acceleration around the Sun?

   (B) What is the gravitational force acting between the Sun and Earth?

   (C) What is the mass of the Sun?

3. An amusement park ride consists of a large cylinder that rotates around its central axis as the passengers stand against the inner wall of the cylinder. Once the passengers are moving at a certain speed, v, the floor on which they are standing is lowered. Each passenger feels pinned against the wall of the cylinder as it rotates. Let r be the inner radius of the cylinder.

   (A) Draw and label all the forces acting on a passenger of mass m as the cylinder rotates with the floor lowered.

   (B) Describe what conditions must hold to keep the passengers from sliding down the wall of the cylinder.

   (C) Compare the conditions discussed in part (b) for an adult passenger of mass m and a child passenger of mass $m/2$.

4. A curved section of a highway has a radius of curvature of r. The coefficient of friction between standard automobile tires and the surface of the highway is μ_s.

   (A) Draw and label all the forces acting on a car of mass m traveling along this curved part of the highway.

   (B) Compute the maximum speed with which a car of mass m could make it around the turn without skidding in terms of μ_s, r, g, and m.

   City engineers are planning to bank this curved section of highway at an angle of θ to the horizontal.

   (C) Draw and label all of the forces acting on a car of mass m traveling along this banked turn. Do not include friction.

   (D) The engineers want to be sure that a car of mass m traveling at a constant speed v (the posted speed limit) could make it safely around the banked turn even if the road were covered with ice (that is, essentially frictionless). Compute this banking angle θ in terms of r, v, g, and m.

Summary

- For objects undergoing uniform circular motion, the centripetal acceleration is given by $a_c = \frac{v^2}{r}$ and the centripetal force is given by $F_c = \frac{mv^2}{r}$.
- For any two masses in the universe, there is a gravitational attraction given by

$$F_G = \frac{Gm_1m_2}{r^2} \text{ where}$$

$$G = 6.67 \times 10^{-11} \frac{\text{N} \cdot \text{m}^2}{\text{kg}^2}$$

- The acceleration due to gravity on any planet is given by

$$g_{planet} = \frac{Gm_{planet}}{r^2}$$

- Many times, universal gravitation is linked up with circular motion (because planetary orbits are very nearly circular). Therefore, it is useful to keep the following equations for circular motion mentally linked for those questions that include orbits:

$$v = \frac{2\pi r}{T} \qquad a_c = \frac{v^2}{r} \qquad a_c = \frac{4\pi^2 r}{T^2}$$

$$F = ma_c \qquad F_c = \frac{mv^2}{r} \qquad F_c = \frac{4\pi^2 mr}{T^2}$$

Chapter 7
Energy

"Energy cannot be created or destroyed: it can only be changed from one form to another."

—Albert Einstein

Kinematics and dynamics are about change. Simple observations of our environment show us that change is occurring all around us. But what is needed to make an object change, and where did that change go to? It wasn't until more than 100 years after Newton that the idea of energy became incorporated into physics, but today it permeates every branch of the subject.

ENERGY: AN OVERVIEW

So, what is **energy**? How do we determine the energy of a system? These are not easy questions. It is difficult to give a precise definition of energy; there are different forms of energy as a result of different kinds of forces. Energy can come as a result of gravitational force, the speed of an object, stored energy in springs, heat loss to nuclear energy. But one truth remains the same for all of them: the Law of Conservation of Energy. Energy cannot just appear out of nowhere, nor can it disappear in a closed system; it must always take on another form. Force is the agent for this change, energy is the measure of that change, and work is the method of transferring energy from one system to another.

WORK

When you lift a book from the floor, you exert a force on it over a distance, and when you push a crate across a floor, you also exert a force on it over a distance. The application of force over a distance and the resulting change in energy of the system that the force acted on, give rise to the concept of **work**.

Work Units

The unit for work, the newton-meter (N·m), is renamed a joule and abbreviated as J.

Named after English physicist James Prescott Joule, one joule is the work required to produce one watt of power for one second.

Definition. If a constant force **F** acts over a distance **d** and **F** is parallel to **d**, then the work done by **F** is the product of force and distance:

$$W = \mathbf{Fd}$$

Notice that, although work depends on two vectors (**F** and **d**), work itself is *not* a vector. *Work is a scalar quantity.* However, even being a scalar quantity there exists positive, negative, and zero work.

Example 1 You slowly lift a book of mass 2 kg at constant velocity a distance of 3 m. How much work did you do on the book?

Solution. In this case, the force you exert must balance the weight of the book (otherwise the velocity of the book wouldn't be constant), so $F = mg = (2 \text{ kg})(10 \text{ m/s}^2) = 20$ N. Since this force is straight upward and the displacement of the book is also straight upward, **F** and **d** are parallel, so the work done by your lifting force is $W = Fd = (20 \text{ N})(3 \text{ m}) = 60$ N·m, or 60 J.

When the Formula for Work Works

$W = Fd \cos \theta$ works only when the forces do not change as the object moves. This also means a constant acceleration is delivered on the mass.

Work at an Angle

The previous formula works only when work is done completely parallel to the intended distance of travel. What happens when the force is done at an angle? The formula becomes:

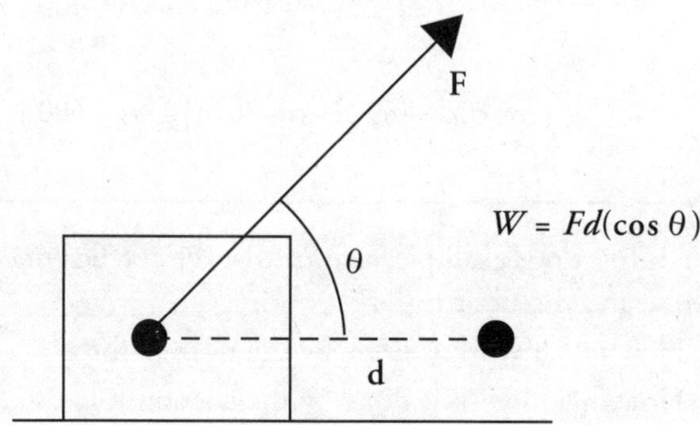

$$W = Fd\,(\cos\theta)$$

Let's compare the work done in a few instances with a force being applied between certain angles:

Angle (θ)	$0 \leq \theta < 90$	$\theta = 90$	$90 < \theta \leq 180$
Cos (θ)	positive	ZERO	negative
Work	positive	ZERO	negative
Speed of Object	increases	constant speed	decreases

Note that when we do positive work, we increase the speed of an object; however, when we do negative work, we slow an object down. This will be important to note when we relate work with kinetic energy and potential energy.

Example 2 A 15 kg crate is moved along a horizontal floor by a warehouse worker who's pulling on it with a rope that makes a 30° angle with the horizontal. The tension in the rope is 69 N, and the crate slides a distance of 10 m. How much work is done on the crate by the worker?

Don't Be Tricked!
A force applied perpendicular to the intended direction of motion always does ZERO work!

Solution. The figure on the next page shows that $\mathbf{F}_T$ and $\mathbf{d}$ are not parallel. It's only the component of the force acting along the direction of motion, $F_T \cos\theta$, that does work.

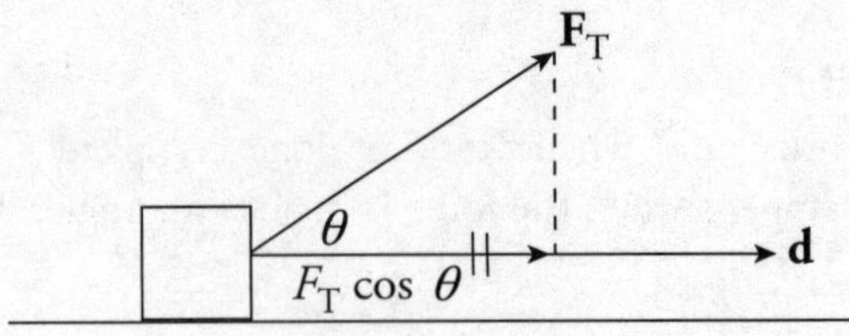

Therefore,

$$W = (F_T \cos\theta)d = (69 \text{ N} \cdot \cos 30°)(10 \text{ m}) = 600 \text{ J}$$

Example 3 In the previous example, assume that the coefficient of kinetic friction between the crate and the floor is 0.4.

(a) How much work is done by the normal force?

(b) How much work is done by the friction force?

Solution.

(a) Clearly, the normal force is not parallel to the motion, so we use the general definition of work. Since the angle between $\mathbf{F}_N$ and **d** is 90° (by definition of *normal*) and cos 90° = 0, the normal force does zero work.

(b) The friction force, $\mathbf{F}_f$, is also not parallel to the motion; it's *antiparallel*. That is, the angle between $\mathbf{F}_f$ and **d** is 180°. Since cos 180° = –1, and since the strength of the normal force is $F_n = F_w - F_{T,y} = mg - F_T \cdot \sin(\theta) = (15 \text{ kg})(10 \text{ m/s}^2) - (69 \text{ N})(1/2) = 115.5 \text{ N}$, the work done by the friction force is

$$W = -F_f d = -\mu_k F_N \cdot d = -(0.4)(115.5 \text{ N})(10 \text{ m}) = -462 \text{ J}$$

Antiparallel
When vectors that are parallel but pointing in opposite directions, if these vectors are joined at the tail, they form an angle of 180 degrees.

Example 4 A box slides down an inclined plane (incline angle = 37°). The mass of the block, m, is 35 kg, the coefficient of kinetic friction between the box and the ramp, μ_k, is 0.3, and the length of the ramp, d, is 8 m.

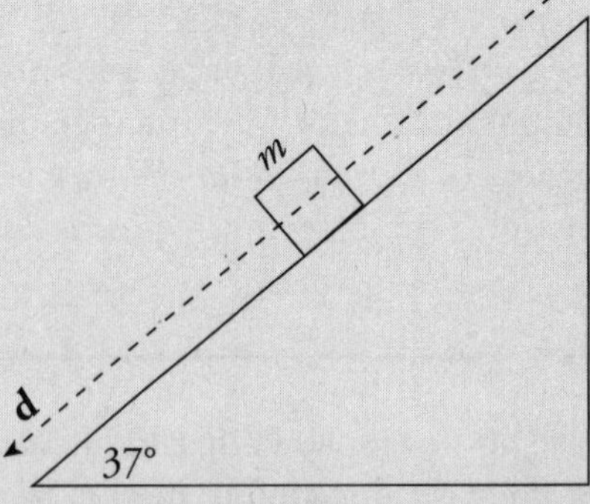

(a) How much work is done by gravity?

(b) How much work is done by the normal force?

(c) How much work is done by friction?

(d) What is the total work done?

Solution.

(a) Recall that the force that's directly responsible for pulling the box down the plane is the component of the gravitational force that's parallel to the ramp: $F_w \sin\theta = mg \sin\theta$ (where θ is the incline angle). This component is parallel to the motion, so the work done by gravity is

$$W_{\text{by gravity}} = (mg\sin\theta)d = (35\text{ kg})(10\text{ N/kg})(\sin 37°)(8\text{ m}) = 1{,}690\text{ J}$$

Note that the work done by gravity is positive, as we would expect it to be, since gravity is helping the motion. Also, be careful with the angle θ. The general definition of work reads $W = (F\cos\theta)d$, where θ is the angle between **F** and **d**. However, the angle between $\mathbf{F}_w$ and **d** is *not* 37° here, so the work done by gravity is not $(mg\cos 37°)d$. The angle θ used in the calculation above is the incline angle.

(b) Since the normal force is perpendicular to the motion, the work done by this force is zero.

(c) The strength of the normal force is $F_w \cos\theta$ (where θ is the incline angle), so the strength of the friction force is $F_f = \mu_k F_N = \mu_k F_w \cos\theta = \mu_k mg\cos\theta$. Since $\mathbf{F}_f$ is antiparallel to **d**, the cosine of the angle between these vectors (180°) is –1, so the work done by friction is

$$\begin{aligned} W_{\text{by friction}} &= -F_f d = -(\mu_k mg\cos\theta)(d) \\ &= -(0.3)(35\text{ kg})(10\text{ N/kg})(\cos 37°)(8\text{ m}) \\ &= -671\text{ J} \end{aligned}$$

Note that the work done by friction is negative, as we expect it to be, since friction is opposing the motion.

(d) The total work done is found simply by adding the values of the work done by each of the forces acting on the box:

$$W_{\text{total}} = \Sigma W = W_{\text{by gravity}} + W_{\text{by normal force}} + W_{\text{by friction}} = 1{,}690 + 0 + (-671) = 1{,}019\text{ J}$$

Zero Work

Part (b) of this question is a typical trick question. Just remember, a force applied perpendicularly to the direction of travel does zero work. Some people think that normal force can never do work, but that's taking things too far. For example, any time you ride up in an elevator, the normal force is doing work.

Work Done by a Variable Force

If a force remains constant over the distance through which it acts, then the work done by the force is simply the product of force and distance. However, if the force does not remain constant, then the work done by the force is given by the area under the curve of a force-versus-displacement graph. In physics language, the term "under the curve" really means between the line itself and zero.

Example 5 A spring exerts a force as shown on the graph below. How much work is done as the spring stretches from 20 to 40 cm?

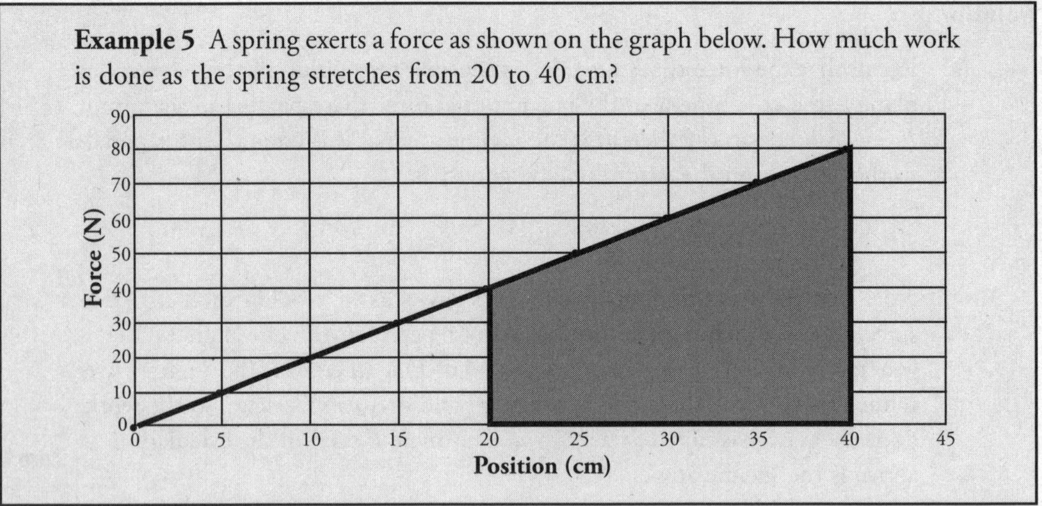

Solution. The area under the curve will be equal to the work done. In this case, we have some choices. You may recognize this shape as a trapezoid (it might help to momentarily rotate your head, or this book, 90 degrees to see this).

$$A = \frac{1}{2}(b_1 + b_2)h$$

$$A = \frac{1}{2}(40\text{N} + 80\text{N})(0.20\text{ m})$$

$$= 12\text{ N} \cdot \text{m or } 12\text{ J}$$

Similar to the previous chapter, units for work and energy should be confined to kg, m, and s, which is why we converted here.

An alternative choice is to recognize this shape as a triangle sitting on top of a rectangle. The total area is simply the area of the rectangle plus the area of the triangle.

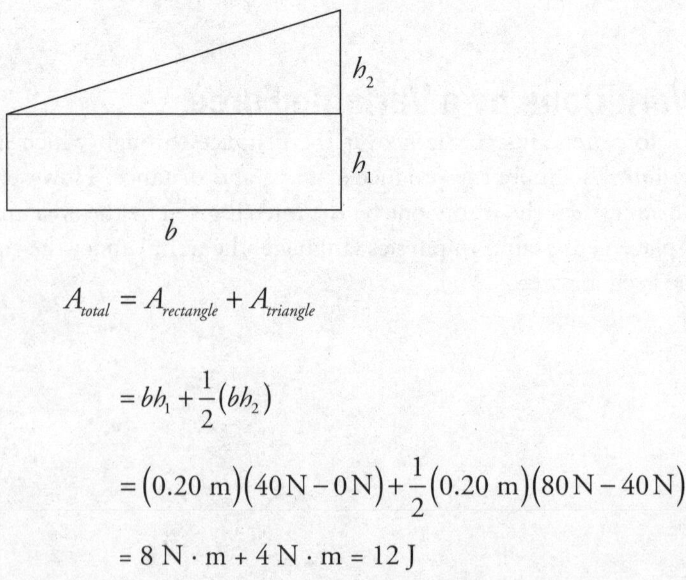

$$A_{total} = A_{rectangle} + A_{triangle}$$

$$= bh_1 + \frac{1}{2}(bh_2)$$

$$= (0.20\text{ m})(40\text{N} - 0\text{N}) + \frac{1}{2}(0.20\text{ m})(80\text{N} - 40\text{N})$$

$$= 8\text{ N} \cdot \text{m} + 4\text{ N} \cdot \text{m} = 12\text{ J}$$

KINETIC ENERGY

Consider an object at rest ($v_0 = 0$), and imagine that a steady force is exerted on it, causing it to accelerate. Let's be more specific; let the object's mass be m, and let **F** be the force acting on the object, pushing it in a straight line. The object's acceleration is $a = F/m$, so after the object has traveled a distance Δx under the action of this force, its final speed, v, is given by Big Five #5:

$$v^2 = v_0^2 + 2a(x - x_0) = 2a(x - x_0) = 2\frac{F}{m}(x - x_0) \quad \Rightarrow \quad F(x - x_0) = \frac{1}{2}mv^2$$

But the quantity $F(x–x_0)$ is the work done by the force, so $W = \frac{1}{2}mv^2$. The work done on the object has transferred energy to it, in the amount of $\frac{1}{2}mv^2$. The energy an object possesses by virtue of its motion is therefore defined as $\frac{1}{2}mv^2$ and is called **kinetic energy**:

$$K = \frac{1}{2}mv^2$$

Potential energy comes from things being in positions they don't want to be in. Two electrons don't want to be near each other. A spring doesn't want to be compressed. When a system moves to a more natural configuration (electrons move apart, spring returns to equilibrium, etc.), potential energy is lost.

THE WORK–ENERGY THEOREM

In the previous section, we derived kinetic energy from Big Five #5. Let's solve it a different way this time:

$v^2 = v_o^2 + 2ad$ Recall: $a = F/m$

$2(F/m)d = v^2 - v_o^2$ Recall: $W = Fd$

$W = (1/2)mv^2 - (1/2)mv_o^2$

$W = K_{\text{final}} - K_{\text{initial}}$

$$W_{\text{total}} = \Delta K$$

One of the questions posed at the beginning of this chapter was "How does a system gain or lose energy?" The **Work–Energy Theorem** begins to answer that question by stating that a system gains or loses kinetic energy by transferring it through work between the environment (forces being introduced into the system) and the system. In basic terms, doing positive work means increasing kinetic energy.

Need for Speed
As we mentioned before, when we do positive work, we increase the speed of an object. As a result, we also increase its kinetic energy. Doing positive work means making a positive change in kinetic energy.

Work Is Less Work
In some situations, using the Big Five Equations is easier. However, in several instances, skipping the Big Five and using work and energy can make problems much simpler. This is because we do not have to worry about the directional part of vectors and deal only with scalar quantities.

Example 6 What is the kinetic energy of a ball (mass = 0.10 kg) moving with a speed of 30 m/s?

Solution. From the definition,

$$K = \frac{1}{2}mv^2 = \frac{1}{2}(0.10 \text{ kg})(30 \text{ m/s})^2 = 45 \text{ J}$$

While the solutions in this book generally present just one method of obtaining the correct answer, that doesn't mean it's the only way. Don't worry about solving a problem the "wrong" way for the free-response questions. As long as your solution is valid, you'll get full credit, even if it isn't the simplest method.

Example 7 A tennis ball (mass = 0.06 kg) is hit straight upward with an initial speed of 50 m/s. How high would it go if air resistance were negligible?

Solution. This could be done using the Big Five, but let's try to solve it using the concepts of work and energy. As the ball travels upward, gravity acts on it by doing negative work. [The work is negative because gravity is opposing the upward motion. F_w and d are in opposite directions, so θ = 180°, which tells us that $W = (F_w \cos \theta)d = -F_w d$.] At the moment the ball reaches its highest point, its speed is 0, so its kinetic energy is also 0. The Work–Energy Theorem says

$$W = \Delta K \quad \Rightarrow \quad -F_w d = 0 - \frac{1}{2}mv_0^2 \quad \Rightarrow \quad d = \frac{\frac{1}{2}mv_0^2}{F_w} = \frac{\frac{1}{2}mv_0^2}{mg} = \frac{\frac{1}{2}v_0^2}{g} = \frac{\frac{1}{2}(50 \text{ m/s})^2}{10 \text{ m/s}^2} = 125 \text{ m}$$

Example 8 Consider the box sliding down the inclined plane in Example 4. If it starts from rest at the top of the ramp, with what speed does it reach the bottom?

Solution. It was calculated in Example 4 that W_{total} = 1,019 J. According to the Work–Energy Theorem,

$$W_{total} = \Delta K \quad \Rightarrow \quad W_{total} = K_f - K_i = K_f = \frac{1}{2}mv^2 \quad \Rightarrow \quad v = \sqrt{\frac{2W_{total}}{m}} = \sqrt{\frac{2(1{,}019 \text{ J})}{35 \text{ kg}}} = 7.6 \text{ m/s}$$

Example 9 A pool cue striking a stationary billiard ball (mass = 0.25 kg) gives the ball a speed of 2 m/s. If the average force of the cue on the ball was 200 N, over what distance did this force act?

Solution. The kinetic energy of the ball as it leaves the cue is

$$K = \frac{1}{2}mv^2 = \frac{1}{2}(0.25 \text{ kg})(2 \text{ m/s})^2 = 0.50 \text{ J}$$

The work (W) done by the cue gave the ball this kinetic energy, so

$$W = \Delta K \quad \Rightarrow \quad W = K_f \quad \Rightarrow \quad Fd = K \quad \Rightarrow \quad d = \frac{K}{F} = \frac{0.50 \text{ J}}{200 \text{ N}} = 0.0025 \text{ m} = 0.25 \text{ cm}$$

POTENTIAL ENERGY

Kinetic energy is the energy an object has by virtue of its motion. Potential energy is independent of motion; it arises from the object's position (or the system's configuration). For example, a ball at the edge of a tabletop has energy that could be transformed into kinetic energy if it falls off. An arrow in an archer's pulled-back bow has energy that could be transformed into kinetic energy if the archer releases the arrow. Both of these examples illustrate the concept of **potential energy**, the energy an object or system has by virtue of its position or configuration. In each case, work was done on the object to put it in the given configuration (the ball was lifted to the tabletop, the bowstring was pulled back), and since work is the means of transferring energy, these things have stored energy that can be retrieved as kinetic energy. Also remember that potential energy can be found in multiple sources such as chemical sources, mechanical sources, and objects in gravitational fields. This is potential energy, denoted by U.

Who Cares About Potential Energy?
What if an object were to never fall? Does it even matter if we calculate its potential energy? Calculating potential energy at a point does not tell us much. We only care about potential energy when we make it change from one point to another point. This in turn translates into doing work (increasing kinetic energy).

Because there are different types of forces, there are different types of potential energy. The ball at the edge of the tabletop provides an example of **gravitational potential energy**, U_g, which is the energy stored by virtue of an object's position in a gravitational field. This energy would be converted to kinetic energy as gravity pulled the ball down to the floor. For now, let's concentrate on gravitational potential energy.

Assume the ball has a mass m of 2 kg and that the tabletop is h = 1.5 m above the floor. How much work did gravity do as the ball was lifted from the floor to the table? The strength of the gravitational force on the ball is $F_w = mg$ = (2 kg)(10 N/kg) = 20 N. The force $\mathbf{F}_w$ points downward, and the ball's motion was upward, so the work done by gravity during the ball's ascent was

$$W_{\text{by gravity}} = -F_w h = -mgh = -(20 \text{ N})(1.5 \text{ m}) = -30 \text{ J}$$

So someone performed +30 J of work to raise the ball from the floor to the tabletop. That energy is now stored, and if the ball were given a push to send it over the edge, by the time the ball reached the floor, it would acquire a kinetic energy of 30 J. We therefore say that the change in the ball's gravitational potential energy in moving from the floor to the table was +30 J. That is,

$$\Delta U_g = -W_{\text{by gravity}}$$

Note that potential energy, like work (and kinetic energy), is expressed in joules.

In general, if an object of mass m is raised a height h (which is small enough that g stays essentially constant over this altitude change), then the increase in the object's gravitational potential energy is

$$\Delta U_g = mgh$$

One way to distinguish between conservative and nonconservative forces is that nonconservative forces do NOT exist in ideal situations.

An important fact that makes the above equation possible is that the work done by gravity as the object is raised does not depend on the path taken by the object. The ball could be lifted straight upward or in some curvy path; it would make no difference. Gravity is said to be a **conservative force** because of this property. Conversely, any work done by a **nonconservative force** is path-dependent. Such is the case for friction and air resistance, as different paths taken may require more or less work to get from the initial to final positions.

If we decide on a reference level to call $h = 0$, then we can say that the gravitational potential energy of an object of mass m at a height h is $U_g = mgh$. In order to use this last equation, it's essential that we choose a reference level for height. For example, consider a passenger in an airplane reading a book. If the book is 1 m above the floor of the plane, then, to the passenger, the gravitational potential energy of the book is mgh, where $h = 1$ m. However, to someone on the ground looking up, the floor of the plane may be, say, 9,000 m above the ground. So, to this person, the gravitational potential energy of the book is mgH, where $H = 9{,}001$ m. What both would agree on, though, is that the difference in potential energy between the floor of the plane and the position of the book is $mg \times (1 \text{ m})$, since the airplane passenger would calculate the difference as $mg \times (1 \text{ m} - 0 \text{ m})$, while the person on the ground would calculate it as $mg \times (9{,}001 \text{ m} - 9{,}000 \text{ m})$. Differences, or changes, in potential energy are unambiguous, but values of potential energy are relative.

Example 10 A stuntwoman (mass = 60 kg) scales a 40-meter-tall rock face. What is her gravitational potential energy (relative to the ground)?

Solution. Calling the ground $h = 0$, we find

$$U_g = mgh = (60 \text{ kg})(10 \text{ m/s}^2)(40 \text{ m}) = 24{,}000 \text{ J}$$

Example 11 If the stuntwoman in the previous example were to jump off the cliff, what would be her final speed as she landed on a large, air-filled cushion lying on the ground?

Solution. The gravitational potential energy would be transformed into kinetic energy. So

$$U \rightarrow K \quad \Rightarrow \quad U \rightarrow \frac{1}{2}mv^2 \quad \Rightarrow \quad v = \sqrt{\frac{2 \cdot U}{m}} = \sqrt{\frac{2(24{,}000 \text{ J})}{60 \text{ kg}}} = 28 \text{ m/s}$$

Go Online!
For answers to test-prep questions for all your tests and additional test taking tips, subscribe to our YouTube channel at www.youtube.com/ThePrincetonReview

One of the requirements for using the equation $\Delta U_g = mgh$ is that h is small enough that g is essentially constant over the altitude change. But what about larger altitude changes that might be experienced by objects such as satellites? For those problems, we need the more general expression for gravitational potential energy

$$U_G = -\frac{Gm_1m_2}{r}$$

Unlike problems using $\Delta U_g = mgh$ in which we could set $h = 0$ wherever it was convenient, this equation for gravitational potential energy defines $U = 0$ at $r = \infty$. The gravitational potential energy is always negative. This just means that energy has to be added to bring an object of mass m that is bound to the gravitational field of M to a point very far from M, at which $U = 0$.

Example 12 A payload on a rocket is launched from the surface of the Earth and travels to an altitude of 3.0×10^5 m. The radius of the Earth is 6.37×10^6 m, and the mass of the Earth is 5.98×10^{24} kg. If the payload has a mass of 1.3×10^5 kg, what is its change in gravitational potential energy during in this ascent?

Solution. The change in gravitational potential energy of the payload is

$$U_G = -\frac{Gm_1m_2}{r_2} + \frac{Gm_1m_2}{r_1} = Gm_1m_2\left(\frac{1}{r_1} - \frac{1}{r_2}\right)$$

The initial distance r_1 is the radius of the Earth. Since altitude is the height above the surface of the Earth, r_2 is the final altitude plus the radius of the Earth.

$$U_G = Gm_1m_2\left(\frac{1}{r_1}-\frac{1}{r_2}\right)$$

$$= (6.67 \times 10^{-11})(5.98 \times 10^{24})(1.3 \times 10^{5})\left(\frac{1}{6.37\times10^{6}}-\frac{1}{6.37\times10^{6}+3.0\times10^{5}}\right)$$

$$= 3.7 \times 10^{11}\ \text{J}$$

CONSERVATION OF MECHANICAL ENERGY

A Note About Kinetic and Potential Energy
Kinetic energy and potential energy have an inverse relationship. As kinetic energy increases, potential energy decreases, and vice-versa.

We have seen energy in its two basic forms: kinetic energy (K) and potential energy (U). The sum of an object's kinetic and potential energies is called its **total mechanical energy**, E.

$$E = K + U$$

(Note that because U is relative, so is E.)

Assuming that no nonconservative forces (friction, for example) act on an object or system while it undergoes some change, then mechanical energy is conserved. Mechanical energy, that is the sum of potential and kinetic energies, is dissipated or converted into other energy forms such as heat, by nonconservative forces. So, if there are no nonconservative forces acting on the system, the initial mechanical energy, E_i, is equal to the final mechanical energy, E_f, or

$$K_i + U_i = K_f + U_f$$

This is the simplest form of the **Law of Conservation of Total Energy**.

Let's evaluate a situation in which an object, initially at rest, is falling from a building and calculate the total mechanical energy in each situation:

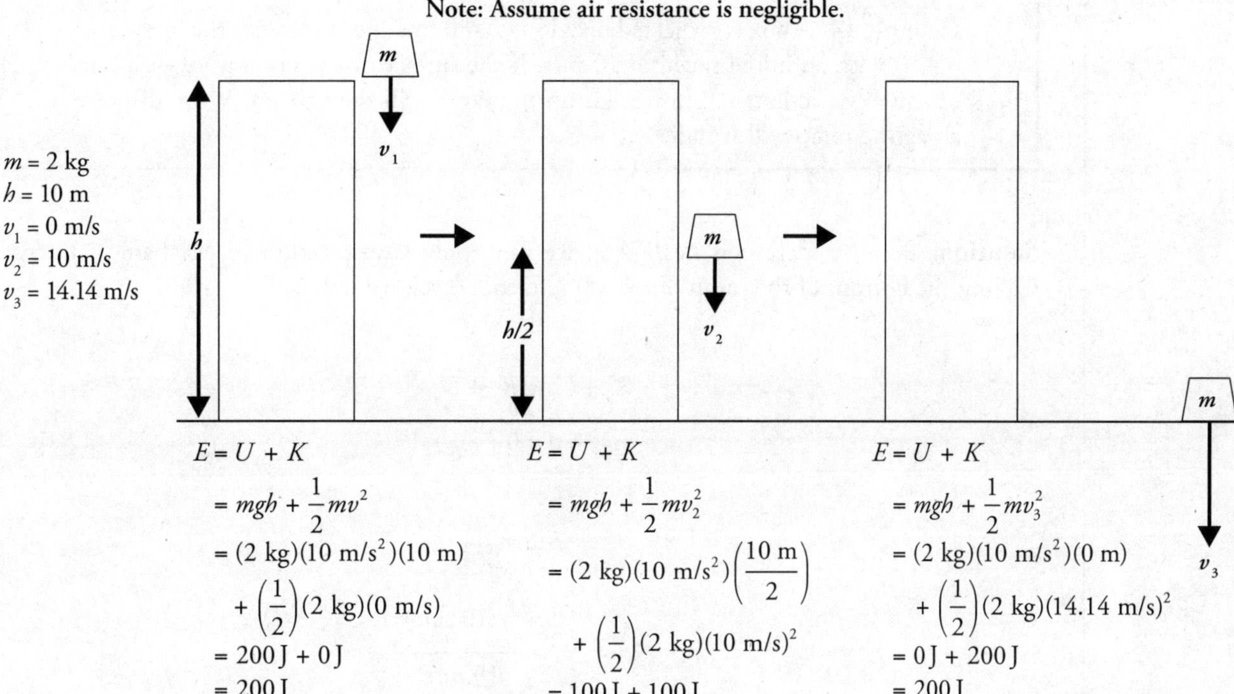

Note: Assume air resistance is negligible.
m = 2 kg
h = 10 m
v1 = 0 m/s
v2 = 10 m/s
v3 = 14.14 m/s
m
v1
h
h/2
v2
v3
E = U + K
= mgh + 1/2 mv^2
= (2 kg)(10 m/s^2)(10 m) + (1/2)(2 kg)(0 m/s)
= 200 J + 0 J
= 200 J
E = U + K
= mgh + 1/2 mv2^2
= (2 kg)(10 m/s^2)(10 m/2) + (1/2)(2 kg)(10 m/s)^2
= 100 J + 100 J
= 200 J
E = U + K
= mgh + 1/2 mv3^2
= (2 kg)(10 m/s^2)(0 m) + (1/2)(2 kg)(14.14 m/s)^2
= 0 J + 200 J
= 200 J

What can you say about the calculated total mechanical energy for each situation? They all are equal. An independent system obeys the Law of Conservation of Energy. Since the total mechanical energy remains the same throughout its travel, sometimes using energy to solve problems is a lot simpler than using kinematics.

Energy In = Energy Out

The Law of Conservation of Energy states that all the energy that was added to the universe during the Big Bang is conserved and energy cannot be added or removed from the universe. This means that all energy added to any system must come out in some form. For example, all of the energy used to bring a roller coaster to the top of the first hill is equal to all of the energy required to bring the roller coaster through all of the twists and loops and back to the starting point. Energy in this system is converted into other forms, but the total input of energy will always equal the total output.

Example 13 A ball of mass 2 kg is gently pushed off the edge of a tabletop that is 5.0 m above the floor. Find the speed of the ball as it strikes the floor.

Solution. Ignoring the friction due to the air, we can apply Conservation of Mechanical Energy. Calling the floor our $h = 0$ reference level, we write

$$
\begin{aligned}
K_i + U_i &= K_f + U_f \\
0 + mgh &= \frac{1}{2}mv^2 + 0 \\
v &= \sqrt{2gh} \\
&= \sqrt{2(10 \text{ m/s}^2)(5.0 \text{ m})} \\
&= 10 \text{ m/s}
\end{aligned}
$$

Note that the ball's potential energy decreased, while its kinetic energy increased. This is the basic idea behind Conservation of Mechanical Energy: one form of energy decreases while the other increases.

Example 14 A box is projected up a long ramp (incline angle with the horizontal = 37°) with an initial speed of 10 m/s. If the surface of the ramp is very smooth (essentially frictionless), how high up the ramp will the box go? What distance along the ramp will it slide?

Solution. Because friction is negligible, we can apply Conservation of Mechanical Energy. Calling the bottom of the ramp our h = 0 reference level, we write

$$\begin{aligned} K_{\text{i}} + U_{\text{i}} &= K_{\text{f}} + U_{\text{f}} \\ \frac{1}{2}mv_0^2 + 0 &= 0 + mgh \\ h &= \frac{\frac{1}{2}v_0^2}{g} \\ &= \frac{\frac{1}{2}(10\text{ m/s})^2}{10\text{ m/s}^2} \\ &= 5\text{ m} \end{aligned}$$

Since the incline angle is θ = 37°, the distance d it slides up the ramp is found in this way:

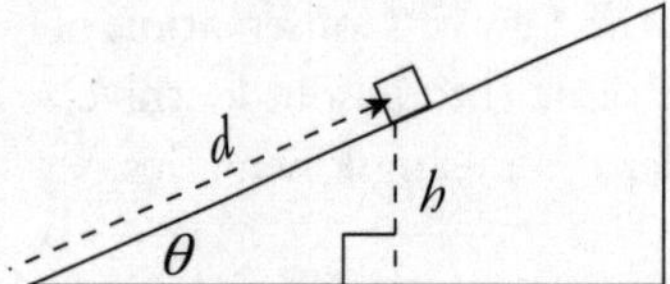

$$h = d\sin\theta$$

$$d = \frac{h}{\sin\theta} = \frac{5\text{ m}}{\sin 37°} = \frac{25}{3}\text{ m} = 8.3\text{ m}$$

Example 15 A skydiver jumps from a hovering helicopter that's 3,000 meters above the ground. If air resistance can be ignored, how fast will he be falling when his altitude is 2,000 m?

Solution. Ignoring air resistance, we can apply Conservation of Mechanical Energy. Calling the ground our h = 0 reference level, we write

$$\begin{aligned} K_{\text{i}} + U_{\text{i}} &= K_{\text{ff}} + U \\ 0 + mgH &= \frac{1}{2}mv^2 + mgh \\ v &= \sqrt{2g(H-h)} \\ &= \sqrt{2(10\text{ m/s}^2)(3{,}000\text{ m} - 2{,}000\text{ m})} \\ &= 140\text{ m/s} \end{aligned}$$

Example 16 With what minimum speed must an object of mass m be launched in order to escape the Earth's gravitational field? (This is called escape speed, v_{esc}.)

Solution. When launched, the object is at the surface of Earth, so r_i is the radius of the Earth, R_E. To escape the Earth's gravitational field, the object must travel far from the Earth, so that $U_f = 0$. To find the *minimum* launch speed, the object's final speed should be zero when it gets to this distant location. So, by Conservation of Energy,

$$K_i + U_i = K_f + U_f$$

$$\frac{1}{2}mv_i^2 + \frac{-GM_E m}{r_E} = 0 + 0$$

Solving for v_i,

$$\frac{1}{2}mv_i^2 = \frac{GM_E m}{r_E}$$

$$v_i = \sqrt{\frac{2GM_E}{r_E}}$$

Substituting the known numerical values for G, M_E, and r_E gives:

$$v_i = v_{esc} = \sqrt{\frac{2\left(6.67\times10^{-11}\right)\left(5.98\times10^{24}\right)}{\left(6.37\times10^{6}\right)}} = 1.12\times10^{4}\text{ m/s}$$

CONSERVATION OF ENERGY WITH NONCONSERVATIVE FORCES

The equation $K_i + U_i = K_f + U_f$ holds if no nonconservative forces are doing work. However, if work is done by such forces during the process under investigation, then the equation needs to be modified to account for this work as follows:

$$K_i + U_i + W_{other} = K_f + U_f$$

Why on the Left and Not the Right?
Work done by non-conservative forces is placed on the initial energy side because the final energy accounts for both the initial energy plus the energy that is dissipated by the object as it overcomes nonconservative forces.

Example 17 Wile E. Coyote (mass = 40 kg) falls off a 50-meter-high cliff. On the way down, the force of air resistance has an average strength of 100 N. Find the speed with which he crashes into the ground.

Solution. The force of air resistance opposes the downward motion, so it does negative work on the coyote as he falls: $W_r = -F_r h$. Calling the ground $h = 0$, we find that

$$K_i + U_i + W_r = K_f + U_f$$
$$0 + mgh + (-F_r h) = \frac{1}{2}mv^2 + 0$$
$$v = \sqrt{2h(g - F_r/m)} = \sqrt{2(50)(10 - 100/40)} = 27 \text{ m/s}$$

> **Example 18** A skier starts from rest at the top of a 20° incline and skis in a straight line to the bottom of the slope, a distance d (measured along the slope) of 400 m. If the coefficient of kinetic friction between the skis and the snow is 0.2, calculate the skier's speed at the bottom of the run.

Working Backward to Go Forward
Working with energy does not take vectors into account. With many problems, we can solve them backward with Conservation of Energy in order to find out the initial energy, work, or final energy; and in turn, we can solve for other useful information such as height or speed.

Solution. The magnitude of the friction force on the skier is $F_f = \mu_k F_N = \mu_k(mg\cos\theta)$, so the work done by friction is $-F_f d = \mu_k(mg \cos\theta)\cdot d$. The vertical height of the slope above the bottom of the run (which we designate the $h = 0$ level) is $h = d\sin\theta$. Therefore, Conservation of Mechanical Energy (including the negative work done by friction) gives

$$K_i + U_i + W_{\text{friction}} = K_f + U_f$$
$$0 + mgh + (-\mu_k mg\cos\theta\cdot d) = \tfrac{1}{2}mv^2 + 0$$
$$mg(d\sin\theta) + (-\mu_k mg\cos\theta\cdot d) = \tfrac{1}{2}mv^2$$
$$gd(\sin\theta - \mu_k\cos\theta) = \tfrac{1}{2}v^2$$
$$v = \sqrt{2gd(\sin\theta - \mu_k\cos\theta)}$$
$$= \sqrt{2(10)(400)[\sin 20° - (0.2)\cos 20°]}$$
$$= 35 \text{ m/s}$$

Time Is of No Difference (And Many Others)
Energy is almost always the easier approach than kinematics. We can throw out so many variables with energy.

So far, any of the problems we have solved in this chapter could have been solved using the kinematics equations and Newton's laws. The truly powerful thing about energy is that in a closed system, changes in energy are independent of the path you take. This allows you to solve many problems you would not otherwise be able to solve. With many energy problems, you do not need to measure time with a stopwatch, you do not need to know the mass of the object, you do not need a constant acceleration (remember that is required for our Big Five equations from kinematics), and you do not need to know the path the object takes.

Example 19 A roller coaster at an amusement park is at rest on top of a 30 m hill (point A). The car starts to roll down the hill and reaches point B, which is 10 m above the ground, and then rolls up the track to point C, which is 20 m above the ground.

(a) A student assumes no energy is lost, and solves for how fast the car is moving at point C using energy arguments. What answer does he get?

(b) If the final speed at C is actually measured to be 2 m/s, where did the lost energy go?

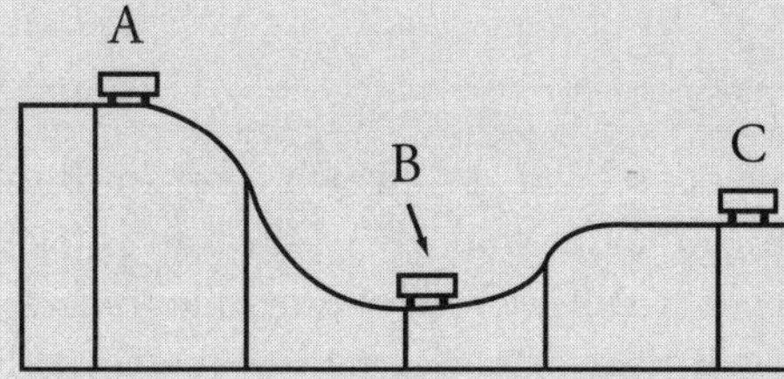

Solution.

(a) Our standard energy equation states

$$K_i + U_i = K_f + U_f$$

or

$$\frac{1}{2}mv_i^2 + mgh_i = \frac{1}{2}mv_f^2 + mgh_f$$

Canceling the mass, setting $v_i = 0$ m/s, and rearranging terms, we get

$$v_f = \sqrt{2g\Delta h}$$

$$v_f = \sqrt{21(10 \text{ m/s}^2)(30 \text{ m} - 20 \text{ m})}$$

$$v_f = \sqrt{200 \text{ m}^2/\text{s}^2} = 10\sqrt{2} \text{ m/s}$$

(b) The lost energy was likely lost as heat.

POWER

Simply put, **power** is the rate at which work gets done. In other words, it is the rate at which energy is transferred into, out of, or within a system. Suppose Scott and Jean each do 1,000 J of work, but Scott does the work in 2 minutes, while Jean does it in 1 minute. They both did the same amount of work, but Jean did it more quickly; thus, Jean was more powerful. Here's the definition of power:

$$\text{Power} = \frac{\text{Work}}{\text{time}} \quad \text{— in symbols} \rightarrow \quad P = \frac{W}{\Delta t} = \frac{\Delta E}{\Delta t}$$

The unit of power is the joule per second (J/s), which is renamed the **watt**, and symbolized W (not to be confused with the symbol for work, *W*). One watt is 1 joule per second: 1 W = 1 J/s. Here in the United States, which still uses older units like inches, feet, yards, miles, ounces, pounds, and so forth, you still hear of power ratings (particularly of engines) expressed in horsepower. One horsepower is defined as 1 hp = 746 W.

Note that this conversion will be provided on the test.

$P = W/t$ Recall: $W = Fd$

$P = Fd/t$ Recall: $v = d/t$

This equation only applies for a constant force parallel to a constant velocity. Remember to check that your equation fits the given circumstances!

$$P = W/t = Fd/t = Fv$$

Example 20 A mover pushes a large crate (mass $m = 75$ kg) from the inside of the truck to the back end (a distance of 6 m), exerting a steady push of 300 N. If he moves the crate this distance in 20 s, what is his power output during this time?

Solution. The work done on the crate by the mover is $W = Fd = (300\text{ N})(6\text{ m}) = 1{,}800$ J. If this much work is done in 20 s, then the power delivered is $P = W/t = (1{,}800\text{ J})/(20\text{ s}) = 90$ W.

Example 21 What must be the power output of an elevator motor that can lift a total mass of 1,000 kg and give the elevator a constant speed of 8.0 m/s?

Solution. The equation $P = Fv$, with $F = mg$, yields

$$P = mgv = (1{,}000 \text{ kg})(10 \text{ N/kg})(8.0 \text{ m/s}) = 80{,}000 \text{ W} = 80 \text{ kW}$$

Study Break
Phew! You have been working hard through these content chapters! Wrap up Chapter 7 and then give yourself a break—go for a walk, take a dance break, get a snack—before you continue on.

CHAPTER 7 KEY TERMS

energy
work
kinetic energy
Work–Energy Theorem
potential energy
gravitational potential energy
conservative force
nonconservative force
total mechanical energy
Law of Conservation of Total Energy
power
watt (W)

Chapter 7 Review Questions

Answers and explanations can be found in Chapter 11.

Section I: Multiple Choice

1. A force **F** of strength 20 N acts on an object of mass 3 kg as it moves a distance of 4 m. If **F** is perpendicular to the 4 m displacement, the work it does is equal to

   (A) 0 J
   (B) 60 J
   (C) 80 J
   (D) 600 J

2. Under the influence of a force, an object of mass 4 kg accelerates from 3 m/s to 6 m/s in 8 s. How much work was done on the object during this time?

   (A) 27 J
   (B) 54 J
   (C) 72 J
   (D) 96 J

3. A box of mass m slides down a frictionless inclined plane of length L and vertical height h. What is the change in its gravitational potential energy?

   (A) $-mgL$
   (B) $-mgh$
   (C) $-mgL/h$
   (D) $-mgh/L$

4. While a person lifts a book of mass 2 kg from the floor to a tabletop, 1.5 m above the floor, how much work does the gravitational force do on the book?

   (A) −30 J
   (B) −15 J
   (C) 0 J
   (D) 15 J

5. A satellite is currently orbiting Earth in a circular orbit of radius R; its kinetic energy is K_1. If the satellite is moved and enters a new circular orbit of radius $2R$, what will be its kinetic energy?

   (A) $\frac{K_1}{4}$
   (B) $\frac{K_1}{2}$
   (C) $2K_1$
   (D) $4K_1$

6. An object of mass m is traveling at constant speed v in a circular path of radius r. How much work is done by the centripetal force during one-half of a revolution?

   (A) πmv^2
   (B) 0
   (C) πmv^2r
   (D) $2\pi mv^2r$

7. A block of mass m slides from rest down an inclined plane of length s and height h. If F is the magnitude of the force of kinetic friction acting on the block as it slides, then the kinetic energy of the block when it reaches the bottom of the incline will be equal to

   (A) mgh
   (B) $mgs - Fh$
   (C) $mgh - Fs$
   (D) $mgs - Fs$

8. As a rock of mass 4 kg drops from the edge of a 40-meter-high cliff, it experiences air resistance, whose average strength during the descent is 20 N. At what speed will the rock hit the ground?

   (A) 10 m/s
   (B) 12 m/s
   (C) 16 m/s
   (D) 20 m/s

9. An astronaut drops a rock from the top of a crater on the Moon. When the rock is halfway down to the bottom of the crater, its speed is what fraction of its final impact speed?

(A) $\frac{1}{4}$

(B) $\frac{1}{2\sqrt{2}}$

(C) $\frac{1}{2}$

(D) $\frac{1}{\sqrt{2}}$

10. A force of 200 N is required to keep an object sliding at a constant speed of 2 m/s across a rough floor. How much power is being expended to maintain this motion?

(A) 50 W
(B) 100 W
(C) 200 W
(D) 400 W

Section II: Free Response

1. A box of mass m is released from rest at Point A, the top of a long, frictionless slide. Point A is at height H above the level of Points B and C. Although the slide is frictionless, the horizontal surface from Point B to C is not. The coefficient of kinetic friction between the box and this surface is μ_k, and the horizontal distance between Point B and C is x.

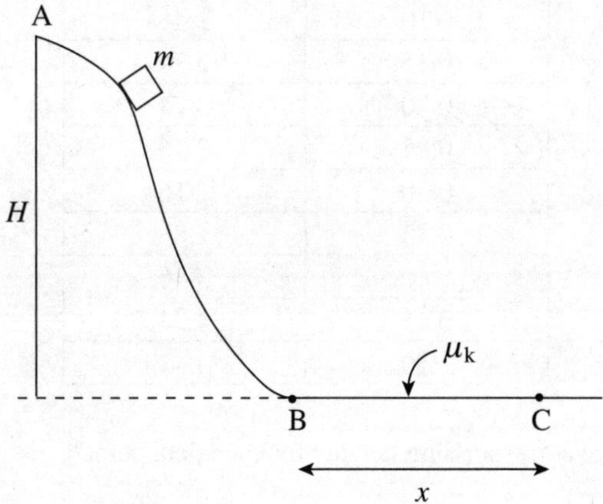

(A) Find the speed of the box when its height above Point B is $\frac{1}{2}H$.

(B) Find the speed of the box when it reaches Point B.

(C) Determine the value of μ_k so that the box comes to rest at Point C.

(D) Now assume that Points B and C were not on the same horizontal level. In particular, assume that the surface from B to C had a uniform upward slope so that Point C were still at a horizontal distance of x from B but now at a vertical height of y above B. Answer the question posed in part (C).

(E) If the slide were not frictionless, determine the work done by friction as the box moved from Point A to Point B if the speed of the box as it reached Point B were half the speed calculated in part (B).

2. A student uses a digital camera and computer to collect the following data about a ball as it slides down a curved frictionless track. The initial release point is 1.5 meters above the ground and the ball is released from rest. He prints up the following data and then tries to analyze it.

time (s)	velocity (m/s)
0.00	0.00
0.05	1.41
0.10	2.45
0.15	3.74
0.20	3.74
0.25	3.46
0.30	3.16
0.35	2.83
0.40	3.46
0.45	4.24
0.50	4.47

(A) Based on the data, what are the corresponding heights for each data point?

(B) What time segment experiences the greatest acceleration, and what is the value of this acceleration?

(C) How would the values change if the ball were replaced by an identical ball with double the mass?

3. A car with a mass of 800 kg is traveling with an initial speed of 10 m/s. When the brakes of the car are applied, the car starts to skid, and it experiences a frictional force with $\mu_k = 0.2$.

(A) What is the skidding distance of the car?

(B) How would the skidding distance change if the initial speed of the car were doubled?

(C) How would the skidding distance change if the initial mass of the car were doubled?

Summary

- Work is force applied across a displacement. Work can cause a change in energy. Positive work puts energy into a system, while negative work takes energy out of a system. Basic equations for work include

$$W = Fd\cos\theta$$

$$\text{Work} = \Delta KE$$

$$W = \text{area under an } F\text{-versus-}d \text{ graph}$$

- Energy is a conserved quantity. By that we mean the total initial energy is equal to the total final energy. Basic equations with energy include

$$K = \frac{1}{2}mv^2$$

$$U_g = mgh \text{ (for small changes to height at the surface of the Earth)}$$

$$U_G = -\frac{Gm_1m_2}{r} \text{ (for large changes in height)}$$

- Often, we limit ourselves to mechanical energy with no heat lost or gained. In this case,

$$K_i + U_i \pm W = K_f + U_f$$

- Power is the rate at which one does work and is given by

$$P = \frac{W}{\Delta t} = \frac{\Delta E}{\Delta t} \text{ or } P = Fv$$

Chapter 8
Momentum

"Nothing will happen until something moves."

—Albert Einstein

Previously we discovered the nature of objects, the reason why objects move the way they do, and the energy required to make objects move. Now we will predict the nature of objects when they interact with other objects and the resulting outcome.

INTRODUCTION TO MOMENTUM

A pool stick hitting the cue ball, a car collision, the Death Star exploding—physics is about the interaction of objects. A collision is a complex interaction between two objects. But what happens after two objects interact with each other? If we use Newton's Second Law, it will prove to be quite a tedious challenge. The forces occurring during a car collision are unimaginably complex. And not all collisions end with the same results: sometimes they move away from each other, sometimes they move in the same direction, and sometimes the objects stick together. If we were to measure their speeds before the collision and their speeds after the collision, we would also get different results. Using our observations and a general understanding of Newton's Second Law, we could predict that the smaller car in a collision (head on and both cars traveling with the same speed) would end up having the greater speed in the end. How can such a complex interaction give rise to a simple outcome?

WHAT IS MOMENTUM?

When Newton first expressed his second law, he didn't write $\mathbf{F}_{net} = m\mathbf{a}$. Instead, he expressed the law in the words, *The alteration of motion is...proportional to the...force impressed....* By "motion," he meant the product of mass and velocity, a vector quantity known as **linear momentum** and denoted by **p**:

Momentum Units
$p = mv$, so the unit for momentum is kg × m/s. There is no special term for this unit.

$$\mathbf{p} = m\mathbf{v}$$

So Newton's original formulation of the second law read $\Delta\mathbf{p} \propto \mathbf{F}$, or, equivalently, $\mathbf{F} \propto \Delta\mathbf{p}$. But a large force that acts for a short period of time can produce the same change in linear momentum as a small force acting for a greater period of time. Knowing this, we can turn the proportion above into an equation, if we take the average force that acts over the time interval Δt:

Momentum is a Vector
Remember that a vector has magnitude and direction. In collision problems, be aware of orientation and assign negative values to negative velocities and positive values to positive velocities.

$$\mathrm{F} = \frac{\Delta \mathrm{p}}{\Delta t}$$

This equation becomes $\mathbf{F} = ma$, since $\Delta\mathbf{p}/\Delta t = \Delta(m\mathbf{v})/\Delta t = m\,(\Delta\mathbf{v}/\Delta t) = ma$.

Reminder!
Remember that momentum and kinetic energy are not the same thing.

Example 1 A golfer strikes a golf ball of mass 0.05 kg, and the time of impact between the golf club and the ball is 1 ms. If the ball acquires a velocity of magnitude 70 m/s, calculate the average force exerted on the ball.

Solution. Using Newton's Second Law, we find

$$\mathbf{F}_{avg} = \frac{\Delta \mathbf{p}}{\Delta t} = \frac{\Delta(m\mathbf{v})}{\Delta t} = m\frac{\mathbf{v} - 0}{\Delta t} = (0.05\text{ kg})\frac{70\text{ m/s}}{10^{-3}\text{ s}} = 3{,}500\text{ N}$$

IMPULSE

The product of force and the time during which it acts is known as **impulse**; it's a vector quantity that's denoted by **J**:

$$\mathbf{J} = \mathbf{F}\Delta t$$

Newton's First Law
Objects naturally resist changes in their motion. In order for us to change an object's speed or direction, we must induce some kind of force over a period of time.

Impulse is equal to change in linear momentum. In terms of impulse, Newton's Second Law can be written in yet another form:

$$\mathbf{J} = \Delta \mathbf{p}$$

Sometimes this is referred to as the **Impulse–Momentum Theorem**, but it's just another way of writing Newton's Second Law. The impulse delivered to an object may be found by taking the area under a force-versus-time (also referred to as an *F*-versus-*t*) graph.

The Impulse–Momentum Theorem basically states that an impulse that is delivered on an object changes its momentum. The momentum "after" the collision is equal to the momentum "before" the collision added in with the impulse required to get your final outcome. In equation form, this is

$$\mathbf{p}_{final} = \mathbf{p}_{initial} + \mathbf{J}$$

Impulse eliminates the need to use $\mathbf{F} = m\mathbf{a}$ and simplifies the intricate forces delivered into mass, initial, and final velocities.

Example 2 A football team's kicker punts the ball (mass = 0.4 kg) and gives it a launch speed of 30 m/s. Find the impulse delivered to the football by the kicker's foot and the average force exerted by the kicker on the ball, given that the impact time is 8 ms.

Solution. As we know, impulse is equal to change in linear momentum, so

$$\mathbf{J} = \Delta \mathbf{p} = \mathbf{p}_f - \mathbf{p}_i = \mathbf{p}_f = m\mathbf{v} = (0.4\text{ kg})(30\text{ m/s}) = 12\text{ kg}\cdot\text{m/s}$$

Using the equation $\mathbf{F}_{avg} = \mathbf{J}/\Delta t$, we find that the average force exerted by the kicker is

$$\mathbf{F}_{avg} = \mathbf{J}/\Delta t = (12 \text{ kg} \cdot \text{m/s} / (8 \times 10^{-3} \text{ s}) = 1{,}500 \text{ N}$$

Example 3 An 80 kg stuntman jumps out of a window that's 45 m above the ground.

(a) How fast is he falling when he reaches ground level?

(b) He lands on a large, air-filled target, coming to rest in 1.5 s. What average force does he feel while coming to rest?

(c) What if he had instead landed on the ground (impact time = 10 ms)?

Solution.

(a) His gravitational potential energy turns into kinetic energy: $mgh = \frac{1}{2}mv^2$, so

$$v = \sqrt{2gh} = \sqrt{2(10)(45)} = 30 \text{ m/s}$$

(You could also have answered this question using Big Five #5.)

(b) Using $\mathbf{F} = \Delta\mathbf{p}/\Delta t$, we find that

$$\mathbf{F} = \frac{\Delta\mathbf{p}}{\Delta t} = \frac{\mathbf{p}_f - \mathbf{p}_i}{\Delta t} = \frac{0 - m\mathbf{v}_i}{\Delta t} = \frac{-(80 \text{ kg})(30 \text{ m/s})}{1.5 \text{ s}} = -1{,}600 \text{ N} \Rightarrow F = 1{,}600 \text{ N}$$

(c) In this case,

$$\mathbf{F} = \frac{\Delta\mathbf{p}}{\Delta t} = \frac{\mathbf{p}_f - \mathbf{p}_i}{\Delta t} = \frac{0 - m\mathbf{v}_i}{\Delta t} = \frac{-(80 \text{ kg})(30 \text{ m/s})}{10\times10^{-3} \text{ s}} = -240{,}000 \text{ N} \Rightarrow F = 240{,}000 \text{ N}$$

The negative signs in the vector answers to (b) and (c) simply tell you that the forces are acting in the opposite direction of motion and will cause the object to slow down. This force is equivalent to about 27 tons (!), more than enough to break bones and cause fatal brain damage. Notice how crucial impact time is: increasing the slowing-down time reduces the acceleration and the force, ideally enough to prevent injury. This is the purpose of air bags in cars, for instance.

Go Online!
For any new news about AP Physics 1, information and advice about college admissions, and much more, head to your online Student Tools!

Example 4 A small block of mass m = 0.07 kg, initially at rest, is struck by an impulsive force F of duration 10 ms whose strength varies with time according to the following graph:

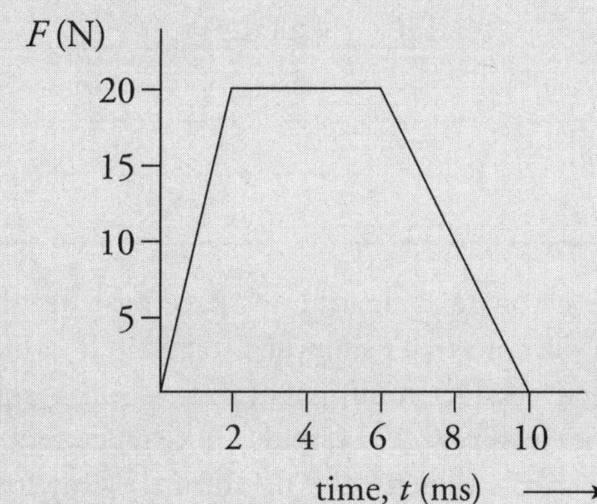

What is the resulting speed of the block?

Curved Graphs?
Calculus will be needed to find the true area under curved graphs. But the AP Physics 1 Exam does not require you to know calculus. You can approximate areas under curved graphs by using basic shapes you have studied before in geometry.

Solution. The impulse delivered to the block is equal to the area under the F-versus-t graph. The region is a trapezoid, so its area, $\frac{1}{2}$(base$_1$ + base$_2$) × height, can be calculated as follows:

$$J = \frac{1}{2}[(10\text{ ms} - 0) + (6\text{ ms} - 2\text{ ms})] \times (20\text{ N}) = 0.14\text{ N}\cdot\text{s}$$

Now, by the Impulse–Momentum Theorem,

$$\mathbf{J} = \Delta\mathbf{p} = \mathbf{p}_f - \mathbf{p}_i = m\mathbf{v} \quad\Rightarrow\quad \mathbf{v}_f = \frac{J}{m} = \frac{0.14\text{ N}\cdot\text{s}}{0.07\text{ kg}} = 2\text{ m/s}$$

I Found the Area, but What Does It Mean?
Whenever you find the area under a curve, you're essentially multiplying the quantities on the two axes. In this example, the area represents the product of force and time. That's how we know we're dealing with impulse.

CONSERVATION OF LINEAR MOMENTUM

Newton's Third Law states that when one object exerts a force on a second object, the second object exerts an equal but opposite force on the first. Newton's Third Law combines with Newton's Second Law when two objects interact with each other.

In the previous section, we redefined Newton's Second Law as Impulse–Momentum Theorem, **J** = Δ**p**. If we combine the laws and interpret them in terms of momentum, it states that two interacting objects experience equal but opposite momentum changes (assuming we have an isolated system, meaning no external forces).

> The total linear momentum of an isolated system remains constant.

No Exceptions!
Just like Conservation of Energy, the Law of Conservation of Momentum is a universal law. There are no special circumstances that allow for violations. While it won't always be helpful to consider that momentum is being conserved, it is nevertheless true in every situation.

The momentum "before" equals the momentum "after." This is the Law of Conservation of Momentum, which states

$$\text{total } \mathbf{p}_{\text{initial}} = \text{total } \mathbf{p}_{\text{final}}$$

Example 5 An astronaut is floating in space near her shuttle when she realizes that the cord that's supposed to attach her to the ship has become disconnected. Her total mass (body + suit + equipment) is 91 kg. She reaches into her pocket, finds a 1 kg metal tool, and throws it out into space with a velocity of 9 m/s directly away from the ship. If the ship is 10 m away, how long will it take her to reach it?

Solution. Here, the astronaut + tool are the system. Because of Conservation of Linear Momentum,

$$\begin{aligned} m_{\text{astronaut}}\mathbf{v}_{\text{astronaut}} + m_{\text{tool}}\mathbf{v}_{\text{tool}} &= 0 \\ m_{\text{astronaut}}\mathbf{v}_{\text{astronaut}} &= -m_{\text{tool}}\mathbf{v}_{\text{tool}} \\ \mathbf{v}_{\text{astronaut}} &= -\frac{m_{\text{tool}}}{m_{\text{astronaut}}}\mathbf{v}_{\text{tool}} \\ &= -\frac{1\text{ kg}}{90\text{ kg}}(-9\text{ m/s}) = +0.1\text{ m/s} \end{aligned}$$

Using *distance* = *rate* × *time*, we find

$$t = \frac{d}{v} = \frac{10\text{ m}}{0.1\text{ m/s}} = 100\text{ s}$$

COLLISIONS

Conservation of Linear Momentum is routinely used to analyze **collisions**. The objects whose collision we will analyze form the *system*, and although the objects exert forces on each other during the impact, these forces are only *internal* (they occur within the system), and the system's total linear momentum is conserved.

Let's break down the collision types:

Two objects collide with one another.

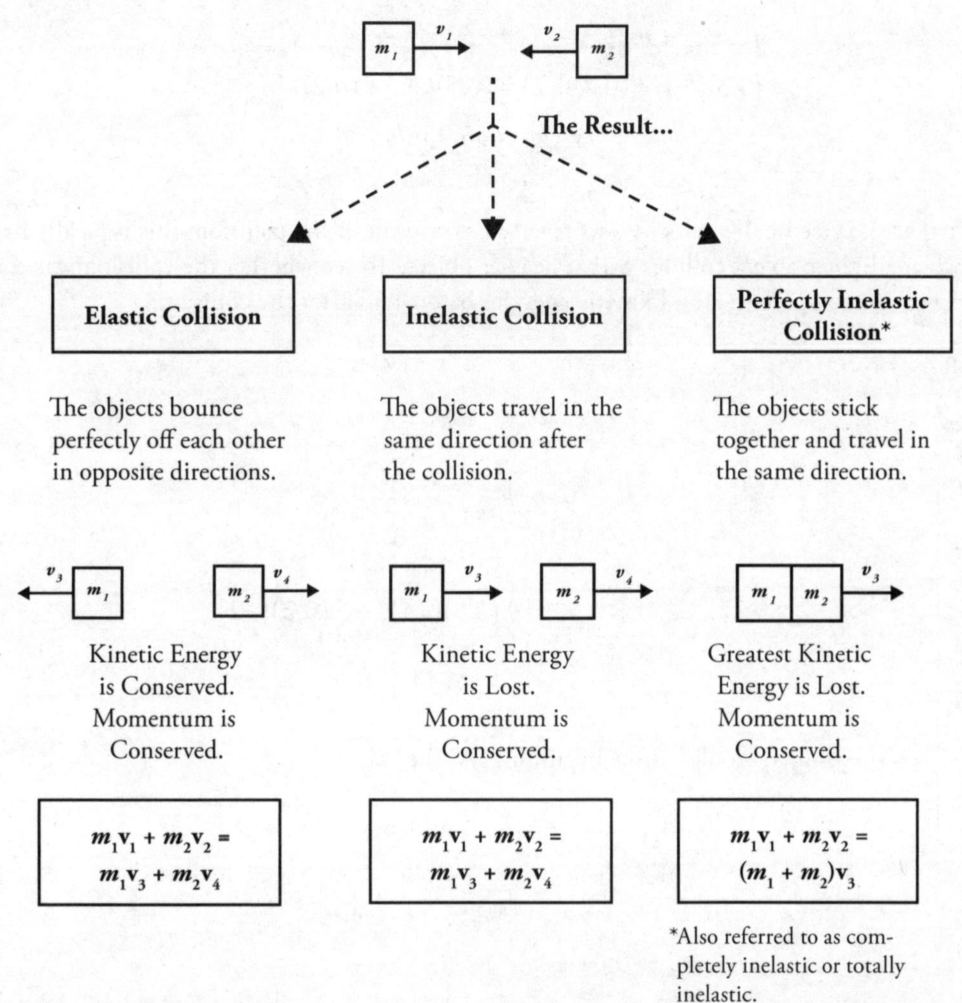

m_1 v_1 v_2 m_2
The Result...
Elastic Collision
Inelastic Collision
Perfectly Inelastic Collision*
The objects bounce perfectly off each other in opposite directions.
The objects travel in the same direction after the collision.
The objects stick together and travel in the same direction.
v_3 m_1 m_2 v_4
m_1 v_3 m_2 v_4
m_1 m_2 v_3
Kinetic Energy is Conserved. Momentum is Conserved.
Kinetic Energy is Lost. Momentum is Conserved.
Greatest Kinetic Energy is Lost. Momentum is Conserved.
$m_1v_1 + m_2v_2 = m_1v_3 + m_2v_4$
$m_1v_1 + m_2v_2 = m_1v_3 + m_2v_4$
$m_1v_1 + m_2v_2 = (m_1 + m_2)v_3$
*Also referred to as completely inelastic or totally inelastic.

Example 6 Two balls roll toward each other. The red ball has a mass of 0.5 kg and a speed of 4 m/s just before impact. The green ball has a mass of 0.2 kg and a speed of 2 m/s. After the head-on collision, the red ball continues forward with a speed of 2 m/s. Find the speed of the green ball after the collision. Was the collision elastic?

Solution. First, remember that momentum is a vector quantity, so the direction of the velocity is crucial. Since the balls roll toward each other, one ball has a positive velocity while the other has a negative velocity. Let's call the red ball's velocity before the collision positive; then $v_{red} = +4$ m/s, and $v_{green} = -2$ m/s.

Remember Vectors!
Momentum is a vector. If you set an object traveling to the right as the positive momentum, you must give an object going to the left a negative momentum.

Using a prime to denote *after the collision*, Conservation of Linear Momentum gives us the following:

$$\begin{aligned} \text{total } \mathbf{p}_{\text{before}} &= \text{total } \mathbf{p}_{\text{after}} \\ m_{\text{red}}\mathbf{v}_{\text{red}} + m_{\text{green}}\mathbf{v}_{\text{green}} &= m_{\text{red}}\mathbf{v}'_{\text{red}} + m_{\text{green}}\mathbf{v}'_{\text{green}} \\ (0.5)(+4) + (0.2)(-2) &= (0.5)(+2) + (0.2)\mathbf{v}'_{\text{green}} \\ \mathbf{v}'_{\text{green}} &= +3.0 \text{ m/s} \end{aligned}$$

Notice that the green ball's velocity was reversed as a result of the collision; this typically happens when a lighter object collides with a heavier object. To see whether the collision was elastic, we need to compare the total kinetic energies before and after the collision.

Initially,

$$\begin{aligned} K_t &= K_1 + K_2 \\ &= \frac{1}{2}m_1 v_{i1}^2 + \frac{1}{2}m_2 v_{i2}^2 \\ &= \frac{1}{2}(0.5)(+4)^2 + \frac{1}{2}(0.2)(-2)^2 \\ &= 4.4 \text{ J} \end{aligned}$$

Elastic Versus Inelastic
All collisions conserve momentum. In order to determine whether an unknown collision was elastic or inelastic, we must check if kinetic energy was lost. If kinetic energy was lost, then the collision was inelastic.

There are 4.4 joules at the beginning. At the end,

$$\begin{aligned} K_t &= K_1 + K_2 \\ &= \frac{1}{2}m_1 v_{f1}^2 + \frac{1}{2}m_2 v_{f2}^2 \\ &= \frac{1}{2}(0.5)(+2)^2 + \frac{1}{2}(0.2)(3)^2 \\ &= 1.9 \text{ J} \end{aligned}$$

So, there is less kinetic energy at the end compared to the beginning. Kinetic energy was lost (so the collision was inelastic). Most of the lost energy was transferred as heat; the two objects are both slightly warmer as a result of the collision.

Example 7 Two balls roll toward each other. The red ball has a mass of 0.5 kg and a speed of 4 m/s just before impact. The green ball has a mass of 0.3 kg and a speed of 2 m/s. If the collision is completely inelastic, determine the velocity of the composite object after the collision.

Solution. If the collision is completely inelastic, then, by definition, the masses stick together after impact, moving with a velocity, v′. Applying Conservation of Linear Momentum, we find

$$\begin{aligned} \text{total } \mathbf{p}_{\text{before}} &= \text{total } \mathbf{p}_{\text{after}} \\ m_{\text{red}}\mathbf{v}_{\text{red}} + m_{\text{green}}\mathbf{v}_{\text{green}} &= (m_{\text{red}} + m_{\text{green}})\mathbf{v}' \\ (0.5)(4) + (0.3)(-2) &= (0.5 + 0.3)\mathbf{v}' \\ \mathbf{v}' &= +1.8 \text{ m/s} \end{aligned}$$

Example 8 A 500 kg car travels 20 m/s due north. It hits a 500 kg car traveling due west at 30 m/s. The cars lock bumpers and stick together. What is the velocity the instant after impact?

Solution. This problem illustrates the vector nature of numbers. First, look only at the x (east-west) direction. There is only one car moving west and its momentum is given by

$$\mathbf{p} = m\mathbf{v} \Rightarrow (500 \text{ kg})(30 \text{ m/s}) = 15{,}000 \text{ kg} \cdot \text{m/s west}$$

Next, look only at the y (north-south) direction. There is only one car moving north and its momentum is given by

$$\mathbf{p} = m\mathbf{v} \Rightarrow (500 \text{ kg})(20 \text{ m/s}) = 10{,}000 \text{ kg} \cdot \text{m/s north}$$

The total final momentum is the resultant of these two vectors. Use the Pythagorean Theorem:

$$(10{,}000)^2 + (15{,}000)^2 = \mathbf{p}_f^2$$

$$\mathbf{p} = 18{,}027 \text{ kg} \cdot \text{m/s}$$

10,000

p

15,000

To find the total velocity, you need to solve for v using

$$\mathbf{p}_f = \text{m}\boldsymbol{v} \Rightarrow 18{,}027 \text{ kg} \cdot \text{m/s} = (500 \text{ kg} + 500 \text{ kg})(v)$$

$$\boldsymbol{v} = 18 \text{ m/s}$$

However, we are not done yet because velocity has both a magnitude (which we now know is 18 m/s) and a direction. The direction can be expressed using $\tan\theta = 10{,}000/15{,}000$ so $\theta = \tan^{-1}(10{,}000/15{,}000)$, or 33.7 degrees north of west. Again, most of the AP Physics 1 Exam is in degrees, not radians.

Example 9 An object of mass m moves with velocity **v** toward a stationary object of mass $2m$. After impact, the objects move off in the directions shown in the following diagram:

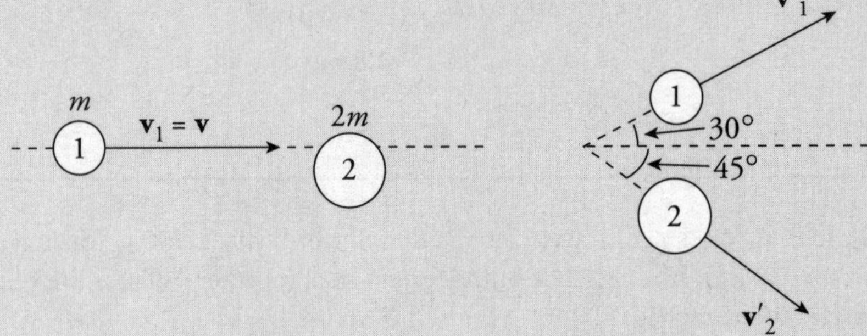

(a) Determine the magnitudes of the velocities after the collision (in terms of v).

(b) Is the collision elastic? Explain your answer.

Solution.

(a) Conservation of Linear Momentum is a principle that establishes the equality of two vectors: $\mathbf{p}_{\text{total}}$ before the collision and $\mathbf{p}_{\text{total}}$ after the collision. Writing this single vector equation as two equations, one for the x-component and one for the y, we have

$$x\text{-component:}\quad mv = mv_1'\cos 30° + 2mv_2'\cos 45° \quad (1)$$

$$y\text{-component:}\quad 0 = mv_1'\sin 30° - 2mv_2'\sin 45° \quad (2)$$

Adding these equations eliminates v_2', because cos 45° = sin 45°.

$$mv = mv'_1(\cos 30° + \sin 30°)$$

and lets us determine v_1':

$$v_1' = \frac{v}{\cos 30° + \sin 30°} = \frac{2v}{1+\sqrt{3}}$$

The math will rarely (if ever) be this intense on the actual exam. Don't stress too much if this problem took you a few minutes to work through. We'd rather you be over-prepared!

Substituting this result into Equation (2) gives us

$$0 = m\frac{2v}{1+\sqrt{3}}\sin 30° - 2mv_2'\sin 45°$$

$$2mv_2'\sin 45° = m\frac{2v}{1+\sqrt{3}}\sin 30°$$

$$v_2' = \frac{\frac{2v}{1+\sqrt{3}}\sin 30°}{2\sin 45°} = \frac{v}{\sqrt{2}(1+\sqrt{3})}$$

(b) The collision is elastic only if kinetic energy is conserved. The total kinetic energy after the collision, K', is calculated as follows:

$$\begin{aligned} K' &= \frac{1}{2}\cdot m v_1'^2 + \frac{1}{2}\cdot 2m v_2'^2 \\ &= \frac{1}{2}m\left(\frac{2v}{1+\sqrt{3}}\right)^2 + m\left(\frac{v}{\sqrt{2}(1+\sqrt{3})}\right)^2 \\ &= mv^2\left[\frac{2}{(1+\sqrt{3})^2} + \frac{1}{2(1+\sqrt{3})^2}\right] \\ &= \frac{5}{2(1+\sqrt{3})^2}mv^2 \end{aligned}$$

However, the kinetic energy before the collision is just $K = \frac{1}{2}mv^2$, so the fact that

$$\frac{5}{2(1+\sqrt{3})^2} < \frac{1}{2}$$

tells us that K' is less than K, so some kinetic energy is lost; the collision is inelastic.

CHAPTER 8 KEY TERMS

linear momentum
impulse
Impulse–Momentum Theorem
collisions
elastic
inelastic
perfectly inelastic

Chapter 8 Review Questions

Answers and explanations can be found in Chapter 11.

Section I: Multiple Choice

1. An object of mass 2 kg has a linear momentum of magnitude 6 kg · m/s. What is this object's kinetic energy?

   (A) 3 J
   (B) 6 J
   (C) 9 J
   (D) 12 J

2. A ball of mass 0.5 kg, initially at rest, acquires a speed of 4 m/s immediately after being kicked by a force of strength 20 N. For how long did this force act on the ball?

   (A) 0.01 s
   (B) 0.1 s
   (C) 0.2 s
   (D) 1 s

3. A box with a mass of 2 kg accelerates in a straight line from 4 m/s to 8 m/s due to the application of a force whose duration is 0.5 s. What is the average strength of this force?

   (A) 4 N
   (B) 8 N
   (C) 12 N
   (D) 16 N

4. A ball of mass m traveling horizontally with velocity $\mathbf{v}$ strikes a massive vertical wall and rebounds back along its original direction with no change in speed. What is the magnitude of the impulse delivered by the wall to the ball?

   (A) $\frac{1}{2}m\mathbf{v}$
   (B) $m\mathbf{v}$
   (C) $2m\mathbf{v}$
   (D) $4m\mathbf{v}$

5. Two objects, one of mass 3 kg moving with a speed of 2 m/s and the other of mass 5 kg and speed 2 m/s, move toward each other and collide head-on. If the collision is perfectly inelastic, find the speed of the objects after the collision.

   (A) 0.25 m/s
   (B) 0.5 m/s
   (C) 0.75 m/s
   (D) 1 m/s

6. Object 1 moves toward Object 2, whose mass is twice that of Object 1 and which is initially at rest. After their impact, the objects lock together and move with what fraction of Object 1's initial kinetic energy?

   (A) 1/18
   (B) 1/9
   (C) 1/6
   (D) 1/3

7. Two objects move toward each other, collide, and separate. If there was no net external force acting on the objects, but some kinetic energy was lost, then

   (A) the collision was elastic and total linear momentum was conserved
   (B) the collision was elastic and total linear momentum was not conserved
   (C) the collision was not elastic and total linear momentum was conserved
   (D) the collision was not elastic and total linear momentum was not conserved

8. Two frictionless carts (mass = 500 g each) are sitting at rest on a perfectly level table. The teacher taps the release so that one cart pushes off the other. If one of the carts has a speed of 2 m/s, then what is the final momentum of the system (in kg · m/s)?

   (A) 2,000
   (B) 1,000
   (C) 2
   (D) 0

9. A wooden block of mass M is moving at speed V in a straight line.

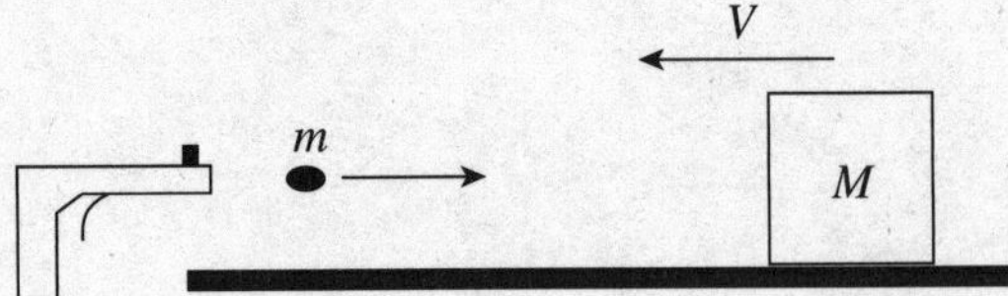

How fast would the bullet of mass m need to travel to stop the block (assuming that the bullet became embedded inside)?

(A) $mV/(m + M)$
(B) mV/M
(C) MV/m
(D) $(m + M)V/m$

10. Which of the following best describes a perfectly inelastic collision free of external forces?

(A) Total linear momentum is never conserved.
(B) Total linear momentum is sometimes conserved.
(C) Kinetic energy is never conserved.
(D) Kinetic energy is always conserved.

11. Object 1 moves with an initial speed of v_0 toward Object 2, which has a mass half that of Object 1. If the final speed of both objects after colliding is 0, what must have been the initial speed of Object 2 (assuming no external forces)?

(A) v_0

(B) $2v_0$

(C) $\frac{1}{2}v_0$

(D) $4v_0$

Section II: Free Response

1. A steel ball of mass m is fastened to a light cord of length L and released when the cord is horizontal. At the bottom of its path, the ball strikes a hard plastic block of mass $M = 4m$, initially at rest on a frictionless surface. The collision is elastic.

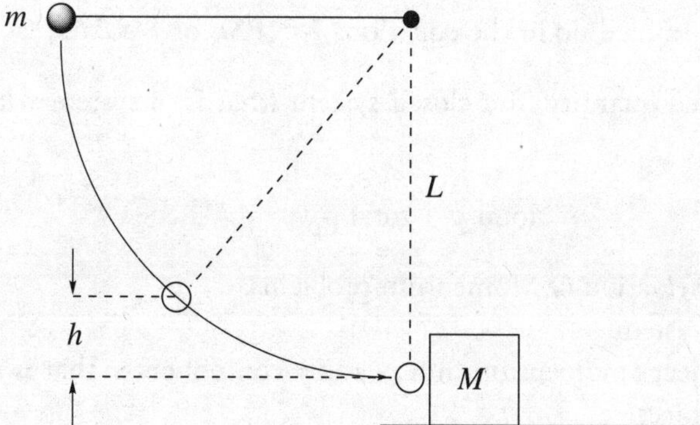

(A) Find the tension in the cord when the ball's height above its lowest position is $\frac{1}{2}L$. Write your answer in terms of m and g.

(B) Find the speed of the block immediately after the collision.

(C) To what height h will the ball rebound after the collision?

2. A *ballistic pendulum* is a device that may be used to measure the muzzle speed of a bullet. It is composed of a wooden block suspended from a horizontal support by cords attached at each end. A bullet is shot into the block, and as a result of the perfectly inelastic impact, the block swings upward. Consider a bullet (mass m) with velocity v as it enters the block (mass M). The length of the cords supporting the block each have length L. The maximum height to which the block swings upward after impact is denoted by y, and the maximum horizontal displacement is denoted by x.

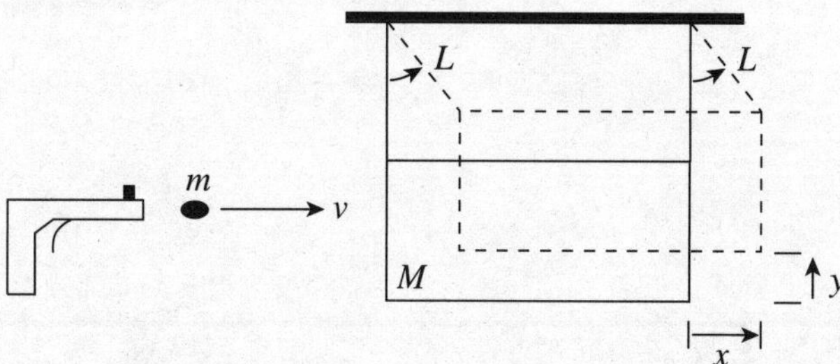

(A) In terms of m, M, g, and y, determine the speed v of the bullet.

(B) What fraction of the bullet's original kinetic energy is lost as a result of the collision? What happens to the lost kinetic energy?

(C) If y is very small (so that y^2 can be neglected), determine the speed of the bullet in terms of m, M, g, x, and L.

(D) Once the block begins to swing, does the momentum of the block remain constant? Why or why not?

Summary

- Momentum is a vector quantity given by $\mathbf{p} = m\mathbf{v}$. If you push on an object for some amount of time, we call that an impulse (**J**). Impulse causes a change in momentum. Impulse is also a vector quantity, and these ideas are summed up in the equations $\mathbf{J} = \mathbf{F}\Delta t$ or $\mathbf{J} = \Delta p$.
- Momentum is a conserved quantity in a closed system (that is, a system with no external forces). That means

$$\text{total } \mathbf{p}_i = \text{total } \mathbf{p}_f$$

- Overall strategy for Conservation of Momentum problems:
  - I. Create a coordinate system.
  - II. Break down each object's momentum into *x*- and *y*-components. That is $p_x = p \cos \theta$ and $p_y = p \sin \theta$ for any given object.
  - III. $\sum p_{xi} = \sum p_{xf}$ and $\sum p_{yi} = \sum p_{yf}$
  - IV. Sometimes you end up rebuilding vectors in the end. Remember total $p = \sqrt{p_x^{\ 2} + p_y^{\ 2}}$ and the angle is given by $\theta = \tan^{-1}\left(\frac{p_y}{p_x}\right)$.

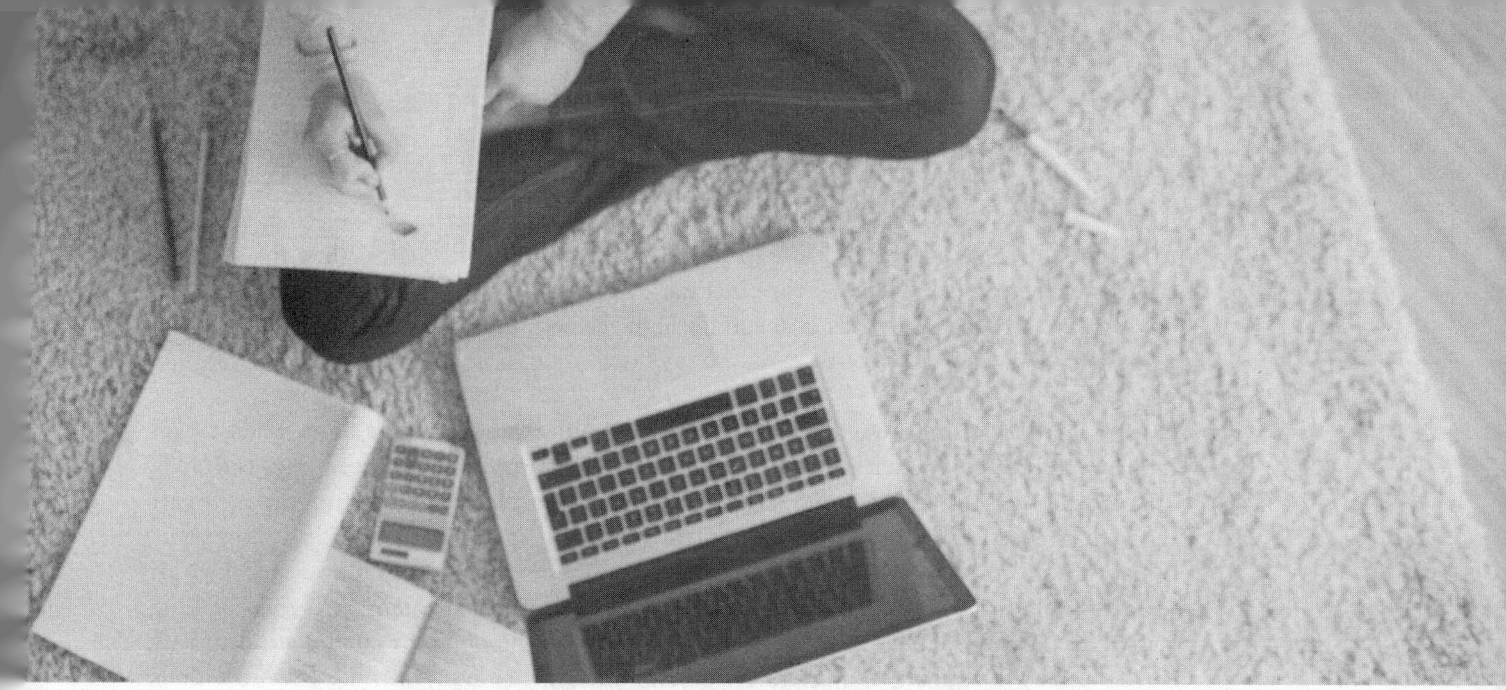

Chapter 9
Simple Harmonic Motion

"There is nothing so small, as to escape our inquiry; hence there is a new visible World discovered to the understanding."

—Robert Hooke

Seventeenth-century British physicist Robert Hooke helped pave the way for simple harmonic motion. Previously, Newton's laws of static equilibrium made it possible to show a relationship between stress and strain for complex objects. Building upon these, he developed a law on simple harmonic motion that is named after him—Hooke's Law. Hooke is credited as the discoverer of the concise mathematical relationship of a spring.

In this section, we will focus on periodic motion that's straightforward and will help to describe oscillations and other simple harmonic motions.

SIMPLE HARMONIC MOTION

In the series of diagrams below is a fixed block on the left side of a wall. When the spring is neither stretched nor compressed (when it sits at its natural length), it is said to be in its equilibrium position. When the block is in equilibrium, the net force on the block is zero. We label this position as $x = 0$.

Let's start with a spring at rest (Diagram 1). First, we pull the block to the right, where it will experience a force pulling back toward equilibrium (Diagram 2). This force brings the block back through the equilibrium position (Diagram 3), and the block's momentum carries it past that point to location $x = -A$. At this point, the block will again be experiencing a force that pushes it toward the equilibrium position (Diagram 4). Once again, the block passes through the equilibrium position, but it's traveling to the right this time (Diagram 5). If this is taking place in ideal conditions (no friction), this back-and-forth motion will continue indefinitely and the block will oscillate from these positions in the same amount of time. The oscillations of the block at the end of this spring provide us with the physical example of simple harmonic motion (often abbreviated SHM).

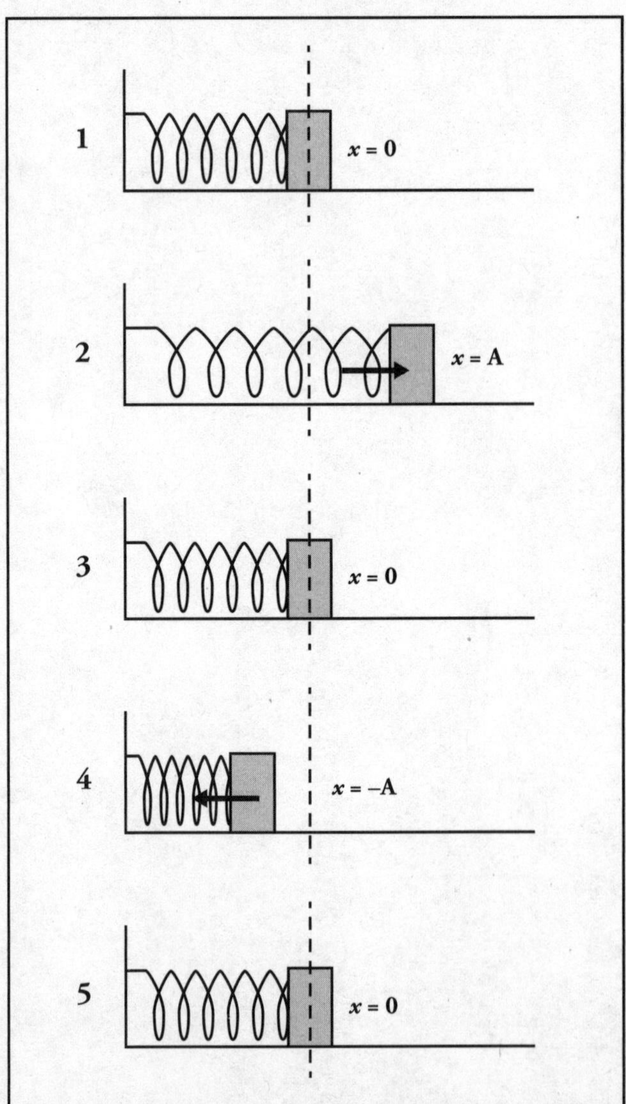

The Dynamics of SHM

Force

Since the block is accelerating and decelerating, there must be some force that is making it do so. In the case of a spring, the spring exerts a force on the block that is proportional to its displacement from its equilibrium point. Setting the equilibrium point as $x = 0$, the force exerted by the spring is

$$F = -kx$$

This is known as Hooke's Law. The proportionality constant, k, is called the **spring constant** and tells us how strong the spring is. The greater the value of k, the stiffer and stronger the spring is. The minus sign in Hooke's Law tells us that the force is a restoring force. A restoring force simply means that the force wants to return the object back to its equilibrium position. Hence, in the previous diagrams, when the block was on the left (when the position is negative), the force exerted was to the right; and when the block was on the right (when the position is positive), the force exerted was to the left. In all cases of the extreme left or right, the spring has a tendency to return to its original length or equilibrium position. This force helps to maintain the oscillations.

Example 1 A 12 cm-long spring has a force constant (k) of 400 N/m. How much force is required to stretch the spring to a length of 14 cm?

Solution. The displacement of the spring has a magnitude of 14 – 12 = 2 cm = 0.02 m so, according to Hooke's Law, the spring exerts a force of magnitude $F = kx = (400 \text{ N/m})(0.02 \text{ m}) = 8 \text{ N}$. Therefore, we'd have to exert this much force to keep the spring in this stretched state.

During the oscillation, the force on the block is zero when the block is at equilibrium (the point we designate as $x = 0$). This is because Hooke's Law says that the strength of the spring's restoring force is given by the equation $F = kx$, so $F = 0$ at equilibrium. The acceleration of the block is also equal to zero at $x = 0$, since $F = 0$ at $x = 0$ and $a = F/m$. At the endpoints of the oscillating region, where the block's displacement, x, has the greatest magnitude, the restoring force and the magnitude of the acceleration are both at their maximums.

Amplitude

The maximum displacement from equilibrium is called the amplitude of oscillation and is denoted by A. So instead of writing $x = x_{max}$, we write $x = A$ ($x = -x_{max}$ is $x = -A$). This number tells us how far to the left and right of equilibrium the block will travel.

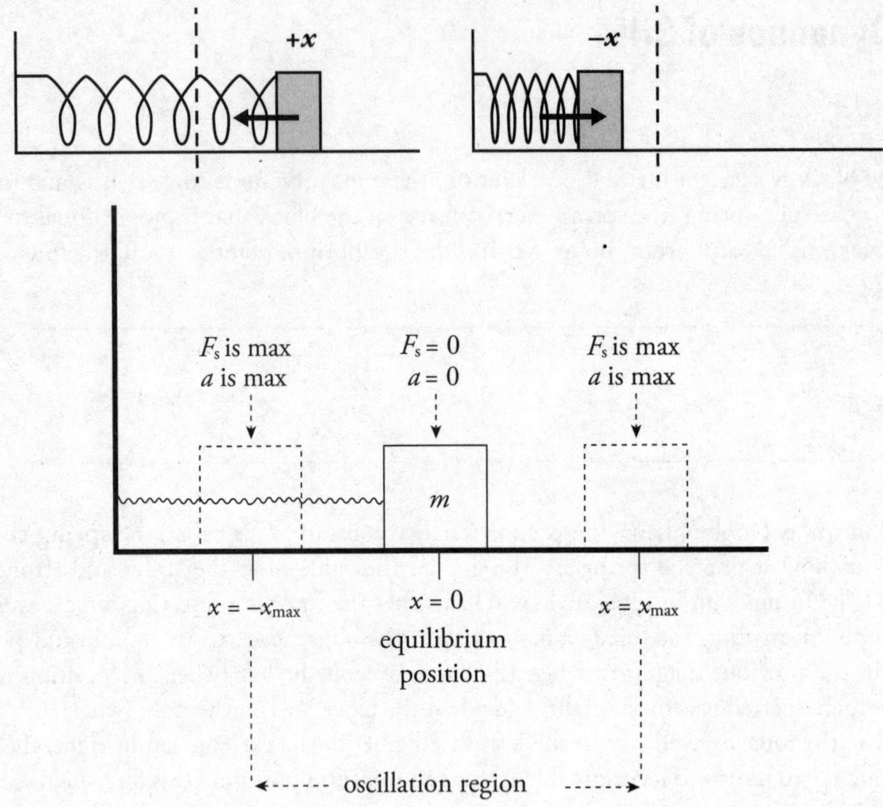

SHM in Terms of Energy

Another way to describe the block's motion is in terms of energy transfers. A stretched or compressed spring stores **elastic potential energy**, which is transformed into kinetic energy (and back again); this shuttling of energy between potential and kinetic causes the oscillations. For a spring with spring constant *k*, the elastic potential energy it possesses—relative to its equilibrium position—is given by the following equation:

$$U_s = \frac{1}{2}kx^2$$

Notice that the farther you stretch or compress a spring, the more work you have to do, and, as a result, the more potential energy that's stored.

Maximum Potential Energy
The maximum potential energy is when $x = A$ or $x = -A$.

In terms of energy transfers, we can describe the block's oscillations as follows. When you initially pull the block out, you increase the elastic potential energy of the system. Upon releasing the block, this potential energy turns into kinetic energy, and the block moves. As it passes through equilibrium, $U_S = 0$, all the energy is kinetic. Then, as the block continues through equilibrium, it compresses the spring, and the kinetic energy is transformed back into elastic potential energy.

By Conservation of Mechanical Energy, the sum $K + U_S$ is a constant. Therefore, when the block reaches the endpoints of the oscillation region (that is, when $x = \pm x_{max}$), U_S is maximized, so K must be minimized; in fact, $K = 0$ at the endpoints. As the block is passing through equilibrium, $x = 0$, so $U_S = 0$ and K is maximized.

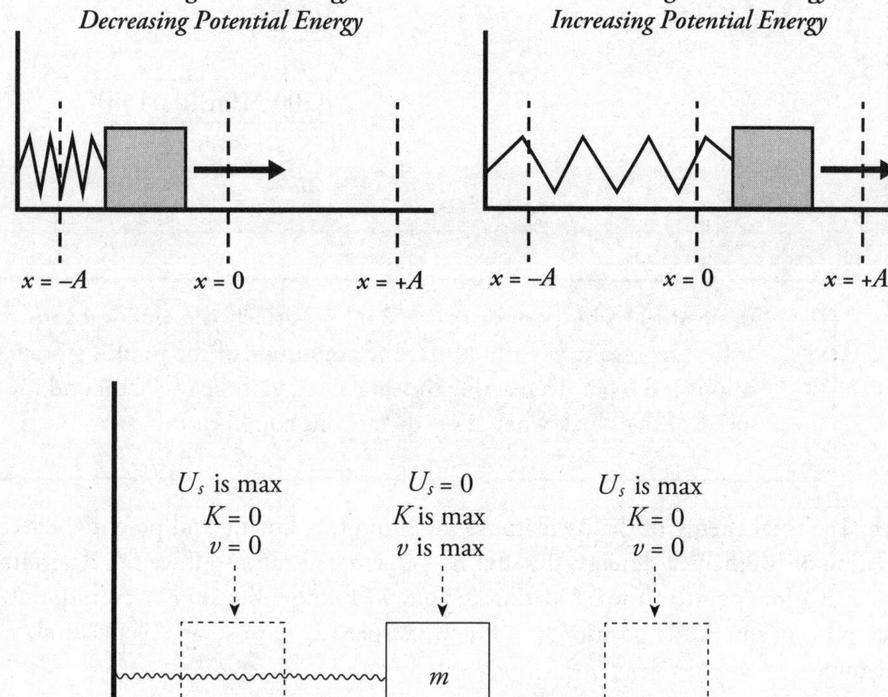

Example 2 A block of mass $m = 0.05$ kg oscillates on a spring whose force constant k is 500 N/m. The amplitude of the oscillations is 4.0 cm. Calculate the maximum speed of the block.

Solution. First, let's get an expression for the maximum elastic potential energy of the system:

$$U_S = \frac{1}{2}kx^2 \quad \Rightarrow \quad U_{S,\,max} = \frac{1}{2}kx_{max}^2 = \frac{1}{2}kA^2$$

When all this energy has been transformed into kinetic energy—which, as we discussed earlier, occurs just as the block is passing through equilibrium—the block will have a maximum kinetic energy and maximum speed of

$$U_{S,\,\max} \rightarrow K_{\max} \quad \Rightarrow \quad \frac{1}{2}kA^2 = \frac{1}{2}mv_{\max}^2$$

$$v_{\max} = \sqrt{\frac{kA^2}{m}}$$

$$= \sqrt{\frac{(500\ \text{N/m})(0.04\ \text{m})^2}{0.05\ \text{kg}}}$$

$$= 4\ \text{m/s}$$

Maximum Velocity
Only the maximum velocity can be calculated in a spring with this method. The AP Physics 1 Exam will not ask you to calculate other velocities at other points because this is not uniform accelerated motion.

Example 3 A block of mass m = 2.0 kg is attached to an ideal spring of force constant k = 500 N/m. The amplitude of the resulting oscillations is 8.0 cm. Determine the total energy of the oscillator and the speed of the block when it's 4.0 cm from equilibrium.

Solution. The total energy of the oscillator is the sum of its kinetic and potential energies. By Conservation of Mechanical Energy, the sum $K + U_S$ is a constant, so if we can determine what this sum is at some point in the oscillation region, we'll know the sum at every point. When the block is at its amplitude position, x = 8 cm, its speed is zero; so at this position, E is easy to figure out:

$$E = K + U_S = 0 + \frac{1}{2}kA^2 = \frac{1}{2}(500\ \text{N/m})(0.08\ \text{m})^2 = 1.6\ \text{J}$$

This gives the total energy of the oscillator at *every* position. At any position x, we have

$$\frac{1}{2}mv^2 + \frac{1}{2}kx^2 = E$$

$$v = \sqrt{\frac{E - \frac{1}{2}kx^2}{\frac{1}{2}m}}$$

so when we substitute in the numbers, we get

$$v = \sqrt{\frac{E - \frac{1}{2}kx^2}{\frac{1}{2}m}} = \sqrt{\frac{(1.6\ \text{J}) - \frac{1}{2}(500\ \text{N/m})(0.04\ \text{m})^2}{\frac{1}{2}(2.0\ \text{kg})}}$$

$$= 1.1\ \text{m/s}$$

Example 4 A block of mass m = 8.0 kg is attached to an ideal spring of force constant k = 500 N/m. The block is at rest at its equilibrium position. An impulsive force acts on the block, giving it an initial speed of 2.0 m/s. Find the amplitude of the resulting oscillations.

Solution. The block will come to rest when all of its initial kinetic energy has been transformed into the spring's potential energy. At this point, the block is at its maximum displacement from equilibrium, at one of its amplitude positions, and

$$K_i + U_i = K_f + U_f$$

$$\frac{1}{2}mv_i^2 + 0 = 0 + \frac{1}{2}kA^2$$

$$A = \sqrt{\frac{mv_i^2}{k}}$$

$$= \sqrt{\frac{(8.0\text{ kg})(2.0\text{ m/s})^2}{500\text{ N/m}}}$$

$$= 0.25\text{ m}$$

Period and Frequency

As you watch the block oscillate, you should notice that it repeats each **cycle** of oscillation in the same amount of time. A cycle is a *round-trip*: for example, from position $x = A$ over to $x = -A$ and back again to $x = A$. The amount of time it takes to complete a cycle is called the **period** of the oscillations, or T. If T is short, the block is oscillating rapidly, and if T is long, the block is oscillating slowly.

Another way of indicating the rapidity of the oscillations is to count the number of cycles that can be completed in a given time interval; the more completed cycles, the more rapid the oscillations. The number of cycles that can be completed per unit time is called the **frequency** of the oscillations, or f, and frequency is expressed in cycles per second. One cycle per second is one **hertz** (abbreviated **Hz**). Do not confuse lowercase f (frequency) with uppercase F (Force).

One of the most basic equations of oscillatory motion expresses the fact that the period and frequency are reciprocals of each other:

$$\text{period} = \frac{\#\text{ seconds}}{\text{cycle}} \quad \text{while} \quad \text{frequency} = \frac{\#\text{ cycles}}{\text{second}}$$

Two Peas in a Pod

If you have the period, you can always get the frequency, and vice versa. Note that the period and the frequency are reciprocals of each other.

Therefore,

$$T = \frac{1}{f} \quad \text{and} \quad f = \frac{1}{T}$$

Example 5 A block oscillating on the end of a spring moves from its position of maximum spring stretch to maximum spring compression in 0.25 s. Determine the period and frequency of this motion.

Solution. The period is defined as the time required for one full cycle. Moving from one end of the oscillation region to the other is only half a cycle. Therefore, if the block moves from its position of maximum spring stretch to maximum spring compression in 0.25 s, the time required for a full cycle is twice as much; $T = 0.5$ s. Because frequency is the reciprocal of period, the frequency of the oscillations is $f = 1/T = 1/(0.5 \text{ s}) = 2$ Hz.

Example 6 A student observing an oscillating block counts 45.5 cycles of oscillation in one minute. Determine its frequency (in hertz) and period (in seconds).

Solution. The frequency of the oscillations, in hertz (which is the number of cycles per second), is

$$f = \frac{45.5 \text{ cycles}}{\text{min}} \times \frac{1 \text{ min}}{60 \text{ s}} = \frac{0.758 \text{ cycles}}{\text{s}} = 0.758 \text{ Hz}$$

Therefore,

$$T = \frac{1}{f} = \frac{1}{0.758 \text{ Hz}} = 1.32 \text{ s}$$

One of the defining properties of the spring-block oscillator is that the frequency and period can be determined from the mass of the block and the force constant of the spring. The equations are as follows:

A useful mnemonic for remembering this equation is to go in alphabetical order (clockwise) for frequency (f to k to m) and reverse alphabetical for period (T to m to k).

$$f = \frac{1}{2\pi}\sqrt{\frac{k}{m}} \quad \text{and} \quad T = 2\pi\sqrt{\frac{m}{k}}$$

Let's analyze these equations. Suppose we had a small mass on a very stiff spring; then intuitively, we would expect that this strong spring would make the small mass oscillate rapidly, with high frequency and short period. Both of these predictions are substantiated by the equations above, because if m is small and k is large, then the ratio k/m is large (high frequency) and the ratio m/k is small (short period).

Example 7 A block of mass $m = 2.0$ kg is attached to a spring whose force constant, k, is 300 N/m. Calculate the frequency and period of the oscillations of this spring-block system.

Solution. According to the equations above,

$$f=\frac{1}{2\pi}\sqrt{\frac{k}{m}}=\frac{1}{2\pi}\sqrt{\frac{300\text{ N/m}}{2.0\text{ kg}}}=1.9\text{ Hz}$$

$$T=2\pi\sqrt{\frac{m}{k}}=2\pi\sqrt{\frac{2.0\text{ kg}}{300\text{ N/m}}}=0.51\text{ s}$$

Notice that $f \approx 2$ Hz and $T \approx 0.5$ s, and that these values satisfy the basic equation $T = 1/f$.

Example 8 A block is attached to a spring and set into oscillatory motion, and its frequency is measured. If this block were removed and replaced by a second block with 1/4 the mass of the first block, how would the frequency of the oscillations compare to that of the first block?

Solution. Since the same spring is used, k remains the same. According to the equation given on the previous page, f is inversely proportional to the square root of the mass of the block: $f \propto 1/\sqrt{m}$. Therefore, if m decreases by a factor of 4, then f increases by a factor of $\sqrt{4} = 2$.

Stay Up to Date!
For late-breaking information about test dates, exam formats, and any other changes pertaining to AP Physics 1, make sure to check the College Board's website at apstudents.collegeboard.org/courses/ap-physics-1-algebra-based/assessment

The equations we saw on the previous page for the frequency and period of the spring-block oscillator do not contain A, the amplitude of the motion. In simple harmonic motion, *both the frequency and the period are independent of the amplitude.* The reason that the frequency and period of the spring-block oscillator are independent of amplitude is that F, the strength of the restoring force, is proportional to x, the displacement from equilibrium, as given by Hooke's Law: $F_S = -kx$.

Example 9 A student performs an experiment with a spring-block simple harmonic oscillator. In the first trial, the amplitude of the oscillations is 3.0 cm, while in the second trial (using the same spring and block), the amplitude of the oscillations is 6.0 cm. Compare the values of the period, frequency, and maximum speed of the block between these two trials.

Solution. If the system exhibits simple harmonic motion, then the period and frequency are independent of amplitude. This is because the same spring and block were used in the two trials, so the period and frequency will have the same values in the second trial as they had in the first. But the maximum speed of the block will be greater in the second trial than in the first. Since the amplitude is greater in the second trial, the system possesses more total energy ($E = \frac{1}{2}kA^2$). So, when the block is passing through equilibrium (its position of greatest speed), the second system has more energy to convert to kinetic, meaning that the block will have a greater speed. In fact, from Example 2, we know that $v_{max} = A\sqrt{k/m}$ so, since A is twice as great in the second trial than in the first, v_{max} will be twice as great in the second trial than in the first.

Example 10 For each of the following arrangements of two springs, determine the **effective spring constant**, k_{eff}. This is the force constant of a single spring that would produce the same force on the block as the pair of springs shown in each case.

(a)

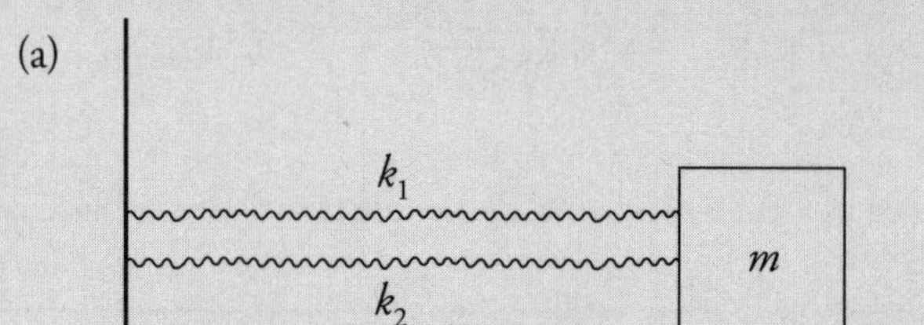

(b)

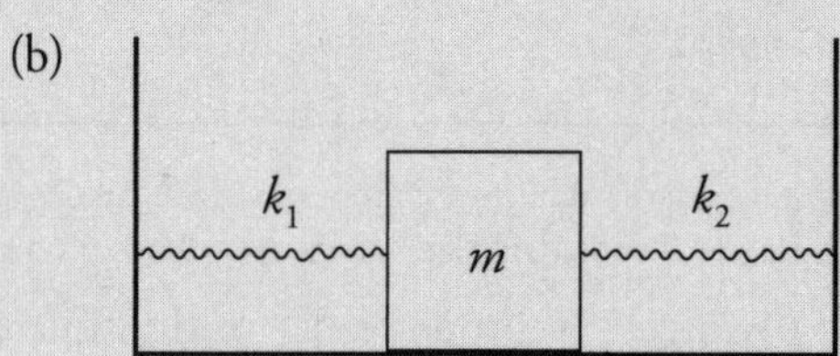

(c)

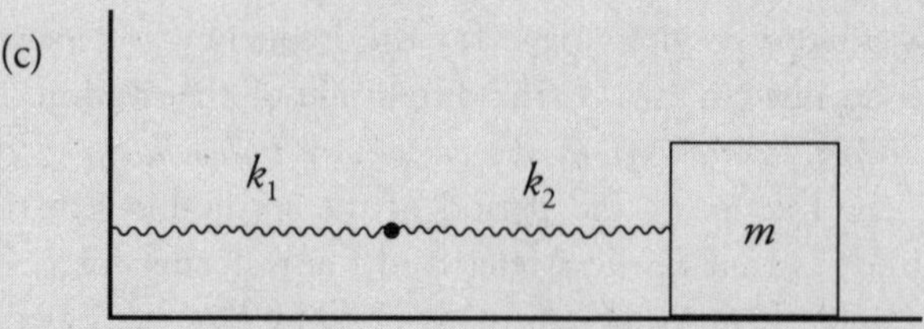

(d) Determine k_{eff} in each of these cases if $k_1 = k_2 = k$.

Solution.

(a) Imagine that the block was displaced a distance x to the right of its equilibrium position. Then the force exerted by the first spring would be $F_1 = -k_1x$ and the force exerted by the second spring would be $F_2 = -k_2x$. The net force exerted by the springs would be

$$F_1 + F_2 = -k_1x + (-k_2x) = -(k_1 + k_2)x$$

Since $F_{eff} = -(k_1 + k_2)x$, we see that $k_{eff} = k_1 + k_2$.

(b) Imagine that the block was displaced a distance x to the right of its equilibrium position. Then the force exerted by the first spring would be $F_1 = -k_1x$ and the force exerted by the second spring would be $F_2 = -k_2x$. The net force exerted by the springs would be

$$F_1 + F_2 = -k_1x + -k_2x = -(k_1 + k_2)x$$

As in part (a), we see that, since $F_{eff} = -(k_1 + k_2)x$, we get $k_{eff} = k_1 + k_2$.

(c) Imagine that the block was displaced a distance x to the right of its equilibrium position. Let x_1 be the distance that the first spring is stretched, and let x_2 be the distance that the second spring is stretched. Then $x = x_1 + x_2$. But $x_1 = -F/k_1$ and $x_2 = -F/k_2$, so

$$\frac{-F}{k_1} + \frac{-F}{k_2} = x$$

$$-F\left(\frac{1}{k_1} + \frac{1}{k_2}\right) = x$$

$$F = -\left(\frac{1}{\frac{1}{k_1} + \frac{1}{k_2}}\right)x$$

$$F = -\frac{k_1 k_2}{k_1 + k_2}x$$

Therefore,

$$k_{\text{eff}} = \frac{k_1 k_2}{k_1 + k_2}$$

(d) If the two springs have the same force constant, that is, if $k_1 = k_2 = k$, then in the first two cases, the pairs of springs are equivalent to one spring that has twice their force constant: $k_{\text{eff}} = k_1 + k_2 = k + k = 2k$. In (c), the pair of springs is equivalent to a single spring with half their force constant:

$$k_{\text{eff}} = \frac{k_1 k_2}{k_1 + k_2} = \frac{kk}{k + k} = \frac{k^2}{2k} = \frac{k}{2}$$

Spring-Block Summary

We can summarize the dynamics of oscillations in this table:

	x = -A	x = 0	x = +A
Magnitude of Restoring Force	MAX	0	MAX
Magnitude of Acceleration	MAX	0	MAX
Potential Energy (U) of Spring	MAX	0	MAX
Kinetic Energy (K) of Block	0	MAX	0
Speed (v) of Block	0	MAX	0

Fortunately, this same table applies to springs, pendulums, and waves. All simple harmonic motion follows this cycle.

Another Approach
Using a force diagram, let's map out the situation described on this page:

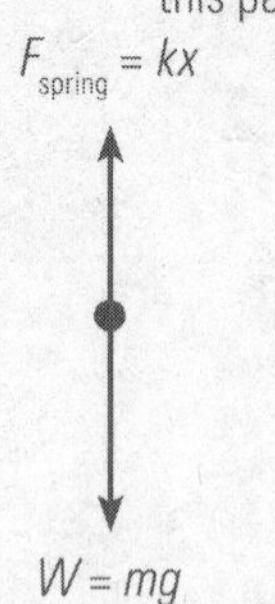

Our net force in this case is zero since it is neither moving up nor moving down. So our force, by the spring, must cancel out with our weight: $kx = mg$.

THE SPRING-BLOCK OSCILLATOR: VERTICAL MOTION

So far, we've looked at a block sliding back and forth on a horizontal table, but the block could also oscillate vertically. The only difference would be that gravity would cause the block to move downward, to an equilibrium position at which, in contrast with the horizontal SHM we've examined, the spring would not be at its natural length. Of course, in calculating energy, the gravitational potential energy (*mgh*) must be included.

Consider a spring of negligible mass hanging from a stationary support. A block of mass *m* is attached to its end and allowed to come to rest, stretching the spring a distance *d*. At this point, the block is in equilibrium; the upward force of the spring is balanced by the downward force of gravity. Therefore,

$$kd = mg \quad \Rightarrow \quad d = \frac{mg}{k}$$

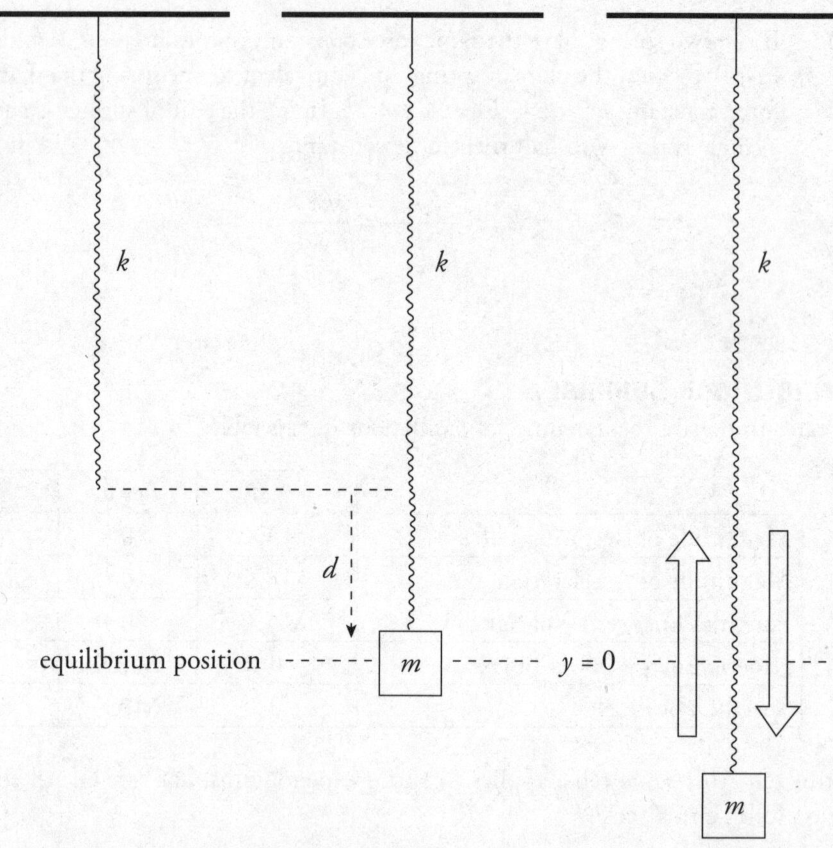

Once you have solved for the distance *d*, simply set this position as your new equilibrium position $x = 0$. At this point, you can treat the vertical spring exactly as you would a horizontal spring. Just be sure to measure distances relative to the new equilibrium when solving for any values (spring force, potential energy, and so on) that have *x* in their formulas.

> **Example 11** A block of mass $m = 1.5$ kg is attached to the end of a vertical spring of force constant $k = 300$ N/m. After the block comes to rest, it is pulled down a distance of 2.0 cm and released.
>
> (a) What is the frequency of the resulting oscillations?
>
> (b) What are the minimum and maximum amounts of stretch of the spring during the oscillations of the block?

New Equilibrium
Normally our equilibrium for the spring (without a block) would be at the position *y*. When we attach the block, our new equilibrium point sits at (*d* + *y*). The only time you need to worry about this is if a question asks about the total length of the spring at a given moment. Use your horizontal spring equations to solve the question normally, and then simply add *d*.

Solution.

(a) The frequency is given by

$$f = \frac{1}{2\pi}\sqrt{\frac{k}{m}} = \frac{1}{2\pi}\sqrt{\frac{300 \text{ N/m}}{1.5 \text{ kg}}} = 2.3 \text{ Hz}$$

(b) Before the block is pulled down, to begin the oscillations, it stretches the spring by a distance calculated as follows:

$$d = \frac{mg}{k} = \frac{(1.5 \text{ kg})(10 \text{ N/kg})}{300 \text{ N/m}} = 0.05 \text{ m} = 5 \text{ cm}$$

Since the amplitude of the motion is 2.0 cm, the spring is stretched a maximum of 5 cm + 2.0 cm = 7 cm when the block is at the lowest position in its cycle, and a minimum of 5 cm – 2.0 cm = 3 cm when the block is at its highest position.

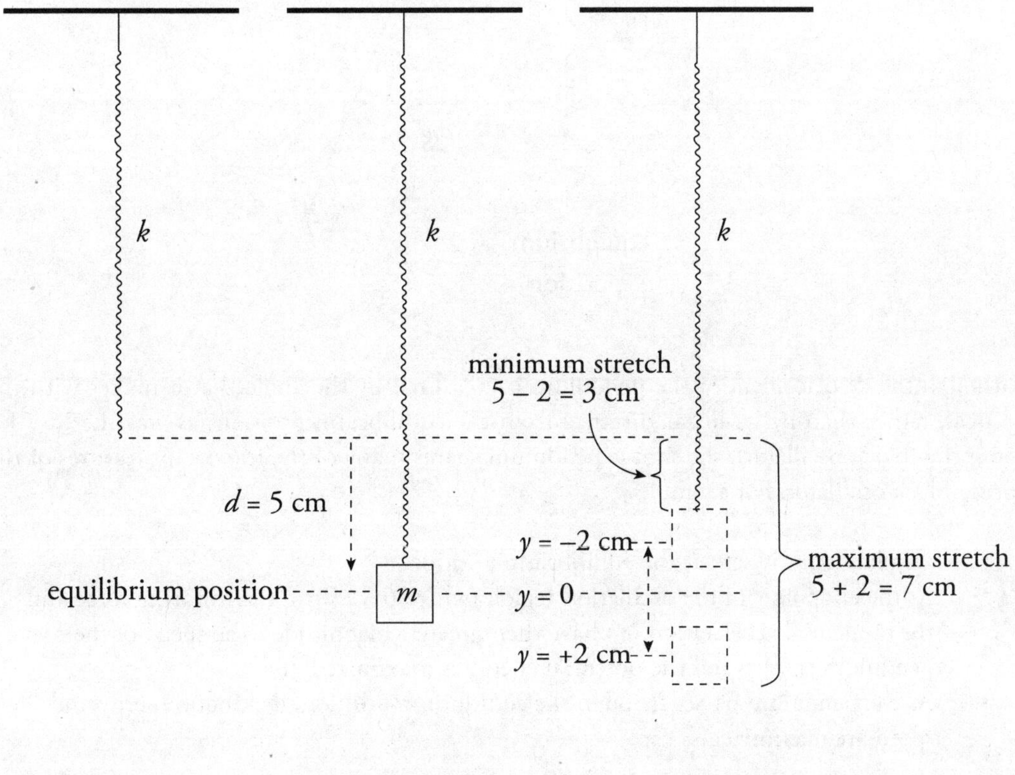

PENDULUMS

A **simple pendulum** consists of a weight of mass *m* attached to a string or a massless rod that swings, without friction, about the vertical equilibrium position. The restoring force is provided by gravity and, as the figure below shows, the magnitude of the restoring force when the bob is θ to an angle to the vertical is given by the equation:

$$F_{restoring} = mg \sin \theta$$

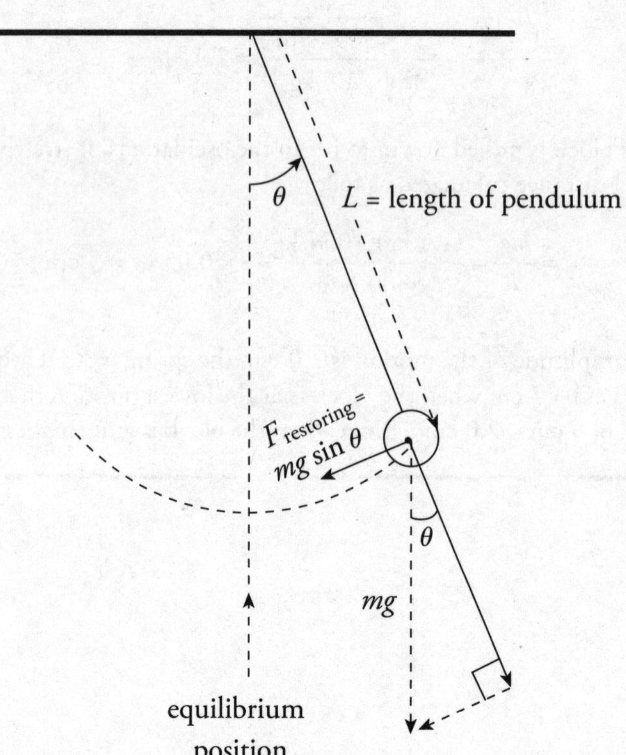

Although the displacement of the pendulum is measured by the angle that it makes with the vertical, rather than by its linear distance from the equilibrium position (as was the case for the spring-block oscillator), the simple pendulum shares many of the important features of the spring-block oscillator. For example,

- Displacement is zero at the equilibrium position.
- At the endpoints of the oscillation region (where $\theta = \pm\theta_{max}$), the restoring force and the tangential acceleration (a_t) have their greatest magnitudes, the speed of the pendulum is zero, and the potential energy is maximized.
- As the pendulum passes through the equilibrium position, its kinetic energy and speed are maximized.

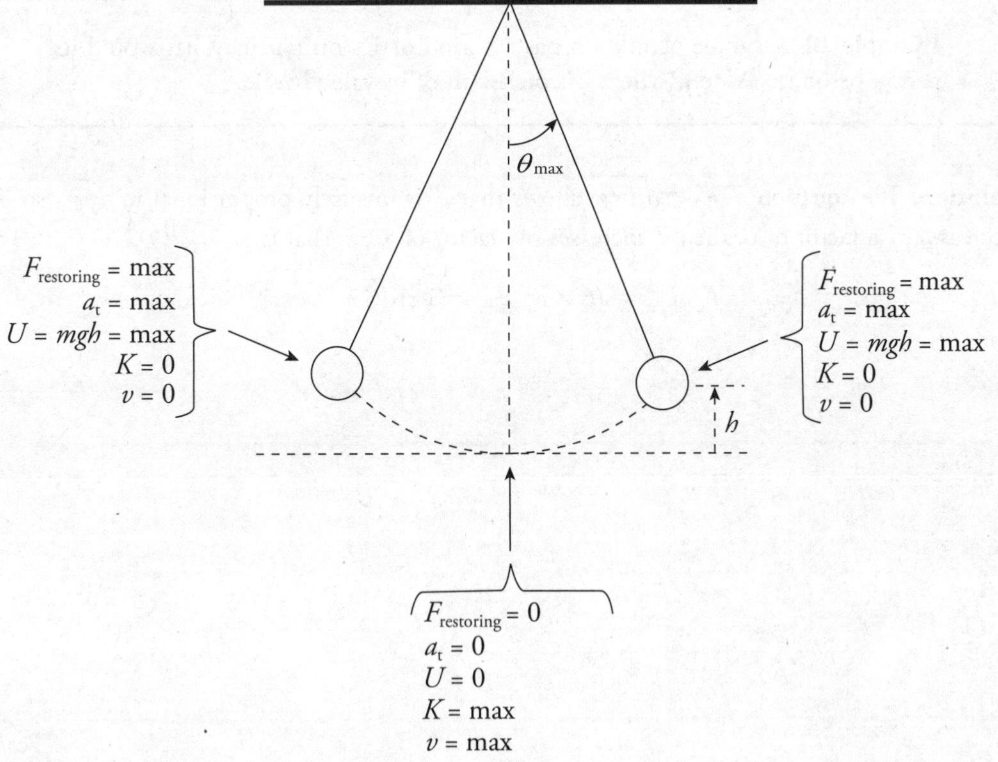

Despite these similarities, there is one important difference. Simple harmonic motion results from a restoring force that has a strength that is proportional to the displacement. The magnitude of the restoring force on a pendulum is $mg \sin \theta$, which is *not* proportional to the displacement (θL, the arc length, with the angle measured in radians). Strictly speaking, the motion of a simple pendulum is not really simple harmonic. However, if θ is small, then $\sin \theta \approx \theta$ (measured in radians) so, in this case, the magnitude of the restoring force is approximately $mg\theta$, which *is* proportional to θ. So if θ_{max} is small, the motion can be treated as simple harmonic.

More so than other topics, pendulum problems are often easier to solve if θ is measured in radians than if it's measured in degrees. If you're not comfortable working in radians, convert to degrees using the conversion 180 degrees = π radians.

If the restoring force is given by $mg\theta$, rather than $mg \sin \theta$, then the frequency and period of the oscillations depend only on the length of the pendulum and the value of the gravitational acceleration, according to the following equations:

$$f = \frac{1}{2\pi}\sqrt{\frac{g}{L}} \quad \text{and} \quad T = 2\pi\sqrt{\frac{L}{g}}$$

The same mnemonic we used for springs can be applied to pendulums. The equation for frequency goes in alphabetical order (clockwise), and the equation for period goes in reverse alphabetical order (clockwise).

Note that neither frequency nor the period depends on the amplitude (the maximum angular displacement, θ_{max}); this is a characteristic feature of simple harmonic motion. Also notice that neither the frequency nor the period depends on the mass of the weight.

Example 12 A simple pendulum has a period of 1 s on Earth. What would its period be on the Moon (where g is one-sixth of its value here)?

Solution. The equation $T = 2\pi\sqrt{L/g}$ shows that T is inversely proportional to $\sqrt{g}$, so if g decreases by a factor of 6, then T increases by factor of $\sqrt{6}$. That is,

$$T_{\text{on Moon}} = \sqrt{6} \times T_{\text{on Earth}} = (1\text{ s})\sqrt{6} = 2.4\text{ s}$$

CHAPTER 9 KEY TERMS

simple harmonic motion
spring constant
amplitude
elastic potential energy
cycle
period
frequency
hertz (Hz)
effective spring constant
simple pendulum

Chapter 9 Review Questions

Answers and explanations can be found in Chapter 11.

Section I: Multiple Choice

1. Which of the following are characteristics of simple harmonic motion? Select two answers.

   (A) The acceleration is constant.
   (B) The restoring force is proportional to the displacement.
   (C) The frequency is independent of the amplitude.
   (D) The period is dependent on the amplitude.

2. A block attached to an ideal spring undergoes simple harmonic motion. The acceleration of the block has its maximum magnitude at the point where

   (A) the speed is the maximum
   (B) the speed is the minimum
   (C) the restoring force is the minimum
   (D) the kinetic energy is the maximum

3. A block attached to an ideal spring undergoes simple harmonic motion about its equilibrium position ($x = 0$) with amplitude A. What fraction of the total energy is in the form of kinetic energy when the block is at position $x = \frac{1}{2}A$?

   (A) $\frac{1}{3}$
   (B) $\frac{1}{2}$
   (C) $\frac{2}{3}$
   (D) $\frac{3}{4}$

4. A student measures the maximum speed of a block undergoing simple harmonic oscillations of amplitude A on the end of an ideal spring. If the block is replaced by one with twice its mass but the amplitude of its oscillations remains the same, then the maximum speed of the block will

   (A) decrease by a factor of 4
   (B) decrease by a factor of 2
   (C) decrease by a factor of $\sqrt{2}$
   (D) increase by a factor of 2

5. A spring-block simple harmonic oscillator is set up so that the oscillations are vertical. The period of the motion is T. If the spring and block are taken to the surface of the Moon, where the gravitational acceleration is 1/6 of its value here, then the vertical oscillations will have a period of

   (A) $\frac{T}{6}$
   (B) $\frac{T}{3}$
   (C) $\frac{T}{\sqrt{6}}$
   (D) T

6. A linear spring of force constant k is used in a physics lab experiment. A block of mass m is attached to the spring and the resulting frequency, f, of the simple harmonic oscillations is measured. Blocks of various masses are used in different trials, and in each case, the corresponding frequency is measured and recorded. If f^2 is plotted versus $1/m$, the graph will be a straight line with slope

   (A) $\frac{4\pi^2}{k^2}$
   (B) $\frac{4\pi^2}{k}$
   (C) $4\pi^2 k$
   (D) $\frac{k}{4\pi^2}$

7. A simple pendulum swings about the vertical equilibrium position with a maximum angular displacement of 5° and period T. If the same pendulum is given a maximum angular displacement of 10°, then which of the following best gives the period of the oscillations?

(A) $\frac{T}{2}$

(B) $\frac{T}{\sqrt{2}}$

(C) T

(D) $2T$

8. A block with a mass of 20 kg is attached to a spring with a force constant $k = 50$ N/m. What is the magnitude of the acceleration of the block when the spring is stretched 4 m from its equilibrium position?

(A) 4 m/s²
(B) 6 m/s²
(C) 8 m/s²
(D) 10 m/s²

9. A block with a mass of 10 kg connected to a spring oscillates back and forth with an amplitude of 2 m. What is the approximate period of the block if it has a speed of 4 m/s when it passes through its equilibrium point?

(A) 1 s
(B) 3 s
(C) 6 s
(D) 12 s

10. A block with a mass of 4 kg is attached to a spring on the wall that oscillates back and forth with a frequency of 4 Hz and an amplitude of 3 m. What would the frequency be if the block were replaced by one with one-fourth the mass and the amplitude of the block is increased to 9 m ?

(A) 4 Hz
(B) 8 Hz
(C) 12 Hz
(D) 24 Hz

Section II: Free Response

1. The figure below shows a block of mass m (Block 1) that is attached to one end of an ideal spring of force constant k and natural length L. The block is pushed so that it compresses the spring to 3/4 of its natural length and is then released from rest. Just as the spring has extended to its natural length L, the attached block collides with another block (also of mass m) at rest on the edge of the frictionless table. When Block 1 collides with Block 2, half of its kinetic energy is lost to heat; the other half of Block 1's kinetic energy at impact is divided between Block 1 and Block 2. The collision sends Block 2 over the edge of the table, where it falls a vertical distance H, landing at a horizontal distance R from the edge.

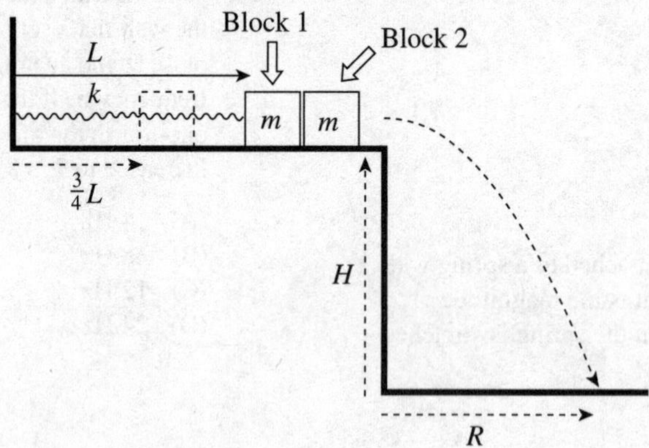

(A) What is the acceleration of Block 1 at the moment it's released from rest from its initial position? Write your answer in terms of k, L, and m.

(B) If v_1 is the velocity of Block 1 just before impact, show that the velocity of Block 1 just after impact is $\frac{1}{2}v_1$.

(C) Determine the amplitude of the oscillations of Block 1 after Block 2 has left the table. Write your answer in terms of L only.

(D) Determine the period of the oscillations of Block 1 after the collision, writing your answer in terms of T_0, the period of the oscillations that Block 1 would have had if it did not collide with Block 2.

(E) Find an expression for R in terms of H, k, L, m, and g.

2. A bullet of mass m is fired from a non-lethal pellet gun horizontally with speed v into a block of mass M initially at rest, at the end of an ideal spring on a frictionless table. At the moment the bullet hits, the spring is at its natural length, L. The bullet becomes embedded in the block, and simple harmonic oscillations result.

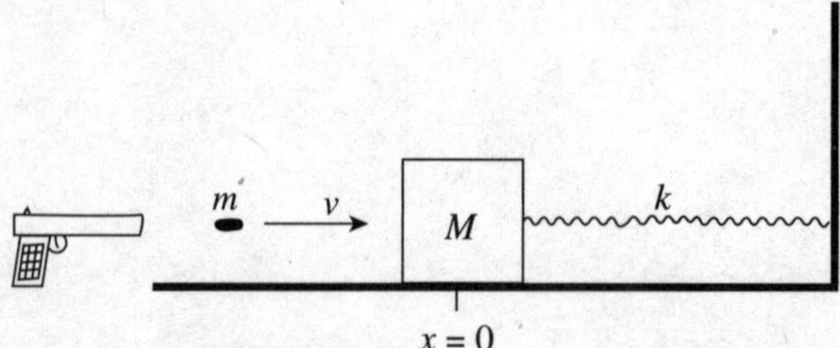

(A) Determine the speed of the block immediately after the impact by the bullet.

(B) Determine the amplitude of the resulting oscillations of the block.

(C) Compute the frequency of the resulting oscillations.

3. A block of mass M oscillating with amplitude A on a frictionless horizontal table is connected to an ideal spring of force constant k. The period of its oscillations is T. At the moment when the block is at position $x = \frac{1}{2}A$ and moving to the right, a ball of clay of mass m dropped from above lands on the block.

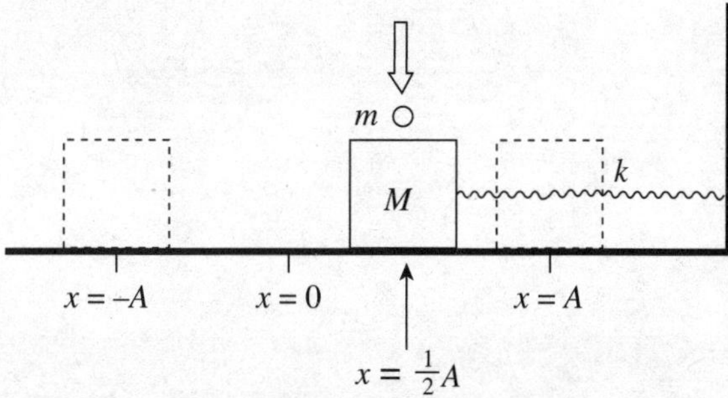

(A) What is the velocity of the block just before the clay hits?

(B) What is the velocity of the block just after the clay hits?

(C) What is the new period of the oscillations of the block?

(D) What is the new amplitude of the oscillations? Write your answer in terms of A, k, M, and m.

(E) Would the answer to part (C) be different if the clay had landed on the block when it was at a different position? Support your answer briefly.

(F) Would the answer to part (D) be different if the clay had landed on the block when it was at a different position? Support your answer briefly.

Summary

- $T = \dfrac{\text{time}}{\#\,\text{cycles}}$

  $f = \dfrac{\#\,\text{cycles}}{\text{time}}$

  $T = \dfrac{1}{f}$

- Hooke's Law holds for most springs. Formulas to keep in mind are the following:

  $F_s = -kx$

  $T = 2\pi\sqrt{\dfrac{m}{k}}$

  $U_s = \dfrac{1}{2}kx^2$

- For small angle of a pendulum swing:

  $T = 2\pi\sqrt{\dfrac{L}{g}}$

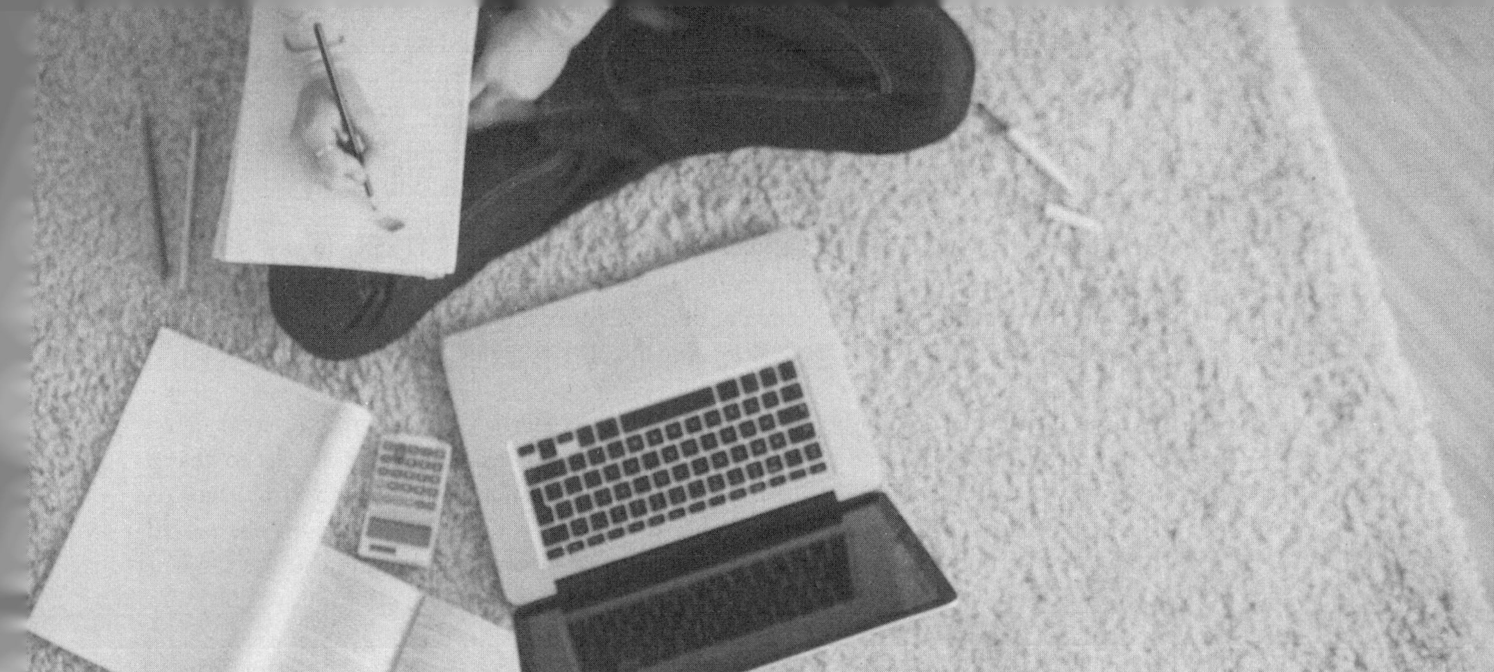

Chapter 10
Torque and Rotational Motion

"Horsepower sells cars, torque wins races."

—Carroll Shelby

So far, we've explored how objects move from one place to another. We first explored how objects moved in one dimension, then we moved on to two dimensions. We described objects moving in a line, a parabola, and even a circle. We're still missing something, though. Motion doesn't always involve going from one place to another. You could spin around in a chair, for instance. So to complete our description of motion, we need to explore rotation: how to describe rotational motion and how to create it with torque.

ROTATIONAL MOTION

Previously, we covered objects that undergo circular motion. This chapter focuses on taking those objects and spinning them. Previous equations involved objects moving in a linear orientation or being manipulated into circular orbit. With rotational motion, we will need to take on a new set of equations that are analogous to the physics of linear motion.

If we recall from before, an object's mass measures its inertia—its resistance to acceleration. The greater the inertia on an object, the harder it is to change its velocity. Harder to change its velocity means the object is harder to deliver an acceleration on—which in turn means the greater the inertia, the greater the force that is required in order for an object to be moved. Comparing two objects, if Object 1 has greater inertia than Object 2 and the same force is applied on both objects, Object 1 will undergo a smaller acceleration.

In the linear model, we put these in terms of force, mass, acceleration, and velocity. When it comes to rotational kinematics, we need to change up a few of these terms:

Linear Kinematics	Rotational Kinematics
Force	Torque (τ)
Mass	Moment of Inertia (I)
Acceleration	Angular Acceleration (α)
$F_{net} = ma$	$\tau_{net} = I\alpha$
Velocity	Angular Velocity (ω)

Rotational Kinematics

Like our linear equations, which are used to determine the distance (x), velocity (v), and linear acceleration, we use rotational equations to determine the same factors. In this section, we will go over angular distance (θ), angular velocity (ω), and angular acceleration (α). Finally, we will explore the relationship between these three rotational parameters and the linear parameters.

It's All Greek to Me
If you haven't seen these symbols before, here's a quick pronunciation guide:

θ = theta
ω = omega
α = alpha

As you work your way through math and physics, you'll eventually learn almost all the Greek letters.

Let's start with some basic definitions.

What is **angular displacement?**

What is **translational displacement?**

If you look at a circle, you can see that 1x around the circle (1 revolution) equals 2π radians, or 360°.

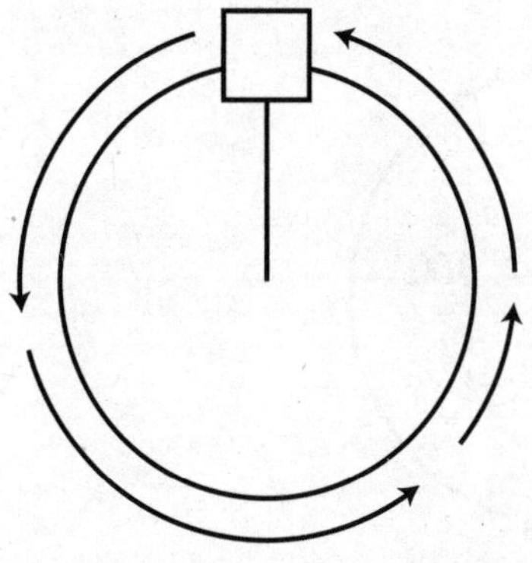

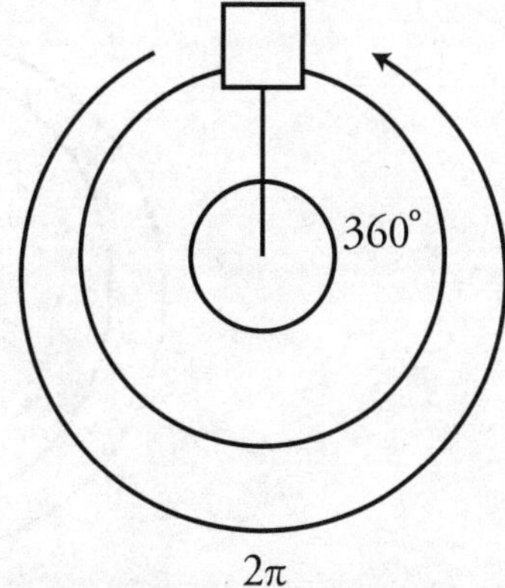

The linear position or physical distance traveled around the circle (Δs) can be related to angular position and the radius of the circle (r) via this equation:

$$\theta = \frac{s}{r}$$

What Is Angular Velocity ($\vec{\omega}$)?

Angular Velocity	Linear Velocity
$\vec{\omega} = \frac{\Delta\theta}{\Delta t}$	$\vec{V} = \frac{\Delta s}{\Delta t}$
Units = $\frac{\text{Rad}}{\text{s}}$ or $\frac{\text{Rev}}{\text{min}}$	Units = m/s
The angular velocity equals change in angular displacement divided by change in time.	The linear velocity equals change in distance (Δx) divided by change in time.

We can relate angular velocity to linear velocity via the following:

$$\vec{V} = r\vec{\omega}$$

Note, just like $\vec{V}$, $\vec{\omega}$ has direction!

The **Right-Hand Rule** states that you must wrap your fingers around the object's path. Let's take the example of a toy car going around a circle.

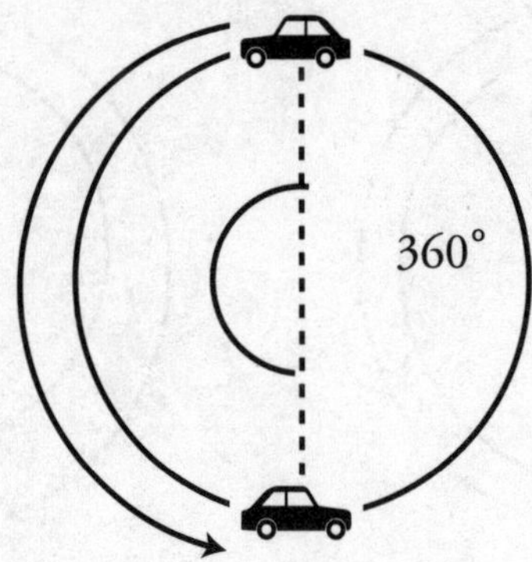

If you follow the car's path, you will find your fingers are wrapping counterclockwise. Your thumb is the direction for angular velocity. In this case, it points out of the page. In physics we write the direction like this ⊙. If it were into the page, we would use this ⊗.

What Is Angular Acceleration (α)?

Angular acceleration, $\alpha = \frac{\Delta\omega}{\Delta t}$

Units: radians/s^2

Many of the rotational kinematics equations reflect linear kinematics equations.

	Rotational Motion	Linear Motion
Big Five #1:	$\Delta\theta = \frac{1}{2}(\omega_0 + \omega)t$	$\Delta x = \frac{1}{2}(v_0 + v)t$
Big Five #2:	$\omega = \omega_0 + \alpha t$	$v = v_0 + at$
Big Five #3:	$\theta = \theta_0 + \omega_0 t + \frac{1}{2}\alpha t^2$	$x = x_0 + v_0 t + \frac{1}{2}at^2$
Big Five #4:	$\theta = \theta_0 + \omega t - \frac{1}{2}\alpha t^2$	$x = x_0 + vt - \frac{1}{2}at^2$
Big Five #5:	$\omega^2 = \omega_0^2 + 2\alpha(\theta - \theta_0)$	$v^2 = v_0^2 + 2a(x - x_0)$

Example 1 Four children climb on a carousel that is initially at rest. If the carousel accelerates to 0.4 radians per second within 10 seconds, what is the angular acceleration? What is its linear rotation 3 m from axis of rotation?

Solution.

$\omega_i = 0$
$\omega_0 = 0.4$ rad/s

$$\alpha = \frac{\omega_o - \omega_i}{t} = \frac{0.4 - 0}{10} = 0.04 \text{ rad/s}^2$$

$\Delta\theta$
$\alpha = ?$
$T = 10$ sec

To find linear acceleration: $a = r\alpha = (3 \text{ m})(0.04 \text{ rad/s}^2) = 0.12 \text{ m/s}^2$

CENTER OF MASS

In the preceding chapters, objects were treated as though they were each a single particle. In many force-diagrams, we have said that all the force is being delivered at a single point on the object. What makes this point the center of mass? And why do we associate all the force being delivered at this single point on the object?

Imagine a series of experiments. We walk into a large room with a hammer and a small light that we can attach to the hammer. In the first experiment, we hold the hammer and we attach a light to the very end of the hammer. Then we turn off the light and throw the hammer across the room. If we trace the path of the hammer, we notice that it gives a weird spiral-shaped path as follows:

You may have heard some people use the term "center of gravity" instead of "center of mass." While there is a technical difference between them, it's not important for this test. The two can be used interchangeably for any AP Physics 1 discussion.

Then we repeat the experiment. This time let's attach the small light to the head of the hammer. Once again we turn off the light, throw the hammer across the room, and trace the path of the hammer. This time we notice that it follows another spiral-shaped path:

If we keep doing the experiment, after countless trials, you will notice that at a specific point, the hammer makes a parabolic path seen in the following drawing:

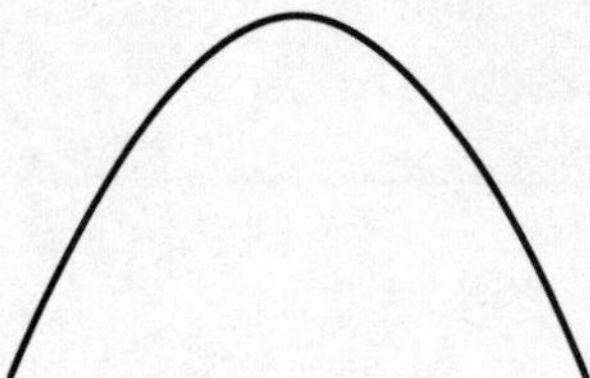

Apparently, there was something important about that specific point. All the other points gave spiraled trajectories, but this one gave a smooth parabolic path. Upon further investigation, if we place that point on our fingers, we notice that the hammer balances nicely and is perfectly horizontal with the floor.

This certain point is the center of mass. Another way of looking at it is to say that the center of mass is the point at which we could consider all the mass of the object to be concentrated.

For a simple object such a sphere, block, or a cylinder, whose density is constant (a term in physics we call homogeneous), the center of mass is at its geometric center.

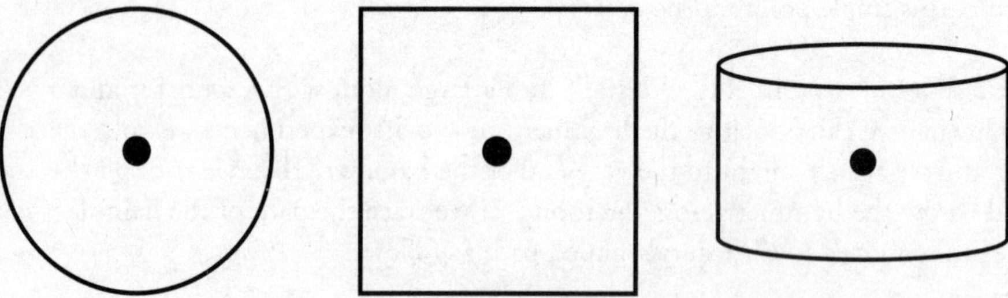

In some cases, the center of mass is not located on the body of the object:

While equations for the location of the center of mass exist, you don't need to know them for the AP Exam. Only a qualitative understanding is required.

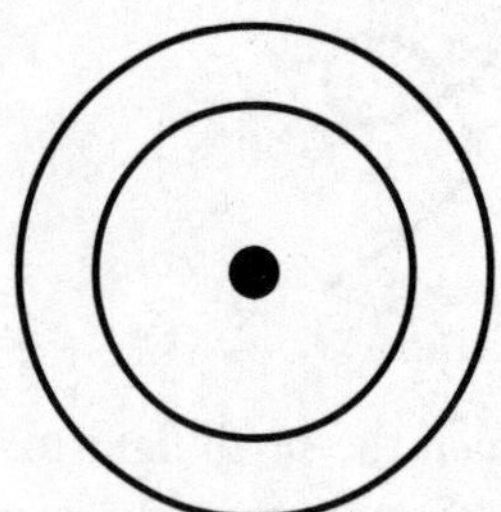

For objects that are irregularly shaped or have different densities throughout, like a hammer, the center of mass is located closer to where most of the mass resides.

Thus, a point close to the head of the hammer would follow a parabolic path when we throw the hammer.

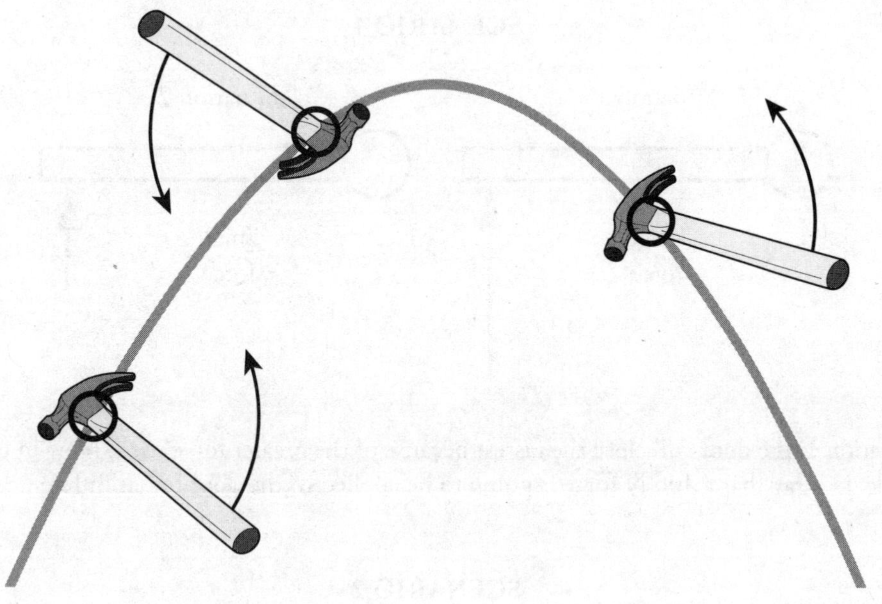

TORQUE

We can tie a ball to a string and make it undergo circular motion, but how would we make that ball itself spin? We could simply palm the ball and rotate our hand, or we could put our hands on opposite sides of the ball and push one hand forward and the other backward. In both cases, in order to make an object's center of mass accelerate, we need to exert a force. In order to make an object spin, we need to exert a torque.

Torque is the measure of a force's effectiveness at making an object spin or rotate. (More precisely, it's the measure of a force's effectiveness at making an object accelerate rotationally.) If an object is initially at rest, and then it starts to spin, something must have exerted a torque. And if an object is already spinning, something would have to exert a torque to get it to stop spinning.

All systems that can spin or rotate have a "center" of turning. This is the point that does not move while the remainder of the object is rotating, effectively becoming the center of the circle. There are many terms used to describe this point, including pivot point and fulcrum.

Torque has always been a topic that students have difficulty understanding. So, let's go through a few examples and ask a few questions before we delve into torque:

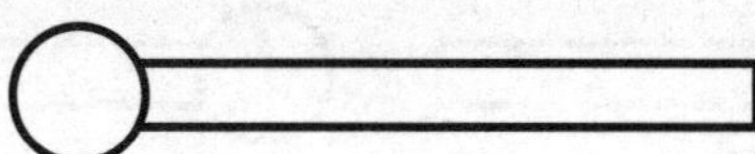

The drawing above shows a door with its hinge (pivot point) located on the left side of the door. You can try some of these examples at home on a door in your house to get a better understanding. Let's pose two different situations for trying to close the door:

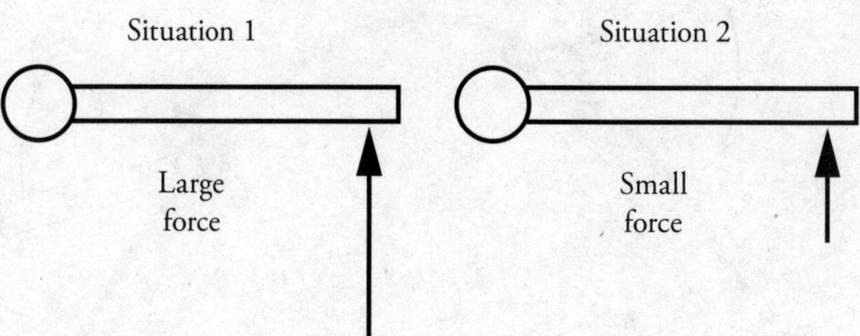

In Situation 1, the door will close the fastest because of the greater force used. Now in the same example, let's say that a 100 N force is going to be applied to the door but at different angles:

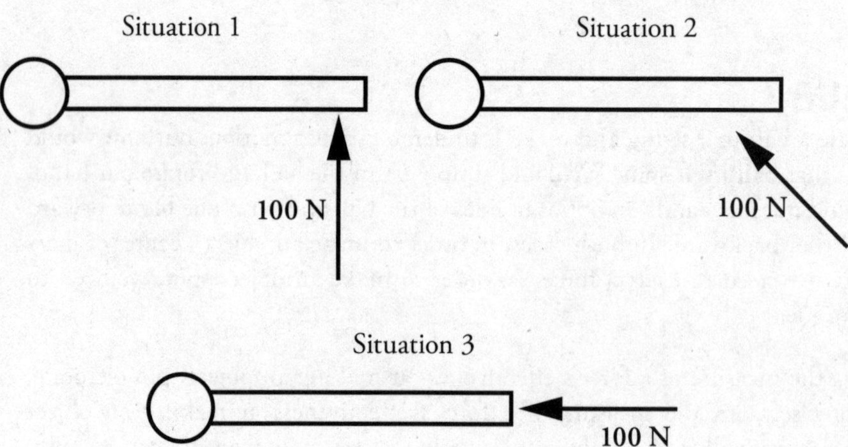

The door will close the fastest in Situation 1 and the second fastest in Situation 2. Situation 3 involves merely pushing on the door to no avail; the door will not close.

Now let's use the same door example and apply the force to different parts of the door:

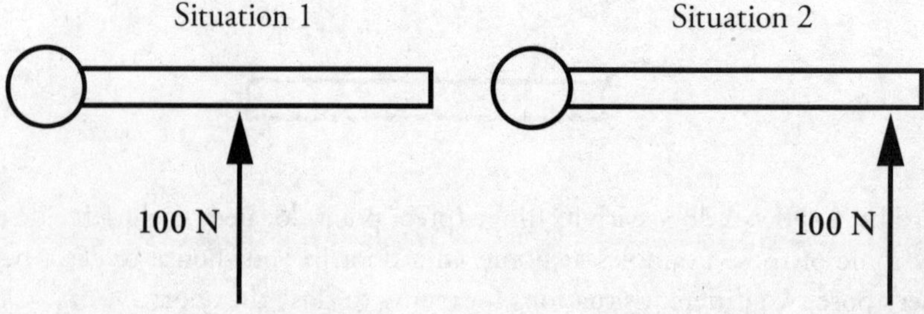

If you try this at home, you will notice that if you push the door as in Situation 2, it will be easier to close the door than if you tried to push the door as in Situation 1.

If you noticed in the three scenarios, there were a few points that mattered when trying to close the door. In Scenario 1, the amount of force used to close the door mattered (magnitude of force). In Scenario 2, the angle in which we pushed the door mattered (angle). And in Scenario 3, the place in which we pushed mattered (radius).

Our force's effectiveness at making something spin or rotate was determined by three factors:

1 the magnitude of force (F)
2. the angle (θ)
3. the radius (r)

These three factors give us our torque equation:

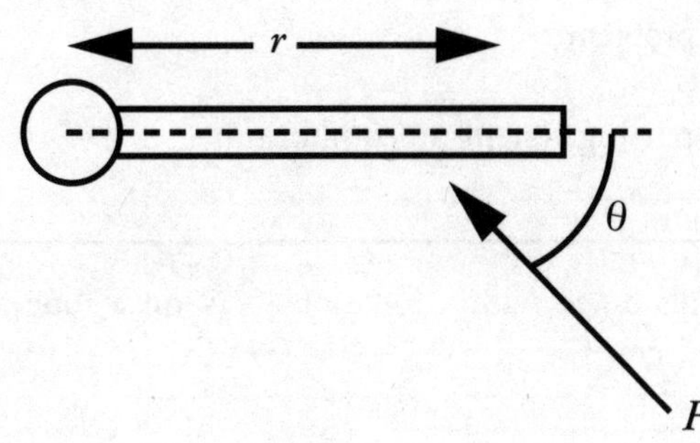

Why Sine?
In math, this is known as a cross product between your force with your radius. It is written as $\tau = (r \times F)$.

$$\tau = rF\sin\theta$$

There's no special name for this unit (torque): it's just called a newton-meter. Because it is not in newtons, torque is NOT a force. (Torque is a measure of how much a force acting on an object causes that object to rotate.) In Scenario 2, Situation 3, a force was being applied straight on to the door directed straight into the pivot point. The magnitude of force did not suddenly disappear, but it was not effective at closing the door. Torque is the rotational equivalent of force in trying to make something accelerate rotationally.

Don't Confuse the Units
In the previous chapter, a newton-meter became a joule. This is not the case with torque.

Another way to calculate torque is using the lever arm. For this method, we consider the line of action of the force, which is the line on which the force vector sits. The lever arm is the perpendicular distance from the pivot to the line of action of the force.

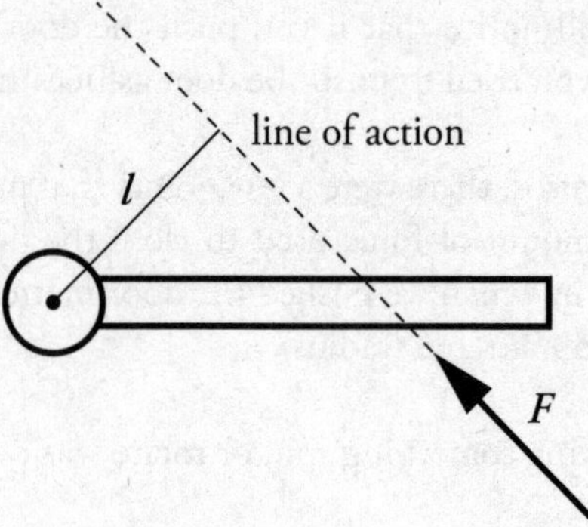

Using the lever arm, torque is defined as

$$\tau = lF$$

If we compare the two methods, we notice that $l = r\sin\theta$, so both methods give the same answer for torque, as we'd expect. Generally, only use the lever arm equation when the lever arm is fairly easy to identify in a problem.

Torque problems usually involve putting systems in equilibrium.

Example 2 A student pulls down with a force of 40 N on a rope that winds around a pulley of radius 5 cm.

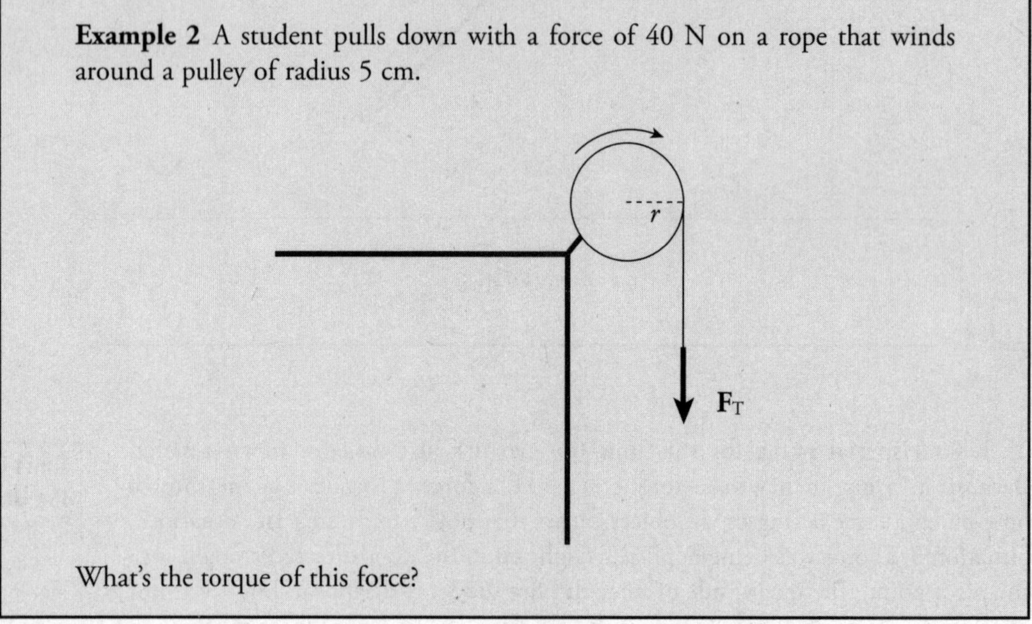

What's the torque of this force?

Solution. Since the tension force, $\mathbf{F}_T$, is tangent to the pulley, it is perpendicular to the radius vector **r** at the point of contact:

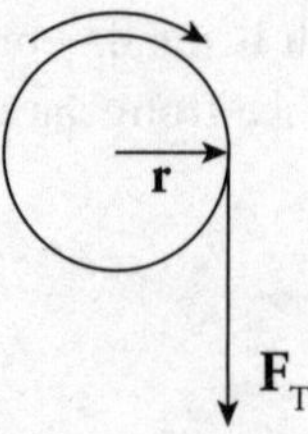

Therefore, the torque produced by this tension force is simply

$$\tau = rF_T = (0.05 \text{ m})(40 \text{ N}) = 2 \text{ N·m}$$

Example 3 What is the net torque on the cylinder shown below, which is pinned at the center?

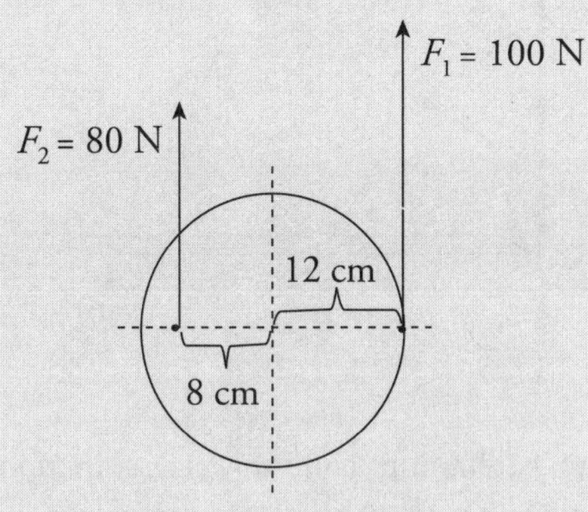

Solution. Each of the two forces produces a torque, but these torques oppose each other. The torque of F_1 is counterclockwise, and the torque of F_2 is clockwise. This can be visualized by imagining the effect of each force, assuming that the other was absent.

The **net torque** is the sum of all the torques. Counting a counterclockwise torque as positive and a clockwise torque as negative, we have

$$\tau_1 = +r_1F_1 = +(0.12 \text{ m})(100 \text{ N}) = +12 \text{ N·m}$$

and

$$\tau_2 = -r_2F_2 = -(0.08 \text{ m})(80 \text{ N}) = -6.4 \text{ N·m}$$

so

$$\tau_{net} = \Sigma\tau = \tau_1 + \tau_2 = (+12 \text{ N·m}) + (-6.4 \text{ N·m}) = +5.6 \text{ N·m}$$

Keep Orientation in Mind
With spinning objects, we will often be referring to directions as clockwise or counter-clockwise.

EQUILIBRIUM

An object is said to be in **translational equilibrium** if the sum of the forces acting on it is zero—that is, if $F_{net} = 0$. Similarly, an object is said to be in **rotational equilibrium** if the sum of the torques acting on it is zero—that is, if $\tau_{net} = 0$. The term *equilibrium* by itself means both translational and rotational equilibrium. A body in equilibrium may be in motion; $F_{net} = 0$ does not mean that the velocity is zero; it only means that the velocity is constant. Similarly, $\tau_{net} = 0$ does not mean that the angular velocity is zero; it only means that it's constant. If an object is at rest, then it is said to be in **static equilibrium**.

It's All About Balance
Equilibrium problems are all about balancing one force's effectiveness at turning something clockwise with another force's effectiveness at turning something counterclockwise.

Example 4 A uniform bar of mass m and length L extends horizontally from a wall. A supporting wire connects the wall to the bar's midpoint, making an angle of 55° with the bar. A sign of mass M hangs from the end of the bar.

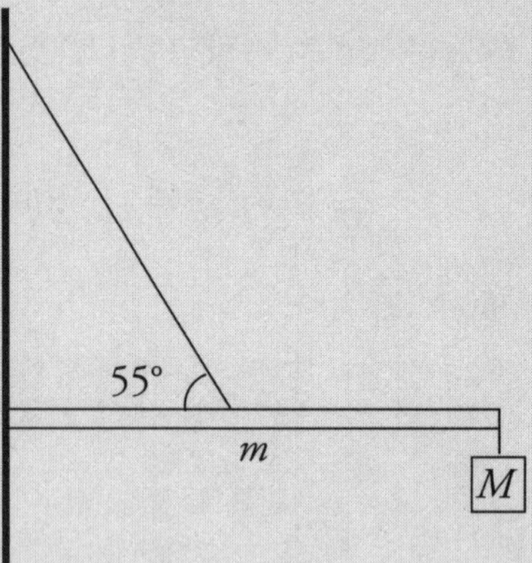

If the system is in static equilibrium and the wall has friction, determine the tension in the wire and the strength of the force exerted on the bar by the wall if m = 8 kg and M = 12 kg.

Choose Your Strategy
Some problems are easier when using center-of-mass, whereas some problems are easier using torque. Choose which one is easier for you.

Solution. Let F_C denote the (contact) force exerted by the wall on the bar. In order to simplify our work, we can write F_C in terms of its horizontal component, F_{Cx}, and its vertical component, F_{Cy}. Also, if F_T is the tension in the wire, then $F_{Tx} = F_T \cos 55°$ and $F_{Ty} = F_T \sin 55°$ are its components. This gives us the following force diagram:

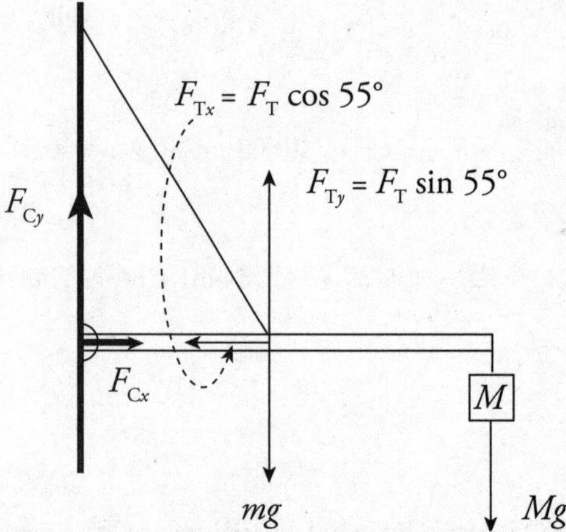

The first condition for equilibrium requires that the sum of the horizontal forces is zero and the sum of the vertical forces is zero:

$$\Sigma F_x = 0: \qquad F_{Cx} - F_T \cos 55° = 0 \qquad (1)$$

$$\Sigma F_y = 0: \qquad F_{Cy} + F_T \sin 55° - mg - Mg = 0 \qquad (2)$$

We notice immediately that we have more unknowns (F_{Cx}, F_{Cy}, F_T) than equations, so this system cannot be solved as is. The second condition for equilibrium requires that the sum of the torques about any point is equal to zero. Choosing the contact point between the bar and the wall as our pivot, only three of the forces in the diagram above produce torque: $\mathbf{F}_{Ty}$ produces a counterclockwise torque, and both $m\mathbf{g}$ and $M\mathbf{g}$ produce clockwise torques, and the sum of the three torques must equal zero. From the definition $\tau = rF \sin \theta$, and taking counterclockwise torque as positive and clockwise torque as negative, we have

$$\Sigma\tau = 0: \qquad \tfrac{L}{2} rF_{Ty} - \tfrac{L}{2} mgr - LMgr = 0 \qquad (3)$$

Note that you can divide out the r term here, so r will not factor into the final equation. The above equation contains only one unknown and can be solved immediately:

$$\begin{aligned} \tfrac{L}{2} F_{Ty} &= \tfrac{L}{2} mg + LMg \\ F_{Ty} &= mg + 2Mg = (m + 2M)g \end{aligned}$$

Since $F_{Ty} = F_T \sin 55°$, we can find that

$$\begin{aligned} F_T \sin 55° = (m + 2M)g \quad \Rightarrow \quad F_T &= \frac{(m+2M)g}{\sin 55°} \\ &= \frac{(8 + 2 \cdot 12)(10)}{\sin 55°} \\ &= 390 \text{ N} \end{aligned}$$

Substituting this result into Equation (1) gives us F_{Cx}:

$$F_{Cx} = F_T \cos 55° = \frac{(m+2M)g}{\sin 55°} \cos 55° = (8 + 2 \cdot 12)(10) \cot 55° = 220 \text{ N}$$

Note: We use cot 55° here, because $\cot 55° = \dfrac{\cos 55°}{\sin 55°}$.

And finally, from Equation (2), we get

$$\begin{aligned} F_{Cy} &= mg + Mg - F_T \sin 55° \\ &= mg + Mg - \frac{(m+2M)g}{\sin 55°} \sin 55° \\ &= -Mg \\ &= -(12)(10) \\ &= -120 \text{ N} \end{aligned}$$

The fact that F_{Cy} turned out to be negative simply means that in our original force diagram, the vector $\mathbf{F}_{Cy}$ points in the direction opposite to how we drew it. That is, $\mathbf{F}_{Cy}$ points downward. Therefore, the magnitude of the total force exerted by the wall on the bar is

$$F_C = \sqrt{(F_{Cx})^2 + (F_{Cy})^2} = \sqrt{220^2 + 120^2} = 250 \text{ N}$$

ROTATIONAL INERTIA

Now that we've studied torque and rotation, we can finally put together the pieces of making an object spin. An object's rotational inertia (also known as the moment of inertia) is defined as the tendency of an object in motion to rotate until acted upon by an outside force. Think of mass as translational inertia, since it measures an object's resistance to translational acceleration, given by a in $F = ma$. Then, just as translational inertia tells us how resistant an object is to translation acceleration, an object's rotational inertia, I, tells us how resistant the object is to rotational acceleration.

Rotational acceleration or angular acceleration is the same as translational acceleration except we are taking a ball at rest and speeding up its angular velocity. In order to achieve this acceleration, a force is required. In terms of rotational inertia, torque is required to produce rotational acceleration. We need to apply a force that is effective in generating acceleration. Torque, in terms of rotational inertia, is

$$\tau = I\alpha$$

This equation makes some key relationships. The larger the rotational inertia (moment of inertia) is, the smaller the value will be for a given torque on an object. If Object 1 has a greater rotational inertia than Object 2, then it will be more difficult to rotate Object 1 than Object 2. More precisely, a greater torque will be required to give Object 1 the same rotational acceleration as Object 2, or, equivalently, if the same torque is applied to both objects, Object 1 will undergo a smaller rotational acceleration.

Study Break

You're almost done with all of the content chapters! Way to go! Wrap up Chapter 10 and then give yourself a break—go for a walk,. take a dance break, get a snack—before you continue on and tackle a practice test. You deserve it!

So, how do we find the rotational inertia of an object? It depends on the object's mass, but there is more to it than that. Two objects can have the same mass but different rotational inertias. Rotational inertia is also dependent on how the mass is distributed in an object with respect to the axis that it rotated around.

The farther away the mass is from the axis of rotation, the greater the rotational inertia will be.

Imagine a barbell with a weight near each end and an identical barbell with the weights pushed near the middle of the bar. These two barbells have the same mass, but their rotational inertias are different. Notice that the first barbell has its attached masses farther away from the axis of rotation than the second one. If we wanted to rotate each bar around its midpoint, we would find it more difficult to rotate the first barbell than the second one. The first barbell has a greater value of I.

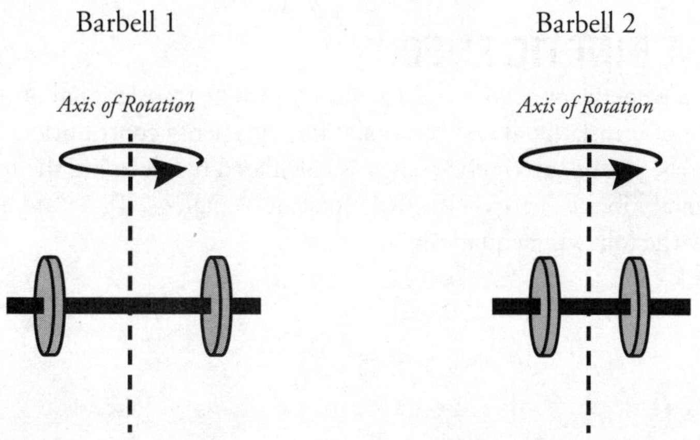

Barbell 1

Axis of Rotation

Barbell 2

Axis of Rotation

ANGULAR MOMENTUM

The same way that objects moving in a straight line have linear momentum, objects experiencing rotation have angular momentum. Angular momentum is a vector quantity denoted by **L**. Similar to the other rotational vectors we've covered so far, angular momentum has only two possible directions: clockwise and counterclockwise. Furthermore, if you remember what we learned about linear momentum back in Chapter 8, then understanding its rotational counterpart is pretty straightforward. All you have to do is replace the linear terms in the equations you already know with their analogous rotational terms.

For example, linear momentum is defined as the product of mass and velocity, and angular momentum is given by the following equation:

$$\mathbf{L} = I\omega$$

which shows *angular* momentum is the product of the rotational inertia and angular velocity.

Additionally, we previously defined the change in linear momentum as the product of force and the change in time. A similar equation exists for finding the change in angular momentum:

$$\Delta\mathbf{L} = \tau\Delta t$$

Finally, in addition to the above equations, it's important to know that, just like linear momentum, angular momentum is conserved for any isolated system.

> **Example 5** A man is standing halfway between the center and one edge of a rotating plank. If he wants to slow down the rotational speed, should he walk toward the center or the edge?

Solution. Because this is an isolated system, we know angular momentum will be conserved. Since $\mathbf{L} = I\omega$, that means our desired decrease to rotational velocity can be achieved by increasing the rotational inertia. Earlier, we learned that rotational inertia is greater when the mass is farther from the axis of rotation. In this case, that means the man should walk toward the edge.

ROTATIONAL KINETIC ENERGY

Kinetic energy is the energy of motion. Since motion can be translational or rotational, kinetic energy can consist of translational and rotational kinetic energy contributions. Similar to other rotational quantities, rotational kinetic energy is calculated by replacing the terms in the equation for translational kinetic energy with their rotational analogs. Therefore, rotational kinetic energy is given by the following equation:

$$K_r = \frac{1}{2}I\omega^2$$

Rotational kinetic energy is just another form of kinetic energy. Therefore, when considering concepts like total mechanical energy, the kinetic energy can be broken down into translational and rotational terms.

$$E = K + U = (K_t + K_r) + U$$

For instance, as a yo-yo travels downward on its string, its gravitational potential energy is converted in part to the translational kinetic energy associated with its motion downward and in part to the rotational kinetic energy associated with the spinning of the yo-yo.

One special scenario is when an object with a circular cross section rolls without slipping. In this case, the object's angular speed is related to its translational speed and its radius, $\omega = v / r$. Therefore, the object's rotational kinetic energy is proportional to its translational kinetic energy.

Example 6 A ball with a rotational inertia of $\frac{2}{3}MR^2$ rolls without slipping with a constant speed v. What fraction of its kinetic energy is rotational?

Solution. For a ball that rolls without slipping, the rotational kinetic energy of the ball is

$$K_r = \frac{1}{2}I\omega^2 = \frac{1}{2}\left(\frac{2}{3}MR^2\right)\left(\frac{v}{R}\right)^2 = \frac{2}{3}\left(\frac{1}{2}Mv^2\right) = \frac{2}{3}K_t$$

Therefore, the fraction of its total kinetic energy that is rotational is

$$\frac{K_r}{K_r + K_t} = \frac{\frac{2}{3}K_t}{\frac{2}{3}K_t + K_t} = \frac{\frac{2}{3}}{\frac{5}{3}} = \frac{2}{5}$$

CHAPTER 10 KEY TERMS

angular displacement
translational displacement
angular velocity
Right-Hand Rule
angular acceleration
center of mass
torque
net torque
translational equilibrium
rotational equilibrium
static equilibrium
rotational inertia
angular momentum
rotational kinetic energy

Chapter 10 Review Questions

Answers and explanations can be found in Chapter 11.

Section I: Multiple Choice

1. A disc has a radius of 5 cm. If the disc rotates about its central axis at an angular speed of 8 rev/s, what is the linear speed of a point on the rim of the disc?

    (A) 0.4 m/s
    (B) 2.5 m/s
    (C) 3.9 m/s
    (D) 50 m/s

2. An object, initially at rest, begins spinning under a constant angular acceleration. In 15 s, it completes an angular displacement of 90 rad. What is the magnitude of the angular acceleration it experiences?

    (A) 0.4 rad/s^2
    (B) 0.8 rad/s^2
    (C) 6 rad/s^2
    (D) 12 rad/s^2

3. A wheel initially rotating at 12 rad/s decelerates uniformly to rest in 0.4 s. If the wheel has a rotational inertia of 0.5 kg·m^2, what is the magnitude of the torque causing this deceleration?

    (A) 1.5 N·m
    (B) 15 N·m
    (C) 30 N·m
    (D) 38 N·m

4. A grinding wheel with a radius of 12 cm is being used to sharpen a knife. The side of the knife is pressed perpendicularly into the wheel's surface with a force of 16 N. The coefficient of kinetic friction between the knife and the wheel is 0.28. What is the torque the grinding wheel's motor must supply to maintain its rotation rate of 50 rad/s?

    (A) 0.54 N·m
    (B) 0.72 N·m
    (C) 4.5 N·m
    (D) 54 N·m

5. A uniform 10-m ladder with a mass of 20 kg rests against a wall such that it makes a 50° angle with the ground. If the friction between the ladder and the wall is negligible, what is the force the ladder exerts on the wall?

    (A) 82 N
    (B) 98 N
    (C) 164 N
    (D) 196 N

6. An object spins with an angular velocity ω. If the object's rotational inertia increases by a factor of 4, without the application of external torque, what will be the object's new angular velocity?

    (A) $\omega/4$
    (B) $\omega/2$
    (C) 2ω
    (D) 4ω

7. A child with a mass of 25 kg is standing at the edge of a 2.0 m radius merry-go-round, and they are rotating without friction at 20 rpm. The child then moves toward the center of the merry-go-round. The merry-go-round has a rotational inertia of 1200 kg·m^2, and the rotational inertia of the child is $I = mr^2$, where r is her distance from the axis of rotation. What is the new angular velocity when she is 0.5 m from the center of the merry-go-round?

    (A) 2.03 rad/s
    (B) 2.09 rad/s
    (C) 2.15 rad/s
    (D) 2.25 rad/s

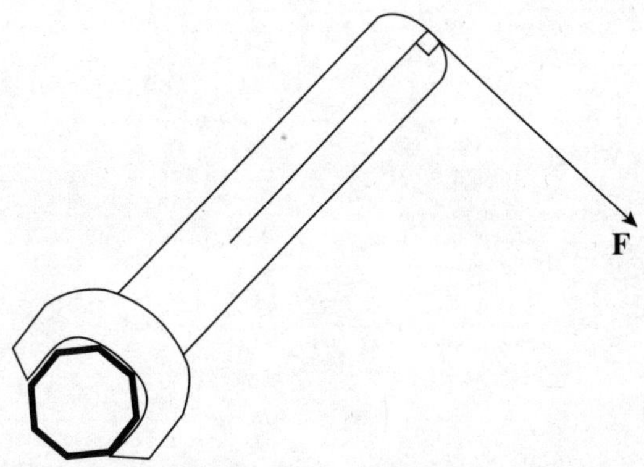

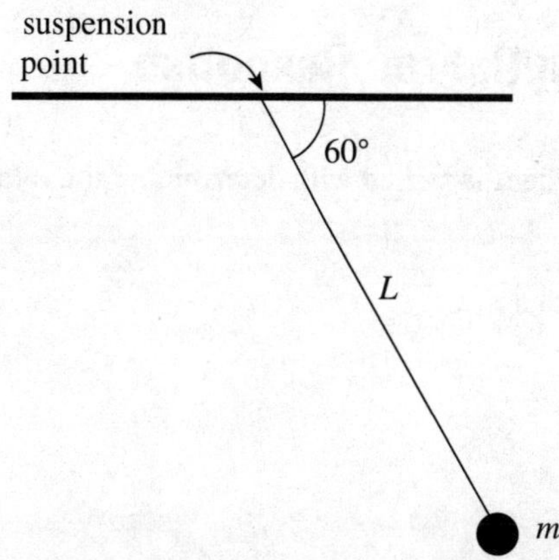

8. In an effort to tighten a bolt, a force **F** is applied as shown in the figure above. If the distance from the end of the wrench to the center of the bolt is 20 cm and $F = 20$ N, what is the magnitude of the torque produced by **F** ?
   (A) 1 N·m
   (B) 2 N·m
   (C) 4 N·m
   (D) 10 N·m

9. In the figure above, what is the torque about the pendulum's suspension point produced by the weight of the bob, given that the length of the pendulum, L, is 80 cm and $m = 0.50$ kg ?
   (A) 0.5 N·m
   (B) 1.0 N·m
   (C) 1.7 N·m
   (D) 2.0 N·m

Section II: Free Response

1. A student is tasked with determining the rotational inertia of a bicycle wheel.

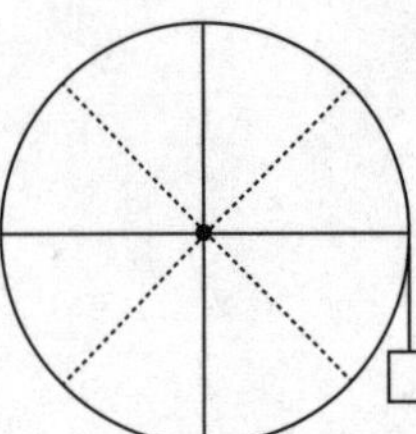

(A) The student plans to mount the wheel on a horizontal axis so that it can spin freely, then hang a mass from a string wrapped around the outer diameter of the wheel.

i. Design an experimental procedure to measure the rotational inertia. Assume equipment usually found in a school physics laboratory is available. In the table below, list the quantities and associated symbols that would be measured in your experiment. Also list the equipment that would be used to measure each quantity. You do not need to fill in every row. If you need additional rows, you may add them to the space just below the table.

Quantity to be Measured	Symbol for Quantity	Equipment for Measurement

ii. Determine an expression for the rotational inertia of the wheel, referring to the quantities listed in the table.

(B) Describe the overall procedure to be used to measure the rotational inertia of the wheel, referring to the table. Provide enough detail so that another student could replicate the experiment, including any steps necessary to reduce experimental uncertainty. As needed, use the symbols defined in the table and/or include a simple diagram of the setup.

(C) The student then decides to make a number of measurements by hanging different masses from the same wheel. On the axes below, sketch a graph of the rotational inertia as a function of the hanging mass.

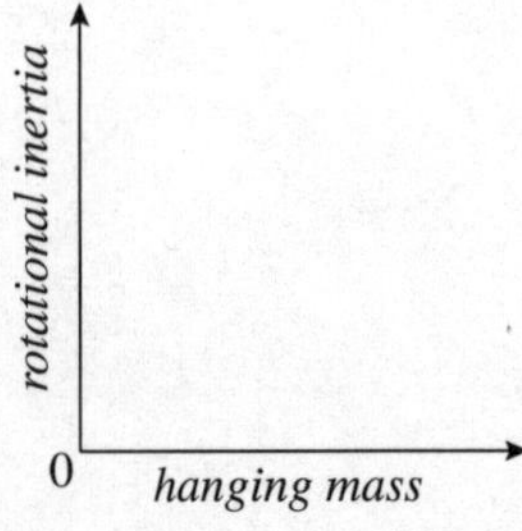

2. A woman is riding her bicycle, which has wheels with a diameter D and a rotational inertia I.

 (A) Assuming that there is negligible friction opposing the rotation of her wheels, derive an expression for the torque, τ_{app}, she applies to one of the wheels to accelerate from rest to a speed of v in a time t in terms of D, I, v, t, and physical constants, as appropriate.

 (B) Suppose there is a kinetic friction force f acting on the wheel at a distance s from the wheel's rotation axis. If the torque the cyclist applies to the wheel remains the same, derive an expression for the time, t, it will take her to attain a speed of v in terms of D, I, v, t, f, s, and physical constants, as appropriate.

3.

Two balls, A and B, of identical mass and radius are placed next to each other at the top of an incline and released from rest. The balls roll without slipping down the incline, along a horizontal surface, and then up another incline. Ball A reaches the bottom of the first incline with a greater translational speed than does Ball B. The balls end up reaching the same height on the second incline.

In terms of energy, explain the different translational speeds at the bottom of the first incline and the same final height of the two balls. Provide your answer in a clear, coherent paragraph-length response that may also contain figures and/or equations.

Summary

- Torque is a property of a force that makes an object rotate. The equation for torque is $\tau = rF \sin \theta$. Torques may be clockwise or counterclockwise. An object is in equilibrium if $\sum F_x = 0$, $\sum F_y = 0$, and $\sum \tau = 0$.
- The rotation of objects is described with rotational analogues to linear kinematic quantities: angular displacement $\Delta\theta$, angular velocity $\vec{\omega} = \frac{\Delta\theta}{\Delta t}$, and angular acceleration $\vec{\alpha} = \frac{\Delta\omega}{\Delta t}$.
- The Big 5 also applies to rotational motion, with the rotational quantities replacing the corresponding linear quantities.
- Torques result from forces applied at a distance from a pivot point, $\tau = rF\sin\theta$. Torques can be clockwise or counterclockwise, and cause angular acceleration depending on the rotational inertia of the object, $\tau = I\alpha$.
- Rotational inertia reflects the distribution of an object's mass relative to an axis of rotation.
- An object is in equilibrium (not accelerating linearly or rotationally) when $\Sigma F_x = 0$, $\Sigma F_y = 0$, and $\Sigma\tau = 0$.
- The angular momentum of a system, $L = I\omega$, is conserved in the absence of outside net torque.
- A net torque applied to a rotating system over some time will change the angular momentum of the system, $\Delta L = \tau\Delta\tau$.
- A rotating object possesses rotational kinetic energy, $K_r = \frac{1}{2}I\omega^2$.
- Any isolated system will have its angular momentum conserved. This is given by the following equation:

  $$\mathbf{L} = I\omega$$

  If outside forces affect the system, then the change to an object's angular momentum can be calculated by the equation:

  $$\Delta\mathbf{L} = \tau\Delta t$$

Chapter 11
Answers and Explanations to the Chapter Review Questions

CHAPTER 3 REVIEW QUESTIONS

Section I: Multiple Choice

1. **B** To add the vectors, draw the first vector and then from the end of the first draw the second. The resultant vector is from the beginning of the first vector to the end of the second:

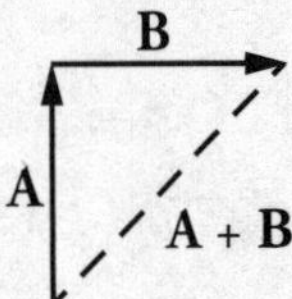

The direction of the resultant vector is therefore northeast, eliminating (C) and (D). Since vectors **A** and **B** are perpendicular to each other and equal in magnitude, the magnitude of the resultant vector can be found using the Pythagorean Theorem:

$$a^2 + b^2 = c^2 \Rightarrow m^2 + m^2 = c^2 \Rightarrow 2m^2 = c^2 \Rightarrow m\sqrt{2} = c$$

This makes (B) correct.

2. **B** To add the three vectors, add their components separately:

$$\mathbf{F}_1 + \mathbf{F}_2 + \mathbf{F}_3 = (0 - 10 + 5)\mathbf{\hat{i}} + (-20 + 0 + 10)\mathbf{\hat{j}} = -5\mathbf{\hat{i}} - 10\mathbf{\hat{j}}$$

3. **C** The magnitude of a vector is given by

$$A = \sqrt{(Ax)^2 + (Ay)^2}$$

If both components of the vector are doubled, the new magnitude, ***A′***, will be

$$A' = \sqrt{(2Ax)^2 + (2Ay)^2} = \sqrt{4(Ax)^2 + 4(Ay)^2} = \sqrt{4[(Ax)^2 + (Ay)^2]} = 2\sqrt{(Ax)^2 + (Ay)^2} = 2A$$

So the magnitude of the vector will also be doubled.

The angle/direction of a vector is given by

$$\theta = \tan^{-1}(A_y / A_x)$$

As can be observed in this equation, doubling the magnitude of both components of the vector will have no effect on the direction of the vector.

4. **D** Subtracting the vector $\boldsymbol{v}_0$ is equivalent to adding the negative of $\boldsymbol{v}_0$ **to** $\boldsymbol{v}_f$. Since $\boldsymbol{v}_f$ and $-\boldsymbol{v}_0$ both point south, adding the two vectors results in a vector that is the sum of their two magnitudes and also points south:

$v_f = 5$ ms

$-v_0 = 15$ ms

This corresponds to (D).

5. **B** The components of the vector must satisfy the equation:

$$A = 10 = \sqrt{(Ax)^2 + (Ay)^2}$$

The only answer choice with components that satisfy this equation is (B): $10 = \sqrt{6^2 + 8^2}$.

6. **B** The vector sum **A** + **B** can be found by adding the individual components:

$$\mathbf{A} + \mathbf{B} = (1 + 4)\hat{\mathbf{i}} + (-2 - 5)\hat{\mathbf{j}} = 5\hat{\mathbf{i}} - 7\hat{\mathbf{j}}$$

The angle that this vector makes with the x-axis is then found with

$$\theta = \tan^{-1}\frac{(\mathbf{A} + \mathbf{B})_y}{(\mathbf{A} + \mathbf{B})_x} = \tan^{-1}\frac{7}{5}$$

This matches (B).

7. **A** If the object travels along the vectors $\boldsymbol{d}_1$ and then $\boldsymbol{d}_2$, then the total distance traveled with be the sum of the two vectors $\boldsymbol{d}_1 + \boldsymbol{d}_2$:

$$\boldsymbol{d}_1 + \boldsymbol{d}_2 = (\mathbf{4} + \mathbf{2})\hat{\mathbf{i}} + (\mathbf{5} + (\mathbf{-3}))\hat{\mathbf{j}}$$
$$= \mathbf{6}\hat{\mathbf{i}} + \mathbf{2}\hat{\mathbf{j}}$$

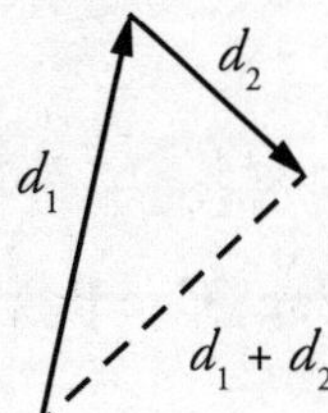

The total distance of the object from the starting position is therefore the magnitude $\boldsymbol{d}_1 + \boldsymbol{d}_2$:

$$|d_1 + d_2| = \sqrt{((d_1 + d_2)_x)^2 + ((d_1 + d_2)_y)^2} = \sqrt{6^2 + 2^2} = \sqrt{40} \approx 6.3 \text{ m}$$

That's (A).

8. **A** Subtracting vector **B** from **A** is equivalent to adding the negative of the second vector to the first:

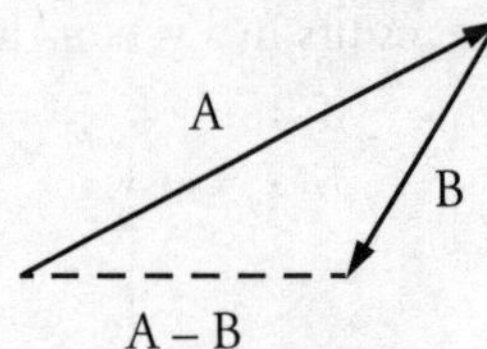

This results in a vector pointing the direction of (A).

9. **C** The x-component of vector **A** can be used to calculate the magnitude of **A:**

$$A_x = A\cos\theta$$
$$42 = A\cos 50°$$
$$A = \frac{42}{\cos 50°}$$

The y-component can then be calculated:

$$A_y = A\sin\theta = \left(\frac{42}{\cos 50°}\right)\sin 50° = 42\tan 50° \approx 50.1$$

which corresponds to (C).

10. **B** If the two vectors were pointed in the same direction, their magnitudes would add, and since magnitudes are positive, the result could not be zero, eliminating (A). If the vectors point in opposite directions, then the magnitude of the resultant vector would be obtained by subtracting the individual magnitudes of the two vectors. The only way for the difference to equal zero is if these two vectors have the same magnitude, eliminating (C) and (D). Choice (B) is therefore correct.

Section II: Free Response

1. (A) The magnitude of A is

$$|A| = \sqrt{(A_x)^2 + (A_y)^2} = \sqrt{3^2 + 6^2} = \sqrt{9+36} = \sqrt{45} \approx 6.7$$

(B) Vector **C** is subtracted from **B** by adding –**C** to **B**:

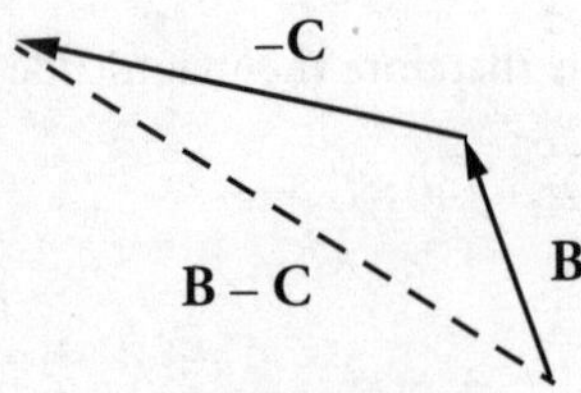

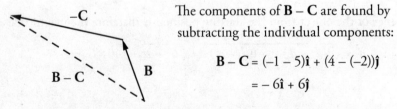

The components of **B** – **C** are found by subtracting the individual components:

$$\mathbf{B} - \mathbf{C} = (-1 - 5)\hat{\imath} + (4 - (-2))\hat{\jmath}$$
$$= -6\hat{\imath} + 6\hat{\jmath}$$

(C) The components of **2B** are found by multiplying each component of **B** by 2:

$$\mathbf{2B} = (2 \times -1)\hat{\mathbf{i}} + (2 \times 4)\hat{\mathbf{j}} = -2\hat{\mathbf{i}} + 8\hat{\mathbf{j}}$$

The components of **A** + **2B** are found by adding the individual components:

$$\mathbf{A} + \mathbf{2B} = (3 + (-2))\hat{\mathbf{i}} + (6 + 8)\hat{\mathbf{j}} = \hat{\mathbf{i}} + 14\hat{\mathbf{j}}$$

(D) The components of **A** – **B** – **C** are found by working with the individual components:

$$\mathbf{A} - \mathbf{B} - \mathbf{C} = (3 - (-1) - 5)\hat{\mathbf{i}} + (6 - 4 - (-2))\hat{\mathbf{j}} = -1\hat{\mathbf{i}} + 4\hat{\mathbf{j}}$$

Let **D** = **A** – **B** – **C** The magnitude of **D** is found using

$$D = \sqrt{(D_x)^2 + (D_y)^2} = \sqrt{(-1)^2 + 4^2} = \sqrt{17} \approx 4.1$$

and the direction of **D** relative to the horizontal is found using

$$\theta = \tan^{-1}\frac{4}{-1} \approx -76°$$

As the resultant vector has components $-1\hat{\mathbf{i}} + 4\hat{\mathbf{j}}$, it is located in Quadrant II. To obtain an angle in Quadrant II, add 180° to get the correct answer of $\theta \approx 104°$ *or* 76° north of west.

2. (A)

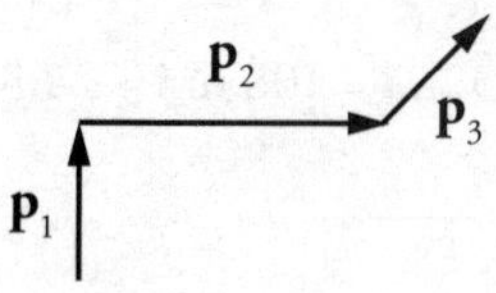

$\mathbf{P}_1$ $\mathbf{P}_2$ $\mathbf{P}_3$

(B) The distance the ant has traveled, *d*, is the magnitude of the sum of the individual vectors in the ant's path: $\mathbf{d} = \mathbf{p}_1 + \mathbf{p}_2 + \mathbf{p}_3$. To do this addition, the components of $\mathbf{p}_3$ must be calculated. If due east is the *x*-direction and due north is the *y*-direction, then northeast is 45° about the horizontal. Thus,

$$p_{3,x} = p_3 \cos\theta = (14 \text{ cm}) \cos 45° \approx 10 \text{ cm}$$
$$p_{3,y} = p_3 \sin\theta = (14 \text{ cm}) \sin 45° \approx 10 \text{ cm}$$

Adding the individual components,

$$\mathbf{d} = \mathbf{p}_1 + \mathbf{p}_2 + \mathbf{p}_3 = (0 + 30 \text{ cm} + 10 \text{ cm})\hat{\mathbf{i}} + (20 \text{ cm} + 0 + 10 \text{ cm})\hat{\mathbf{j}}$$
$$= 40 \text{ cm } \hat{\mathbf{i}} + 30 \text{ cm } \hat{\mathbf{j}}$$

The magnitude of this vector is

$$d = \sqrt{(40^2 + 30^2)} \text{ cm} = \sqrt{2500} \text{ cm} = 50 \text{ cm}$$

(C) The ant's position is $\mathbf{d}$ = 40 cm $\hat{\mathbf{i}}$ + 30 cm $\hat{\mathbf{j}}$. If the ant walks due north, the *x*-component of its position will not change, but the *y*-component will increase. Since its final position vector will have a magnitude of 80 cm, the *y*-component of the final position vector, **f**, can be calculated:

$$\mathbf{f} = \sqrt{(f_x)^2 + (f_y)^2} \Rightarrow 80\text{ cm} = \sqrt{(40\text{ cm})^2 + (f_y)^2} \Rightarrow (80\text{ cm})^2 - (40\text{ cm})^2 = (f_y)^2 \Rightarrow f_y \approx 69.3\text{ cm}$$

Since the ant was already 30 cm north of his original position, he can walk another 69.3 cm – 30 cm = 39.3 cm due north.

3. (A) The components of $\mathbf{p}$ = 140 m directed 10° above the horizontal are given by

$$p_x = p\cos\theta = (140\text{ m})\cos 10° = 138\text{ cm}$$

$$p_y = p\sin\theta = (140\text{ m})\sin 10° = 24.3\text{ cm}$$

thus, $\mathbf{p} = 138\hat{\mathbf{i}} + 24.3\hat{\mathbf{j}}$.

(B) The vector addition $\mathbf{p}' = \mathbf{p} + \mathbf{w}$ is accomplished by starting to draw the tail of the second vector at the end of the first:

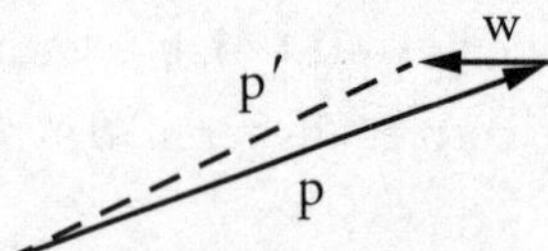

(C) Adding the components of $\mathbf{p} + \mathbf{w}$ yields p′ = (138 m – 30 m)$\hat{\mathbf{i}}$ + 24.3 m $\hat{\mathbf{j}}$ = 108 m $\hat{\mathbf{i}}$ + 24.3 m $\hat{\mathbf{j}}$. The new angle with the horizontal is therefore $\theta = \tan^{-1}\frac{24.3}{108} = 12.7°$.

4. (A) Since **v** depends on **c** and **r**, it makes sense to use **v** as the resultant vector. The diagram shows that **r** and **c** are connected tip-to-tail, indicating that these vectors are being added together, and thus, **v** = **r** + **c**.

(B) If the angle between **v** and **c** is 90°, a right triangle is formed, so that the Pythagorean Theorem can be used to solve for the magnitude of **v** as a function of the other two vectors:

$$v^2 + c^2 = r^2 \Rightarrow v^2 = r^2 - c^2 \Rightarrow v = \sqrt{r^2 - c^2}$$

(C) Let due east be the $\hat{\mathbf{i}}$ direction and due north be the $\hat{\mathbf{j}}$ direction. Then the vectors **c** and **v** are defined as: $\mathbf{c} = 5\text{ m/s}\,\hat{\mathbf{i}}$ and $\mathbf{v} = -10\text{ m/s}\,\hat{\mathbf{j}}$. Solving the vector equation from part (A) for **r** yields

$$\mathbf{r} = \mathbf{v} - \mathbf{c} = (0 - 5\text{ m/s})\hat{\mathbf{i}} + (-10\text{ m/s} - 0)\hat{\mathbf{j}} = -5\text{ m/s}\,\hat{\mathbf{i}} - 10\text{ m/s}\,\hat{\mathbf{j}}$$

Using these components, the magnitude and direction of **r** are given by

$$|r| = \sqrt{(r_x)^2 + (r_y)^2} = \sqrt{(-5)^2 + (-10)^2} = \sqrt{125} = 5\sqrt{5} \approx 11.2\text{ m/s}$$

$$\theta = \tan^{-1}\frac{-10}{-5} = -116.6° = 26.6°\text{ west of south or }63.4°\text{ south of west}$$

CHAPTER 4 REVIEW QUESTIONS

Section I: Multiple Choice

1. **A, D** Traveling once around a circular path means that the final position is the same as the initial position. Therefore, the displacement is zero. The average speed, which is *total* distance traveled divided by elapsed time, cannot be zero. Remember that velocity has direction and magnitude. When the object travels in a circular path, the direction is always changing, so the velocity is always changing. Because the velocity is changing, this means there is an acceleration that is delivered on the object.

2. **D** Section 1 represents a constant positive speed. Section 2 shows an object slowing down, moving in the positive direction. Section 3 represents an object speeding up in the negative direction. Section 4 demonstrates a constant negative speed, and section 5 represents the correct answer: slowing down moving in the negative direction. Though the slope is positive, this corresponds to acceleration, indicating that the direction of acceleration is opposite to the direction of velocity and thus is slowing down. However, the section remains in the negative quadrant, and the velocity becomes slower but is still negative.

3. **B, D** In parabolic motion, a projectile experiences only the constant direction due to gravity, but the velocity does not point in the same direction. Eliminate (A). Zero acceleration means no change in speed (or direction), making (B) and (D) true. An object whose speed remains constant but whose velocity vector is changing direction is accelerating. Eliminate (C).

4. **B** The baseball is still under the influence of Earth's gravity. Its acceleration throughout the *entire* flight is constant, equal to g downward.

5. **A** Use Big Five #3 with $v_0 = 0$:

$$x = x_0 + v_0 t + \frac{1}{2}at^2 = \frac{1}{2}at^2 \quad \Rightarrow \quad t = \sqrt{\frac{2\Delta x}{a}} = \sqrt{\frac{2(200\text{ m})}{5\text{ m/s}^2}} = 9\text{ s}$$

6. **C** Use Big Five #5 with $v_0 = 0$ (calling *down* the positive direction):

$$v^2 = v_0^2 + 2a(x - x_0) = 2a(x - x_0) \quad \Rightarrow \quad (x - x_0) = \frac{v^2}{2a} = \frac{v^2}{2g} = \frac{(30\text{ m/s})^2}{2(10\text{ m/s}^2)} = 45\text{ m}$$

7. **C** Apply Big Five #3 to the vertical motion, calling *down* the positive direction:

$$\Delta y = v_{0y}t + \frac{1}{2}a_y t^2 = \frac{1}{2}a_y t^2 = \frac{1}{2}gt^2 \quad \Rightarrow \quad t = \sqrt{\frac{2\Delta y}{g}} = \sqrt{\frac{2(80\text{ m})}{10\text{ m/s}^2}} = 4\text{ s}$$

Note that the stone's initial horizontal speed (v_{0x} = 10 m/s) is irrelevant.

8. **B** First, determine the time required for the ball to reach the top of its parabolic trajectory (which is the time required for the vertical velocity to drop to zero).

$$v_y \overset{\text{set}}{=} 0 \quad \Rightarrow \quad v_{0y} - gt = 0 \quad \Rightarrow \quad t = \frac{v_{0y}}{g}$$

The total flight time is equal to twice this value:

$$t_t = 2t = 2\frac{v_{0y}}{g} = 2\frac{v_0 \sin\theta_0}{g} = \frac{2(10 \text{ m/s})\sin 30°}{10 \text{ m/s}^2} = 1 \text{ s}$$

9. **C** After 4 seconds, the stone's vertical speed has changed by $\Delta v_y = a_y t = (10 \text{ m/s}^2)(4 \text{ s}) = 40 \text{ m/s}$. Since $v_{0y} = 0$, the value of v_y at $t = 4$ is 40 m/s. The horizontal speed does not change. Therefore, when the rock hits the water, its velocity has a horizontal component of 30 m/s and a vertical component of 40 m/s.

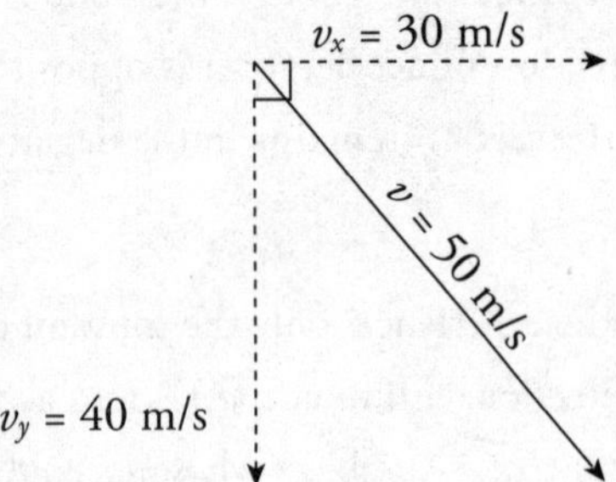

By the Pythagorean Theorem, the magnitude of the total velocity, v, is 50 m/s.

10. **D** Since the acceleration of the projectile is always downward (because of its gravitational acceleration), the vertical speed decreases as the projectile rises and increases as the projectile falls. Choices (A), (B), and (C) are all false.

11. **B** Use Big Five #2:

$$v = v_0 + at = 5 \text{ m/s} + (-10 \text{ m/s}^2)(3 \text{ s}) = -25 \text{ m/s}$$

Because you called up the positive direction, the negative sign for the velocity indicates that the stone is traveling downward.

12. **C** The variables involved in this question are the initial velocity (given in both cases), the acceleration (constant), final velocity (0 in both cases), and displacement (the skidding distance). As the missing variable is time, use Big Five #5 to solve for the displacement:

$$v^2 = v^2 + 2a(x - x_0)$$

$$x - x_0 = \frac{v^2 - v_0^2}{2a}$$

As the initial position, x_0, and the final velocity, v^2, are equal to 0, this equation simplifies to

$$x = \frac{-v_0^2}{2a}$$

When the initial velocity is doubled, the final position quadruples. (Note that the acceleration in this problem is negative, as the brakes cause the car to decelerate.)

Section II: Free Response

1. (A) At time $t = 1$ s, the car's velocity starts to decrease as the acceleration (which is the slope of the given velocity-versus-time graph) changes from positive to negative.

   (B) The average velocity between $t = 0$ and $t = 1$ s is $\frac{1}{2}(v_{t=0} + v_{t=1}) = \frac{1}{2}(0 + 20 \text{ m/s}) = 10 \text{ m/s}$, and the average velocity between $t = 1$ and $t = 5$ is $\frac{1}{2}(v_{t=1} + v_{t=5}) = \frac{1}{2}(20 \text{ m/s} + 0) = 10 \text{ m/s}$. The two average velocities are the same.

   (C) The displacement is equal to the area bounded by the graph and the t-axis, taking areas above the t-axis as positive and those below as negative. In this case, the displacement from $t = 0$ to $t = 5$ s is equal to the area of the triangular region whose base is the segment along the t-axis from $t = 0$ to $t = 5$ s:

   $$\Delta x\ (t = 0 \text{ to } t = 5 \text{ s}) = \frac{1}{2} \times \text{base} \times \text{height} = \frac{1}{2}(5 \text{ s})(20 \text{ m/s}) = 50 \text{ m}$$

   The displacement from $t = 5$ s to $t = 7$ s is equal to the negative of the area of the triangular region whose base is the segment along the t-axis from $t = 5$ s to $t = 7$ s:

   $$\Delta x\ (t = 5 \text{ s to } t = 7 \text{ s}) = -\frac{1}{2} \times \text{base} \times \text{height} = -\frac{1}{2}(2 \text{ s})(10 \text{ m/s}) = -10 \text{ m}$$

   Therefore, the displacement from $t = 0$ to $t = 7$ s is

   $$\Delta x\ (t = 0 \text{ to } t = 5 \text{ s}) + \Delta s\ (t = 5 \text{ s to } t = 7 \text{ s}) = 50 \text{ m} + (-10 \text{ m}) = 40 \text{ m}$$

(D) The acceleration is the slope of the velocity-versus-time graph. The segment of the graph from $t = 0$ to $t = 1$ s has a slope of $a = \Delta v/\Delta t$ = (20 m/s – 0)/(1 s – 0) = 20 m/s^2, and the segment of the graph from $t = 1$ s to $t = 7$ s has a slope of $a = \Delta v/\Delta t$ = (–10 m/s – 20 s)/(7 s – 1 s) = –5 m/s^2. Therefore, the acceleration-versus-time graph is

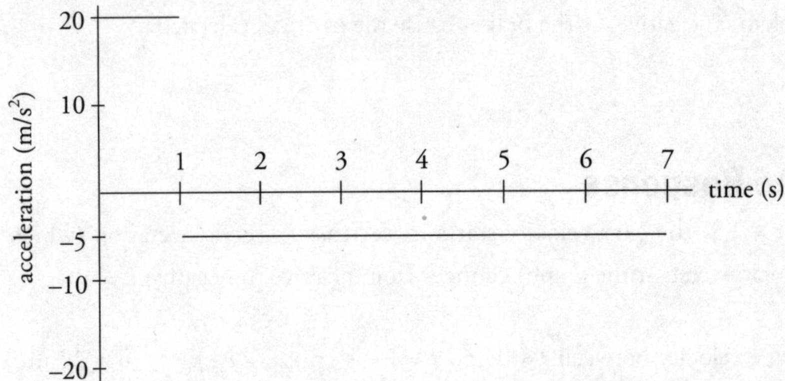

(E)

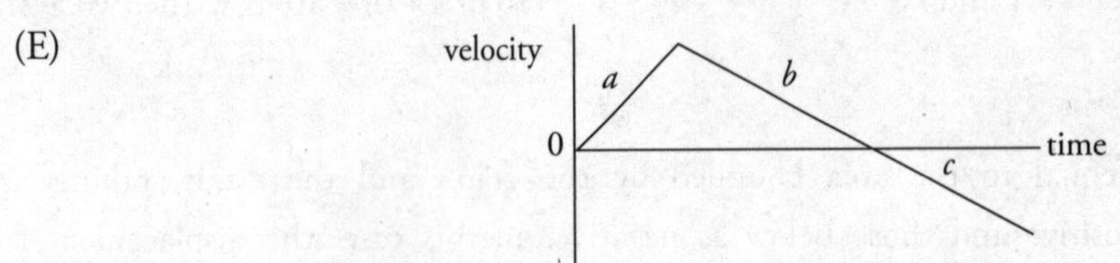

Section a shows that the object is speeding up in the positive direction. Section b shows that the object is slowing down, yet still moving in the positive direction. At five seconds, the object has stopped for an instant. Section c shows that the object is moving in the negative direction and speeding up. The corresponding position-versus-time graph for each section would look like this:

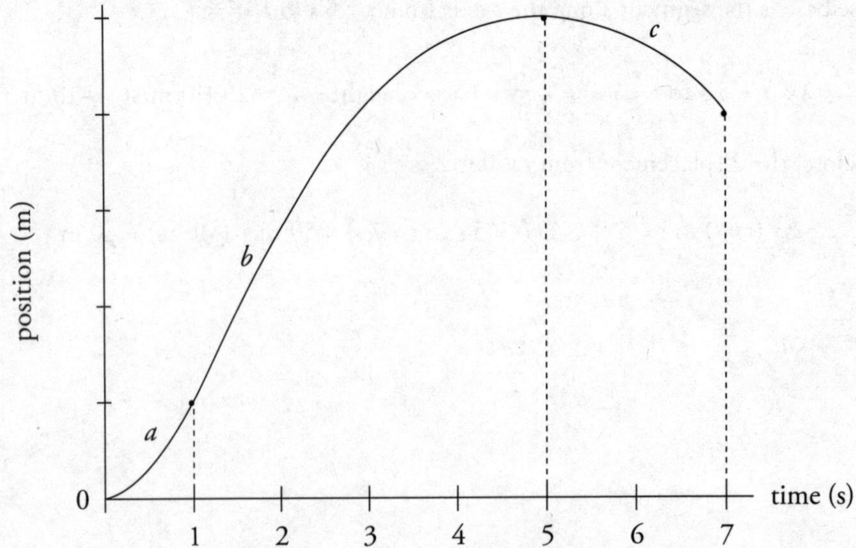

2. (A) The maximum height of the projectile occurs at the time at which its vertical velocity drops to zero:

$$v_y \overset{\text{set}}{=} 0 \quad \Rightarrow \quad v_{0y} - gt = 0 \quad \Rightarrow \quad t = \frac{v_{0y}}{g}$$

The vertical displacement of the projectile at this time is computed as follows:

$$\Delta y = v_{0y}t - \frac{1}{2}gt^2 \quad \Rightarrow \quad H = v_{0y}\frac{v_{0y}}{g} - \frac{1}{2}g\left(\frac{v_{0y}}{g}\right)^2 = \frac{v_{0y}^2}{2g} = \frac{v_0^2 \sin^2 \theta_0}{2g}$$

(B) The total flight time is equal to twice the time computed in part (A):

$$t_t = 2t = 2\frac{v_{0y}}{g}$$

The horizontal displacement at this time gives the projectile's range:

$$\Delta x = v_{0x}t \quad \Rightarrow \quad R = v_{0x}t_t = \frac{v_{0x} \cdot 2v_{0y}}{g} = \frac{2v_0^2 \sin\theta_0 \cos\theta_0}{g} \text{ or } \frac{v_0^2 \sin 2\theta_0}{g}$$

(C) For any given value of v_0, the range,

$$\Delta x = v_{0x}t \quad \Rightarrow \quad R = v_{0x}t_t = \frac{v_{0x} \cdot 2v_{0y}}{g} = \frac{2v_0^2 \sin\theta_0 \cos\theta_0}{g} \text{ or } \frac{v_0^2 \sin 2\theta_0}{g}$$

will be maximized when $\sin 2\theta_0$ is maximized. This occurs when $2\theta_0 = 90°$, that is, when $\theta_0 = 45°$.

(D) Set the general expression for the projectile's vertical displacement equal to h and solve for the two values of t (assuming that $g = +10$ m/s²):

$$v_{0y}t - \frac{1}{2}gt^2 \overset{\text{set}}{=} h \quad \Rightarrow \quad \frac{1}{2}gt^2 - v_{0y}t + h = 0$$

Applying the quadratic formula, find that

$$t = \frac{v_{0y} + \sqrt{(-v_{0y})^2 - 4(\frac{1}{2}g)(h)}}{2(\frac{1}{2}g)} = \frac{v_{0y} \pm \sqrt{v_{0y}^2 - 2gh}}{g}$$

Therefore, the two times at which the projectile crosses the horizontal line at height h are

$$t_1 = \frac{v_{0y} - \sqrt{v_{0y}^2 - 2gh}}{g} \quad \text{and} \quad t_2 = \frac{v_{0y} + \sqrt{v_{0y}^2 - 2gh}}{g}$$

so the amount of time that elapses between these events is

$$\Delta t = t_2 - t_1 = \frac{2\sqrt{v_{0y}^2 - 2gh}}{g}$$

3. (A) The cannonball will certainly reach the wall (which is only 220 m away) since the ball's range is

$$R = \frac{v_0^2 \sin 2\theta_0}{g} = \frac{(50\text{ m/s})^2 \sin 2(40°)}{9.8\text{ m/s}^2} = 251\text{ m}$$

You simply need to make sure that the cannonball's height is less than 30 m at the point where its horizontal displacement is 220 m (so that the ball actually hits the wall rather than flying over it). To do this, find the time at which $x = 220$ m by first writing

$$x = v_{0x}t \quad \Rightarrow \quad t = \frac{x}{v_{0x}} = \frac{x}{v_0 \cos\theta_0} \qquad (1)$$

Thus, the cannonball's vertical position can be written in terms of its horizontal position as follows:

$$\begin{aligned} y = v_{0y}t - \tfrac{1}{2}gt^2 &= v_0 \sin\theta_0 \frac{x}{v_0\cos\theta_0} - \tfrac{1}{2}g\left(\frac{x}{v_0\cos\theta_0}\right)^2 \\ &= x\tan\theta_0 - \frac{gx^2}{2v_0^2\cos^2\theta_0} \end{aligned} \qquad (2)$$

Substituting the known values for x, θ_0, g, and v_0, you get

$$\begin{aligned} y(\text{at } x = 220\text{ m}) &= (220\text{ m})\tan 40° - \frac{(9.8\text{ m/s}^2)(220\text{ m})^2}{2(50\text{ m/s})^2\cos^2 40°} \\ &= 23\text{ m} \end{aligned}$$

This is indeed less than 30 m, as desired.

(B) From Equation (1) derived in part (A),

$$t = \frac{x}{v_0\cos\theta_0} = \frac{220\text{ m}}{(50\text{ m/s})\cos 40°} = 5.7\text{ s}$$

(C) The height at which the cannonball strikes the wall was determined in part (A) to be 23 m.

4. (A) For parabolic trajectories, the total flight time can be determined by doubling the amount of time it takes for the projectile to reach its apex. However, this trajectory is NOT parabolic. As the initial position is 25 m, the final position is 0 m, the initial velocity is $v_0 = v_0 \sin\theta = 40 \sin 30° = 40\left(\frac{1}{2}\right) = 20$ m/s, and the acceleration is –10 m/s²; the missing variable is the final velocity, so the flight time of the cannonball can be computed using Big Five #3:

$$x_y = x_0 + v_0 t + \frac{1}{2}at^2$$

$$0 = 25\text{ m} + 20\text{ m/s} \cdot t + \frac{1}{2}(-10\text{ m/s}^2) \cdot t^2$$

$$0 = 25\text{ m} + 20\text{ m/s} \cdot t - 5\text{ m/s}^2 \cdot t^2$$

$$0 = -5\text{ m/s}^2 \cdot t^2 + 20\text{ m/s} \cdot t + 25\text{ m}$$

Applying the quadratic formula, find that,

$$t = \frac{-20 \pm \sqrt{20^2 - 4(25)(-5)}}{2(-5)} = \frac{-20 \pm \sqrt{400 + 500}}{-10} = \frac{-20 \pm \sqrt{900}}{-10} = \frac{-20 \pm 30}{-10}$$

$$t = -1\text{ s or } t = 5\text{ s}$$

As time cannot be a negative value, the flight time of the cannonball is 5 seconds.

(B) As the horizontal speed of a projectile is constant, the range is given by

$$\Delta x = v_{0x} t = 40\text{ m/s} \cdot \cos 30° \cdot 5\text{ s} = 173\text{ m}$$

(C) The cannonball is fired with an initial horizontal velocity of

$$v_{0y} = 40\text{ m/s} \cdot \cos 30° = 34.6\text{ m/s}$$

As there is no horizontal acceleration, the horizontal speed of the projectile remains constant throughout the entire flight, leading to a flat line with a *y*-value of 34.6 m/s over the flight time of 5 s. The horizontal speed-versus-time graph can then be plotted.

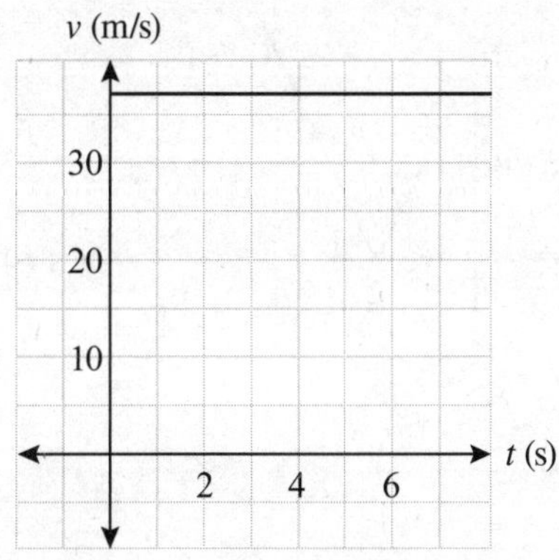

The initial vertical velocity of the cannonball is

$$v_{0y} = 40 \text{ m/s} \cdot \sin 30° = 20 \text{ m/s}$$

In the vertical direction, there is a constant acceleration due to gravity (a = 10 m/s^2). As the slope of the velocity-versus-time graph is equal to the acceleration, the resulting velocity-versus-time graph is a linear line with a negative slope since acceleration due to gravity points downward. As the magnitude of the acceleration of gravity is 10 m/s^2, the vertical velocity of the cannonball thus decreases by 10 m/s every second. With an initial velocity of 20 m/s, this means that after 2 s, the vertical velocity of the projectile is 0 m/s. After a total of 5 s, the vertical velocity of the projectile is –30 m/s. This can be visualized in the vertical velocity-versus-time graph below:

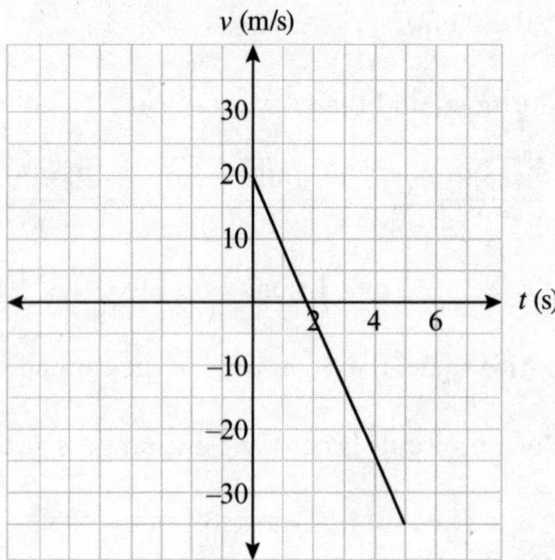

As the problem asks for the graph of the vertical speed versus time, the absolute value of the graph must be drawn to get the correct plot.

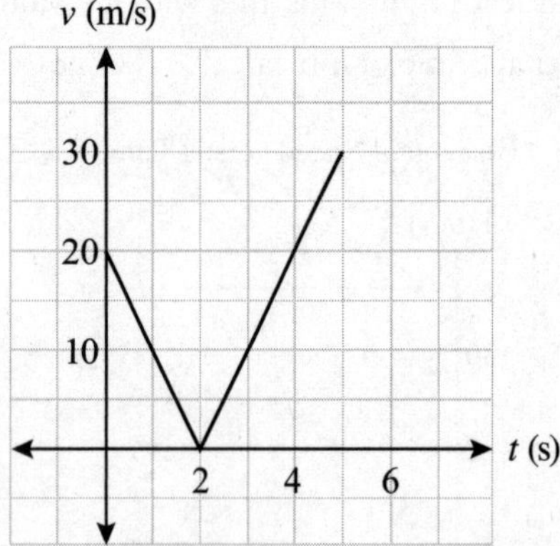

CHAPTER 5 REVIEW QUESTIONS

Section I: Multiple Choice

1. **B** Because the person is not accelerating, the net force he feels must be zero. Therefore, the magnitude of the upward normal force from the floor must balance that of the downward gravitational force. Although these two forces have equal magnitudes, they do not form an action/reaction pair because they both act on the same object (namely, the person). The forces in an action/reaction pair always act on different objects. The correct action/reaction pair in this situation is the Earth's pull on the person (the weight) and the person's pull on the Earth.

2. **D** First, draw a free-body diagram:

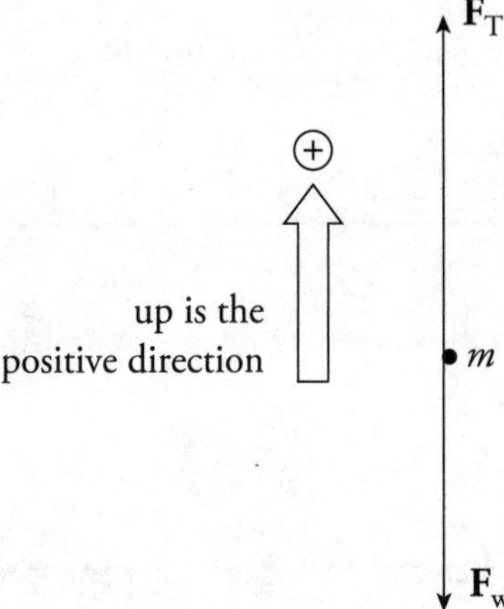

The person exerts a downward force on the scale, and the scale pushes up on the person with an equal (but opposite) force, $\mathbf{F}_N$. Thus, the scale reading is F_N, the magnitude of the normal force. Since $F_N - F_w = ma$, you get $F_N = F_w + ma = (800\text{ N}) + [800\text{ N}/(10\text{ m/s}^2)](5\text{ m/s}^2) = 1{,}200\text{ N}$.

3. **A** The net force that the object feels on the inclined plane is $mg\sin\theta$, the component of the gravitational force that is parallel to the ramp. Since $\sin\theta = (5\text{ m})/(20\text{ m}) = \frac{1}{4}$, you get $F_{net} = (2\text{ kg})(10\text{ N/kg})(\frac{1}{4}) = 5\text{ N}$.

4. **C** The net force on the block is $F - F_f = F - \mu_k F_N = F - \mu_k F_w = (18\text{ N}) - (0.4)(20\text{ N}) = 10\text{ N}$. Since $F_{net} = ma = (F_w/g)a$, you get $10\text{ N} = [(20\text{ N})/(10\text{ m/s}^2)]a$, which gives $a = 5\text{ m/s}^2$.

5. **A** The force pulling the block down the ramp is $mg\sin\theta$, and the maximum force of static friction is $\mu_s F_N = \mu_s mg\cos\theta$. If $mg\sin\theta$ is greater than $\mu_s mg\cos\theta$, then there is a net force down the ramp, and the block will accelerate down. So, the question becomes, "Is $\sin\theta$ greater than $\mu_s\cos\theta$?" Since $\theta = 30°$ and $\mu_s = 0.5$, the answer is "yes."

6. **D** One way to attack this question is to notice that if the two masses happen to be equal, that is, if $M = m$, then the blocks won't accelerate (because their weights balance). The only expression given that becomes zero when $M = m$ is the one given in (D). Draw a free-body diagram:

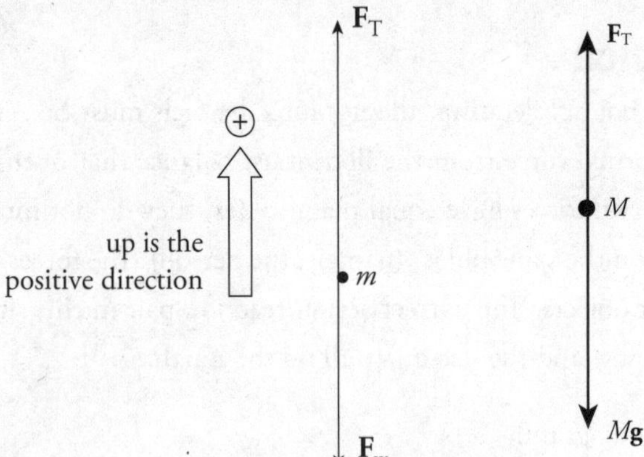

Newton's Second Law gives the following two equations:

$F_T - mg = ma$ (1)

$F_T - Mg = M(-a)$ (2)

Subtracting these equations yields $Mg - mg = ma + Ma = (M + m)a$, so

$$a = \frac{Mg - mg}{M + m} = \frac{M - m}{M + m}g$$

7. **D** If $F_{net} = 0$, then $a = 0$. No acceleration means constant speed (possibly, but not necessarily zero) with no change in direction. Therefore, (B) and (C) are false, and (A) is not necessarily true.

8. **C** The horizontal motion across the frictionless tables is unaffected by (vertical) gravitational acceleration. It would take as much force to accelerate the block across the table on Earth as it would on the Moon. (If friction *were* taken into account, then the smaller weight of the block on the Moon would imply a smaller normal force by the table and hence a smaller frictional force. Less force would be needed on the Moon in this case.)

9. **D** The maximum force which static friction can exert on the crate is $\mu_s F_N = \mu_s F_w = \mu_s mg$ = (0.4)(100 kg)(10 N/kg) = 400 N. Since the force applied to the crate is only 344 N, static friction is able to apply that same magnitude of force on the crate, keeping it stationary. Choice (B) is incorrect because the static friction force is *not* the reaction force to **F**; both **F** and $\mathbf{F}_{f\,(static)}$ act on the same object (the crate) and therefore cannot form an action/reaction pair.

10. **A** With Crate #2 on top of Crate #1, the force pushing downward on the floor is greater, so the normal force exerted by the floor on Crate #1 is greater, which increases the friction force. Choices (B), (C), and (D) are all false.

Section II: Free Response

1. (A) The forces acting on the crate are $\mathbf{F}_T$ (the tension in the rope), $\mathbf{F}_w$ (the weight of the block), $\mathbf{F}_N$ (the normal force exerted by the floor), and $\mathbf{F}_f$ (the force of kinetic friction):

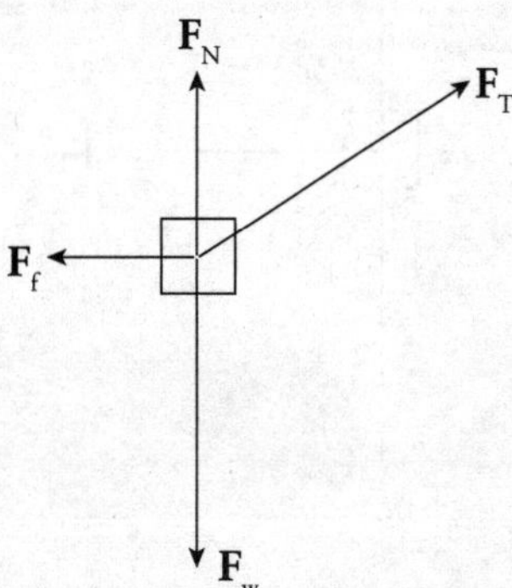

(B) First, break $\mathbf{F}_T$ into its horizontal and vertical components:

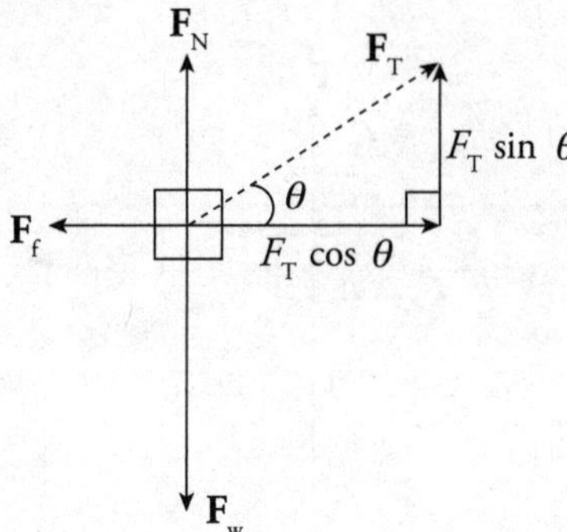

Since the net vertical force on the crate is zero, you get $F_N + F_T \sin\theta = F_w$, so $F_N = F_w - F_T \sin\theta = mg - F_T \sin\theta$.

(C) From part (B), notice that the net horizontal force acting on the crate is

$$F_T \cos\theta - F_f = F_T \cos\theta - \mu F_N = F_T \cos\theta - \mu(mg - F_T \sin\theta)$$

so the crate's horizontal acceleration across the floor is

$$a = \frac{F_{net}}{m} = \frac{F_T \cos\theta - \mu(mg - F_T \sin\theta)}{m}$$

2. (A) The forces acting on Block #1 are $\mathbf{F}_T$ (the tension in the string connecting it to Block #2), $\mathbf{F}_{w1}$ (the weight of the block), and $\mathbf{F}_{N1}$ (the normal force exerted by the tabletop):

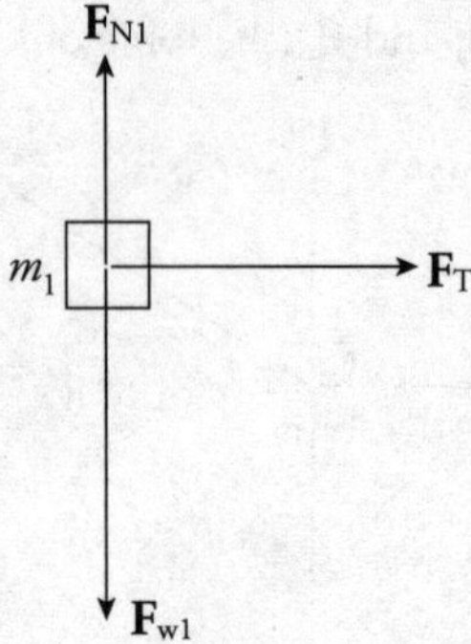

(B) The forces acting on Block #2 are $\mathbf{F}$ (the pulling force), $\mathbf{F}_T$ (the tension in the string connecting it to Block #1), $\mathbf{F}_{w2}$ (the weight of the block), and $\mathbf{F}_{N2}$ (the normal force exerted by the tabletop):

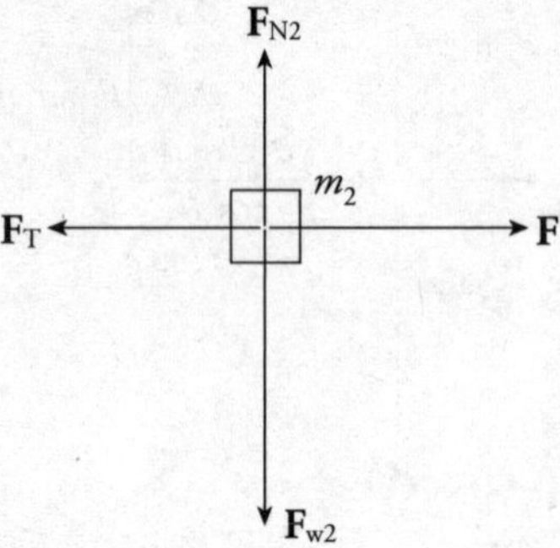

(C) Newton's Second Law applied to Block #2 yields $F - F_T = m_2a$ and applied to Block #1 yields $F_T = m_1a$. Adding these equations, you get $F = (m_1 + m_2)a$, so

$$a = \frac{F}{m_1 + m_2}$$

(D) Substituting the result of part (C) into the equation $F_T = m_1a$, you get

$$F_T = m_1a = \frac{m_1}{m_1 + m_2}F$$

(E) (i) Since the force $\mathbf{F}$ must accelerate all three masses—m_1, m, and m_2—the common acceleration of all parts of the system is

$$a = \frac{F}{m_1 + m + m_2}$$

(ii) Let $\mathbf{F}_{T1}$ denote the tension force in the connecting string acting on Block #1, and let $\mathbf{F}_{T2}$ denote the tension force in the connecting string acting on Block #2. Then, Newton's Second Law applied to Block #1 yields $F_{T1} = m_1 a$ and applied to Block #2 yields $F - F_{T2} = m_2 a$. Therefore, using the value for a computed above, you get

$$\begin{aligned} F_{T2} - F_{T1} &= (F - m_2 a) - m_1 a \\ &= F - (m_1 + m_2)a \\ &= F - (m_1 + m_2)\frac{F}{m_1 + m + m_2} \\ &= F\left(1 - \frac{m_1 + m_2}{m_1 + m + m_2}\right) \\ &= F\frac{m}{m_1 + m + m_2} \end{aligned}$$

3. (A) First, draw free-body diagrams for the two boxes:

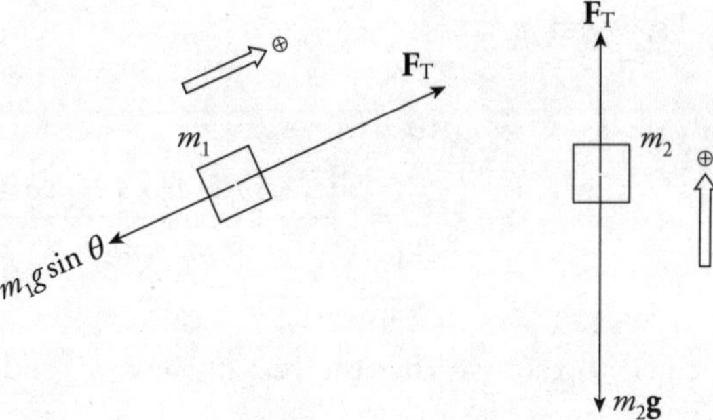

Applying Newton's Second Law to the boxes yields the following two equations:

$F_T - m_1 g \sin\theta = m_1 a$ (1)

$F_T - m_2 g = m_2(-a)$ (2)

Subtract the equations and solve for a:

$$m_2 g - m_1 g \sin\theta = (m_1 + m_2)a$$

$$a = \frac{m_2 - m_1 \sin\theta}{m_1 + m_2} g$$

(i) For a to be positive, you must have $m_2 - m_1 \sin\theta > 0$, which implies that $\sin\theta < m_2/m_1$, or, equivalently, $\theta < \sin^{-1}(m_2/m_1)$.

(ii) For a to be zero, you must have $m_2 - m_1 \sin\theta = 0$, which implies that $\sin\theta = m_2/m_1$, or, equivalently, $\theta = \sin^{-1}(m_2/m_1)$.

(B) Including the force of kinetic friction, the force diagram for m_1 is

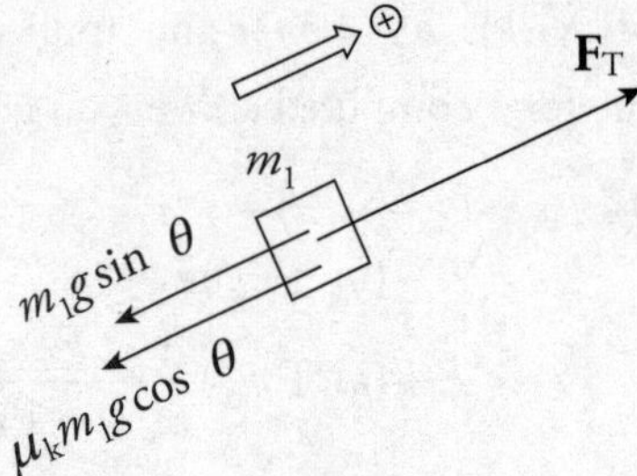

Since $F_f = \mu_k F_N = \mu_k m_1 g \cos\theta$, applying Newton's Second Law to the boxes yields these two equations:

$F_T - m_1 g \sin\theta - \mu_k mg \cos\theta = m_1 a$ (1)

$m_2 g - F_T = m_2 a$ (2)

Add the equations and solve for a:

$$m_2 g - m_1 g \sin\theta - \mu_k mg \cos\theta = (m_1 + m_2)a$$
$$a = \left(\frac{m_2 - m_1(\sin\theta + \mu_k \cos\theta)}{m_1 + m_2}\right)g$$

In order for a to be equal to zero (so that the box of mass m_1 slides up the ramp with constant velocity),

$$m_2 - m_1(\sin\theta + \mu_k \cos\theta) = 0$$
$$\sin\theta + \mu_k \cos\theta = \frac{m_2}{m_1}$$

4. (A) The forces acting on the skydiver are $\mathbf{F}_r$, the force of air resistance (upward), and $\mathbf{F}_w$, the weight of the skydiver (downward):

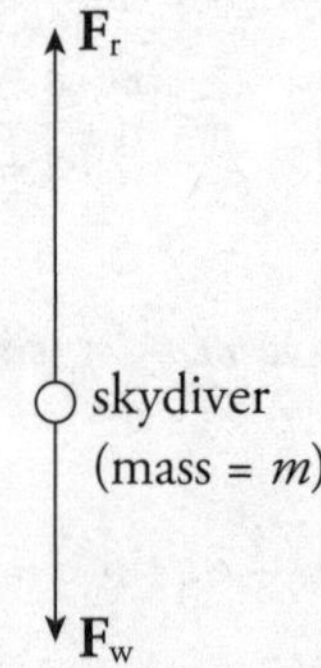

(B) Since $F_{net} = F_w - F_r = mg - kv$, the skydiver's acceleration is

$$a = \frac{F_{net}}{m} = \frac{mg - kv}{m}$$

(C) Terminal speed occurs when the skydiver's acceleration becomes zero, since then the descent velocity becomes constant. Setting the expression derived in part (B) equal to 0, find the speed $v = v_t$ at which this occurs:

$$v = v_t \text{ when } a = 0 \quad \Rightarrow \quad \frac{mg - kv_t}{m} = 0 \quad \Rightarrow \quad v_t = \frac{mg}{k}$$

(D) The skydiver's descent speed is initially v_0 and the acceleration is (close to) g. However, once the parachute opens, the force of air resistance provides a large (speed-dependent) upward acceleration, causing her descent velocity to decrease. The slope of the velocity-versus-time graph (the acceleration) is not constant but instead decreases to zero as her descent speed decreases from v_0 to v_t. Therefore, the graph is not linear.

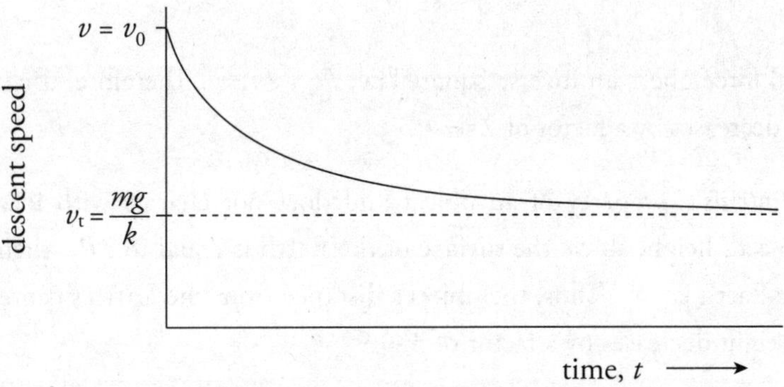

CHAPTER 6 REVIEW QUESTIONS

Section I: Multiple Choice

1. **A, C** In uniform circular motion, the speed is constant, but the velocity is constantly changing because the direction is changing. Because the direction is changing, there is an acceleration. The magnitude of acceleration is constant; however, the acceleration is directed toward the center and perpendicular to the velocity. As a result, since the velocity is always changing, so is the direction of acceleration.

2. **B** When the bucket is at the lowest point in its vertical circle, it feels a tension force $\mathbf{F}_T$ upward and the gravitational force $\mathbf{F}_w$ downward. The net force toward the center of the circle, which is the centripetal force, is $F_T - F_w$. Thus,

$$F_T - F_w = m\frac{v^2}{r} \quad \Rightarrow \quad v = \sqrt{\frac{r(F_T - mg)}{m}} = \sqrt{\frac{(0.60\text{ m})[50\text{ N} - (3\text{ kg})(10\text{ N/kg})]}{3\text{ kg}}} = 2\text{ m/s}$$

3. **C** When the bucket reaches the topmost point in its vertical circle, the forces acting on the bucket are its weight, $\mathbf{F}_w$, and the downward tension force, $\mathbf{F}_T$. The net force, $\mathbf{F}_w + \mathbf{F}_T$, provides the centripetal force. In order for the rope to avoid becoming slack, $\mathbf{F}_T$ must not vanish. Therefore, the cut-off speed for ensuring that the bucket makes it around the circle is the speed at which $\mathbf{F}_T$ just becomes zero; any greater speed would imply that the bucket would make it around. Thus,

$$F_w + F_T = m\frac{v^2}{r} \Rightarrow F_w + 0 = m\frac{v_{\text{cut-off}}^2}{r} \Rightarrow v_{\text{cut-off}} = \sqrt{\frac{rF_w}{m}} = \sqrt{gr}$$
$$= \sqrt{(10 \text{ m/s}^2)(0.60 \text{ m})}$$
$$= 2.4 \text{ m/s}$$

4. **D** Centripetal acceleration is given by the equation $a_c = v^2/r$. Since the object covers a distance of $2\pi r$ in 1 revolution, its speed is $2\pi r$. Therefore,

$$a_c = \frac{v^2}{r} = \frac{(2\pi r)^2}{r} = 4\pi^2 r$$

5. **A** Gravitational force obeys an inverse-square law: $F_{grav} \propto 1/r^2$. Therefore, if r increases by a factor of 2, then F_{grav} decreases by a factor of $2^2 = 4$.

6. **D** Mass is an intrinsic property of an object and does not change with location. This eliminates (B). If an object's height above the surface of the Earth is equal to $2R_E$, then its distance from the center of the Earth is $3R_E$. Thus, the object's distance from the Earth's center increases by a factor of 3, so its weight decreases by a factor of $3^2 = 9$.

7. **C** The gravitational force that the Moon exerts on the planet is equal in magnitude to the gravitational force that the planet exerts on the Moon (Newton's Third Law).

8. **D** The gravitational acceleration at the surface of a planet of mass M and radius R is given by the equation $g = GM/R^2$. Therefore, for the dwarf planet Pluto,

$$g_{\text{Pluto}} = G\frac{M_{\text{Pluto}}}{R_{\text{Pluto}}^2} = G\frac{\frac{1}{500}M_{\text{Earth}}}{\left(\frac{1}{15}R_{\text{Earth}}\right)^2} = \frac{15^2}{500}\cdot G\frac{M_{\text{Earth}}}{R_{\text{Earth}}^2} = \frac{225}{500}(10g \text{ m/s}^2) = \frac{225}{50}\text{m/s}^2$$

9. **D** The gravitational pull by Jupiter provides the centripetal force on its moon:

$$G\frac{Mm}{R^2} = \frac{mv^2}{R}$$

$$G\frac{M}{R} = v^2$$

$$G\frac{M}{R} = \left(\frac{2\pi R}{T}\right)^2$$

$$G\frac{M}{R} = \frac{4\pi^2 R^2}{T^2}$$

$$M = \frac{4\pi^2 R^3}{(GT^2)}$$

10. **D** Let the object's distance from Body A be x; then its distance from Body B is $R - x$. In order for the object to feel no net gravitational force, the gravitational pull by A must balance the gravitational pull by B. Therefore, if you let M denote the mass of the object, then

$$G\frac{m_A M}{x^2} = G\frac{m_B M}{(R-x)^2}$$

$$\frac{m}{x^2} = \frac{4m}{(R-x)^2}$$

$$\frac{(R-x)^2}{x^2} = \frac{4m}{m}$$

$$\left(\frac{R-x}{x}\right)^2 = 4$$

$$\left(\frac{R}{x} - 1\right)^2 = 4$$

$$\frac{R}{x} - 1 = 2$$

$$x = R/3$$

11. **B** Because the planet is spinning clockwise and the velocity is tangent to the circle, the velocity must point down. The acceleration and force point toward the center of the circle.

12. **B, D** If there were no forces or balanced forces, in and out, the satellite would have a net force of zero. If the net force were zero, the satellite would continue in a straight line and not orbit the planet.

The force of gravity produces the centripetal force that keeps the object orbiting (where r is the distance between the center of the planet and the orbiting satellite).

$$F_g \rightarrow F_C$$

$$G\frac{m_{\text{planet}} m_{\text{satellite}}}{r^2} = m_{\text{satellite}} \frac{v^2_{\text{orbit}}}{r}$$

$$v_{\text{orbit}} = \sqrt{G\frac{m_{\text{planet}}}{r}}$$

The mass of the satellite does not determine the orbital speed. Only the mass of the planet and its distance from the center of the planet determine orbital speed.

Section II: Free Response

1. (A) Given a mass of 1 kg, weight $F_g = 5$ N, and diameter $= 8 \times 10^6$ m (this gives you $r = 4 \times 10^6$ m), you can fill in this equation:

$$F_G = \frac{Gm_1m_2}{r^2} \Rightarrow m_1 = \frac{F_G r^2}{Gm_2}$$

This becomes

$$m_1 = \frac{(5\text{ N})(4\times10^6\text{ m})^2}{\left(6.67\times10^{-11}\frac{\text{N}\cdot\text{m}^2}{\text{kg}^2}\right)(1\text{ kg})} = 1.2 \times 10^{24}\text{ kg}$$

(B) $$g = \frac{Gm_1}{r^2} \Rightarrow g = \frac{\left(6.67\times10^{-11}\frac{\text{N}\cdot\text{m}^2}{\text{kg}^2}\right)1.2\times10^{24}\text{ kg}}{(4\times10^6\text{ m})^2} = 5\text{ m/s}^2$$

Note that you could have also observed that, because a 1 kg mass (which normally weighs 10 N on the surface of the Earth) only weighed 5 N, gravity on this planet must be half the Earth's gravity.

If you want to look at it in terms of g, g is 10 m/s² on Earth, so you can simply convert.

$$5\text{ m/s}^2\left(\frac{1g}{10\text{ m/s}^2}\right) = 0.5g$$

(C) Density is given by mass per unit volume ($\rho = \frac{m}{V}$). In addition, use the equation for the volume of a sphere as $V = \frac{4}{3}\pi r^3$ to get

$$\rho = \frac{m}{\frac{4}{3}\pi r^3} \Rightarrow \rho = \frac{3m}{4\pi r^3} \Rightarrow \rho = \frac{3(1.2\times10^{24}\text{ kg})}{4(3.14)(4\times10^6\text{ m})^3} \text{ or}$$

$$\rho = 4{,}480\frac{\text{kg}}{\text{m}^3}$$

2. (A) The centripetal acceleration is given by the equation $a_c = \frac{v^2}{R}$. You also know that for objects traveling in circles (Earth's orbit can be considered a circle), $v = \frac{2\pi R}{T}$. Substituting this v into the previous equation for centripetal acceleration, you get $a_c = \frac{4\pi^2 R}{T^2}$. This becomes

$$a_c = \frac{4\pi^2 1.5\times10^{11}\text{ m}}{(3.15\times10^7\text{ s})^2} = 6.0 \times 10^{-3}\text{ m/s}^2$$

(B) The gravitational force is the force that keeps the Earth traveling in a circle around the Sun. More specifically, $ma_c = (6 \times 10^{24}\text{ kg})(6.0 \times 10^{-3}\text{ m/s}^2) = 3.6 \times 10^{22}$ N.

(C) Use the Universal Law of Gravitation: $F = GmM/r^2$. You know F from the previous question, G is constant, m is the mass of the Earth (given in question), and r is the radius from Earth to Sun (also given in the question). So you can rearrange this equation to solve for M (mass of the Sun) as $M = F \cdot r^2/(G \cdot m)$ and then just plug in the appropriate values.

$F = GmM/r^2 \rightarrow M = Fr^2/(Gm) = (3.6 \times 10^{22}\ \text{N})(1.5 \times 10^{11}\ \text{m})^2/\{[6.67 \times 10^{-11}\ \text{Nm}^2/(\text{kg}^2)][6.0 \times 10^{24}\ \text{kg}]\} = 2.0 \times 10^{30}$

3. (A) The forces acting on a person standing against the cylinder wall are gravity ($\mathbf{F}_w$, downward), the normal force from the wall ($\mathbf{F}_N$, directed toward the center of the cylinder), and the force of static friction ($\mathbf{F}_f$, directed upward):

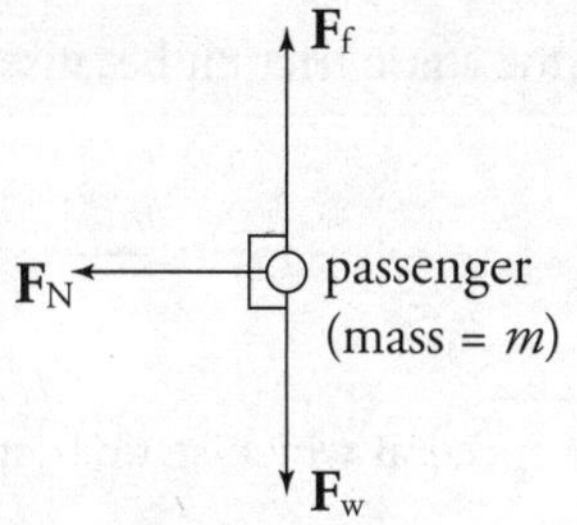

$\mathbf{F}_f$

$\mathbf{F}_N$ passenger (mass = m)

$\mathbf{F}_w$

(B) In order to keep a passenger from sliding down the wall, the maximum force of static friction must be at least as great as the passenger's weight: $F_{f\,(max)} \geq mg$. Since $F_{f\,(max)} = \mu_s F_N$, this condition becomes $\mu_s F_N \geq mg$.

Now, consider the circular motion of the passenger. Neither $\mathbf{F}_f$ nor $\mathbf{F}_w$ has a component toward the center of the path, so the centripetal force is provided entirely by the normal force:

$$F_N = \frac{mv^2}{r}$$

Substituting this expression for F_N into the previous equation, you get

$$\mu_s \frac{mv^2}{r} \geq mg$$

$$\mu_s \geq \frac{gr}{v^2}$$

Therefore, the coefficient of static friction between the passenger and the wall of the cylinder must satisfy this condition in order to keep the passenger from sliding down.

(C) Since the mass m canceled out in deriving the expression for μ_s, the conditions are independent of mass. Thus, the inequality $\mu_s \geq gr/v^2$ holds for both the adult passenger of mass m and the child of mass $m/2$.

4. (A) The forces acting on the car are gravity ($\mathbf{F}_w$, downward), the normal force from the road ($\mathbf{F}_N$, upward), and the force of static friction ($\mathbf{F}_f$, directed toward the center of curvature of the road):

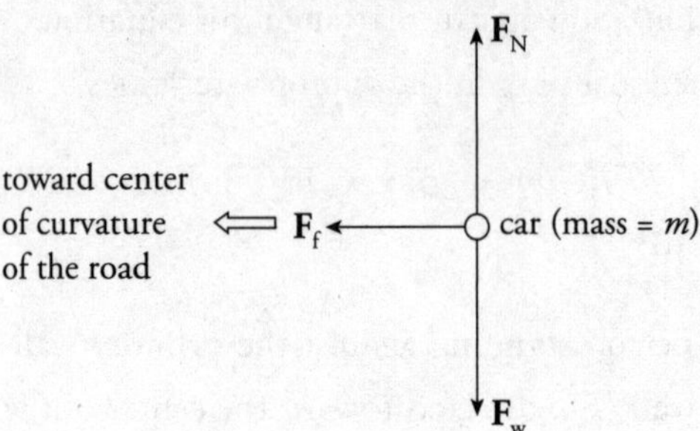

(B) The force of static friction (assume static friction because you *don't* want the car to slide) provides the necessary centripetal force:

$$F_f = \frac{mv^2}{r}$$

Therefore, to find the maximum speed at which static friction can continue to provide the necessary force, write

$$F_{f\,(max)} = \frac{mv_{max}^2}{r}$$

$$\mu_s F_N = \frac{mv_{max}^2}{r}$$

$$\mu_s mg = \frac{mv_{max}^2}{r}$$

$$v_{max} = \sqrt{\mu_s gr}$$

(C) Ignoring friction, the forces acting on the car are gravity ($\mathbf{F}_w$, downward) and the normal force from the road ($\mathbf{F}_N$, which is now tilted toward the center of curvature of the road):

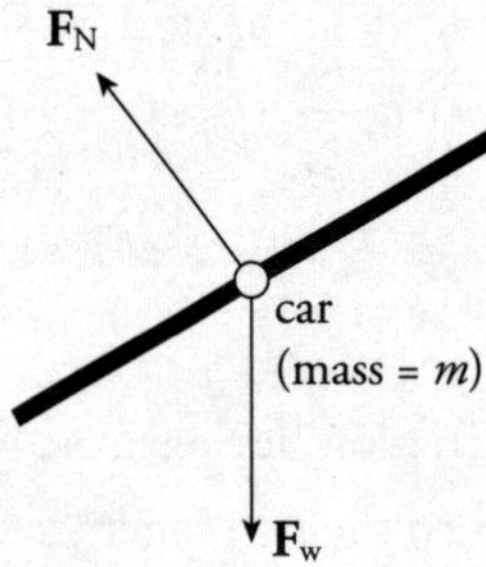

(D) Because of the banking of the turn, the normal force is tilted toward the center of curvature of the road. The component of $\mathbf{F}_N$ toward the center can provide the centripetal force, making reliance on friction unnecessary.

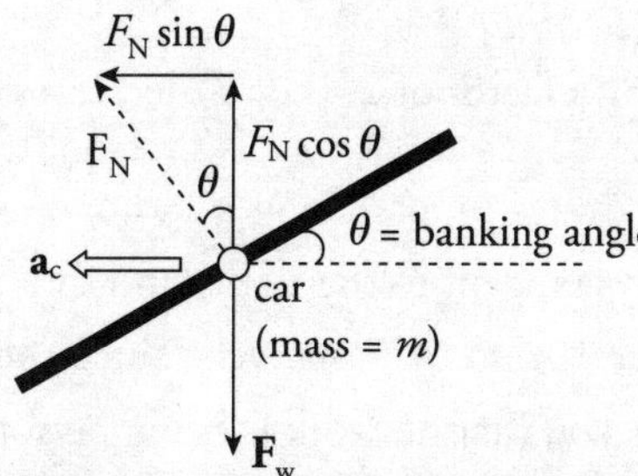

However, there's no vertical acceleration, so there is no net vertical force. Therefore, $F_N \cos\theta = F_w = mg$, so $F_N = mg/\cos\theta$. The component of F_N toward the center of curvature of the turn, $F_N \sin\theta$, provides the centripetal force:

$$F_N \sin\theta = \frac{mv^2}{r}$$

$$\frac{mg}{\cos\theta}\sin\theta = \frac{mv^2}{r}$$

$$g\tan\theta = \frac{v^2}{r}$$

$$\theta = \tan^{-1}\frac{v^2}{gr}$$

CHAPTER 7 REVIEW QUESTIONS

Section I: Multiple Choice

1. **A** Since the force **F** is perpendicular to the displacement, the work it does is zero.

2. **B** By the Work–Energy Theorem,

$$W = \Delta K = \frac{1}{2}m(v^2 - v_0^2) = \frac{1}{2}(4\text{ kg})[(6\text{ m/s})^2 - (3\text{ m/s})^2] = 54\text{ J}$$

3. **B** Since the box (mass m) falls through a vertical distance of h, its gravitational potential energy decreases by mgh. The length of the ramp is irrelevant here. If friction were involved, then the length of the plane would matter.

4. **A** The gravitational force points downward while the book's displacement is upward. Therefore, the work done by gravity is $-mgh = -(2\text{ kg})(10\text{ N/kg})(1.5\text{ m}) = -30\text{ J}$.

5. **B** The gravitational pull by the Earth provides the centripetal force on the satellite, so $\frac{GMm}{R^2} = R^2 = \frac{mc^2}{R}$. This gives $\frac{1}{2}mv^2 = \frac{GMm}{2R}$, so the kinetic energy K of the satellite is inversely proportional to R. Therefore, if R increases by a factor of 2, then K decreases by a factor of 2.

6. **B** Since the centripetal force always points along a radius toward the center of the circle, and the velocity of the object is always tangent to the circle (and thus perpendicular to the radius), the work done by the centripetal force is zero. Alternatively, since the object's speed remains constant, the Work–Energy Theorem tells you that no work is being performed.

7. **C** Since a nonconservative force (namely, friction) is acting during the motion, use the modified Conservation of Mechanical Energy equation.

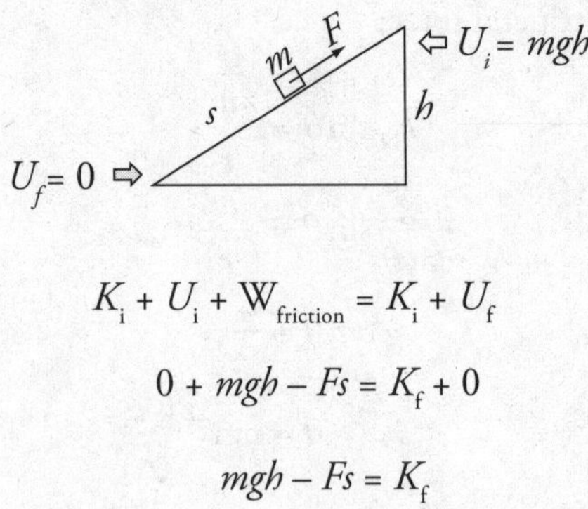

$$K_i + U_i + W_{friction} = K_i + U_f$$

$$0 + mgh - Fs = K_f + 0$$

$$mgh - Fs = K_f$$

8. **D** Apply Conservation of Mechanical Energy (including the negative work done by $\mathbf{F}_r$, the force of air resistance):

$$\begin{aligned}
K_i + U_i + W_r &= K_f + U_f \\
0 + mgh - F_r h &= \frac{1}{2}mv^2 + 0 \\
v &= \sqrt{\frac{2h(mg - F_r)}{m}} \\
&= \sqrt{\frac{2(40\text{ m})[(4\text{ kg})(10\text{ N/kg}) - 20\text{ N}]}{4\text{ kg}}} \\
&= 20\text{ m/s}
\end{aligned}$$

9. **D** Because the rock has lost half of its gravitational potential energy, its kinetic energy at the halfway point is half of its kinetic energy at impact. Since K is proportional to v^2, if $K_{\text{at halfway point}}$ is equal to $\frac{1}{2}K_{\text{at impact}}$, then the rock's speed at the halfway point is $\sqrt{1/2} = 1/\sqrt{2}$ its speed at impact.

10. **D** Using the equation $P = Fv$, find that $P = (200\text{ N})(2\text{ m/s}) = 400$ W.

Section II: Free Response

1. (A) Applying Conservation of Energy,

$$K_A + U_A = K_{\text{at } H/2} + U_{\text{at } H/2}$$
$$0 + mgH = \frac{1}{2}mv^2 + mg(\frac{1}{2}H)$$
$$\frac{1}{2}mgH = \frac{1}{2}mv^2$$
$$v = \sqrt{gH}$$

(B) Applying Conservation of Energy again,

$$K_A + U_A = K_B + U_B$$
$$0 + mgH = \frac{1}{2}mv_B^2 + 0$$
$$v_B = \sqrt{2gH}$$

(C) By the Work–Energy Theorem, you want the work done by friction to be equal (but opposite) to the kinetic energy of the box at Point B:

$$W = \Delta K = \frac{1}{2}m(v_C^2 - v_B^2) = -\frac{1}{2}mv_B^2 = -\frac{1}{2}m(\sqrt{2gH})^2 = -mgH$$

Therefore,

$$W = -mgH \quad \Rightarrow \quad -F_f x = -mgH \quad \Rightarrow \quad -\mu_k mgx = -mgH \quad \Rightarrow \quad \mu_k = H/x$$

(D) Apply Conservation of Energy (including the negative work done by friction as the box slides up the ramp from B to C):

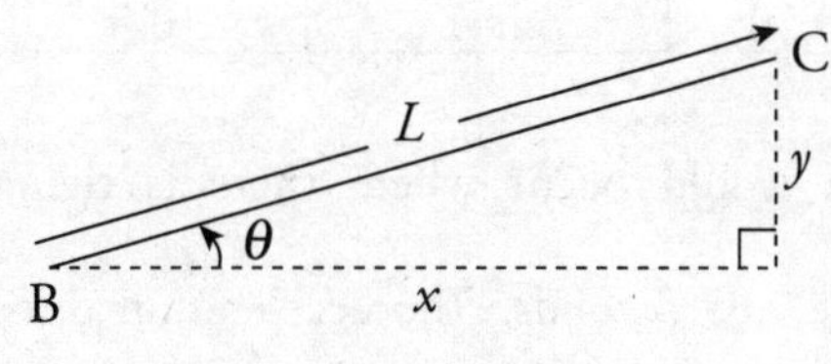

$$K_B + U_B + W_f = K_C + U_C$$
$$\frac{1}{2}m(\sqrt{2gH})^2 + 0 - F_f L = 0 + mgy$$
$$mgH + 0 - F_f L = 0 + mgy$$
$$mg(H - y) - (\mu_k mg\cos\theta)(L) = 0$$
$$\mu_k = \frac{H - y}{L\cos\theta} = \frac{H - y}{x}$$

(E) The result of part (B) reads $v_B = \sqrt{2gH}$. Therefore, by Conservation of Mechanical Energy (with the work done by the frictional force on the slide included), you get

$$K_A + U_A + W_f = K'_B + U_B$$
$$0 + mgH + W_f = \frac{1}{2}m\left(\frac{1}{2}v_B\right)^2 + 0$$
$$mgH + W_f = \frac{1}{2}m\left(\frac{1}{2}\sqrt{2gH}\right)^2$$
$$mgH + W_f = \frac{1}{4}mgH$$
$$W_f = -\frac{3}{4}mgH$$

2. (A) Using Conservation of Energy $K_i + U_i = K_f + U_f$ and $v_i = 0$, this becomes $U_i = K_f + U_f$ or $K_f = U_i - U_f$. This is equivalent to $\frac{1}{2}mv^2 = mgh_i - mgh_f$, which simplifies to $\frac{1}{2}v^2 - gh_i = -gh_f$ or $h_f = h_i - \frac{v^2}{2g}$. Now fill in the table.

Time (s)	Velocity (m/s)	Height (m)
0.00	0.00	1.5
0.05	1.41	1.4
0.10	2.45	1.2
0.15	3.74	0.8
0.20	3.74	0.8
0.25	3.46	0.9
0.30	3.16	1.0
0.35	2.83	1.1
0.40	3.46	0.9
0.45	4.24	0.6
0.50	4.47	0.5

(B) The greatest acceleration would occur where there is the greatest change in velocity. This occurs between 0.00 and 0.05 seconds. The acceleration during that time interval is given by $a = \frac{v_f - v_i}{t_f - t_i} \Rightarrow \frac{1.41 - 0}{0.05 - 0.00}$ or $a = 28\ \text{m/s}^2$.

(C) Changing the mass does not affect the time spent falling or the velocity of the object. Thus, a change in mass will not affect the results.

3. (A) Use the Work–Energy Theorem:

$$W_{total} = \Delta K$$

The force doing work during the motion is provided by the force of friction:

$$W = F_f \cdot d \cdot \cos\theta = \Delta K$$

$$\mu_k F_N \cdot d \cdot \cos\theta = \frac{1}{2}m(v^2 - v_0^2)$$

As the force of friction is antiparallel to the direction of the displacement, $\theta = 180°$.

$$\mu_k F_N \cdot d \cdot \cos\theta = \frac{1}{2}m(v^2 - v_0^2)$$

$$\mu_k mg \cdot d \cdot \cos\theta = \frac{1}{2}m(v^2 - v_0^2)$$

$$\mu_k g \cdot d \cdot \cos\theta = \frac{(v^2 - v_0^2)}{2}$$

$$d = \frac{(v^2 - v_0^2)}{2\mu_k \cdot g \cdot \cos\theta} = \frac{(0 - 10^2)}{2(0.2) \cdot 10 \cdot \cos 180°} = \frac{-100}{-4} = 25 \text{ m}$$

The skidding distance is 25 m.

(B) The final equation for the skidding distance is

$$d = \frac{(v^2 - v_0^2)}{2\mu_k \cdot g \cdot \cos\theta}$$

Since the final velocity is 0, the stopping distance is proportional to the initial speed. So if the initial speed is doubled, the skidding distance is quadrupled. This can also be determined by plugging in an initial velocity of 20 m/s:

$$d = \frac{(v^2 - v_0^2)}{2\mu_k \cdot g \cdot \cos\theta} = \frac{(0 - 20^2)}{2(0.2) \cdot 10 \cdot \cos 180°} = \frac{-400}{-4} = 100 \text{ m}$$

(C) Look back at the equation for the skidding distance:

$$d = \frac{(v^2 - v_0^2)}{2\mu_k \cdot g \cdot \cos\theta}$$

This equation does not include mass, so mass does not affect the skidding distance. (Note: While doubling the mass doubles the initial kinetic energy of the car, it also doubles the normal force and thus the frictional force acting on the car.)

CHAPTER 8 REVIEW QUESTIONS

Section I: Multiple Choice

1. **C** The magnitude of the object's linear momentum is $p = mv$. If $p = 6$ kg · m/s and $m = 2$ kg, then $v = 3$ m/s. Therefore, the object's kinetic energy is $K = \frac{1}{2}mv^2 = \frac{1}{2}(2 \text{ kg})(3 \text{ m/s})^2 = 9$ J.

2. **B** The impulse delivered to the ball, $J = F\Delta t$, equals its change in momentum. Since the ball started from rest, you get

$$F\Delta t = mv \quad \Rightarrow \quad \Delta t = \frac{mv}{F} = \frac{(0.5 \text{ kg})(4 \text{ m/s})}{20 \text{ N}} = 0.1 \text{ s}$$

3. **D** The impulse delivered to the ball, $J = \bar{F}\Delta t$, equals its change in momentum. Thus,

$$\bar{F}\Delta t = \Delta p = p_{\text{fi}} - p = m(v_{\text{fi}} - v) \quad \Rightarrow \quad \bar{F} = \frac{m(v_{\text{fi}} - v)}{\Delta t} = \frac{(2 \text{ kg})(8 \text{ m/s} - 4 \text{ m/s})}{0.5 \text{ s}} = 16 \text{ N}$$

4. **C** The impulse delivered to the ball is equal to its change in momentum. The momentum of the ball was $m\mathbf{v}$ before hitting the wall and $m(-\mathbf{v})$ after. Therefore, the change in momentum is $m(-\mathbf{v}) - m\mathbf{v} = -2m\mathbf{v}$, so the magnitude of the momentum change (and the impulse) is $2mv$.

5. **B** By definition of *perfectly inelastic*, the objects move off together with one common velocity, $\mathbf{v}'$, after the collision. By Conservation of Linear Momentum,

$$\begin{aligned} m_1\mathbf{v}_1 + m_2\mathbf{v}_2 &= (m_1 + m_2)\mathbf{v}' \\ \mathbf{v}' &= \frac{m_1 v_1 + m_2 v_2}{m_1 + m_2} \\ &= \frac{(3 \text{ kg})(2 \text{ m/s}) + (5 \text{ kg})(-2 \text{ m/s})}{3 \text{ kg} + 5 \text{ kg}} \\ &= 0.5 \text{ m/s} \end{aligned}$$

6. **D** First, apply Conservation of Linear Momentum to calculate the speed of the combined object after the (perfectly inelastic) collision:

$$\begin{aligned} m_1v_1 + m_2v_2 &= (m_1 + m_2)v' \\ v' &= \frac{m_1v_1 + m_2v_2}{m_1 + m_2} \\ &= \frac{m_1v_1 + (2m_1)(0)}{m_1 + 2m_1} \\ &= \frac{1}{3}v_1 \end{aligned}$$

Therefore, the ratio of the kinetic energy after the collision to the kinetic energy before the collision is

$$\frac{K'}{K} = \frac{\frac{1}{2}m'v'^2}{\frac{1}{2}m_1v_1^2} = \frac{\frac{1}{2}(m_1 + 2m_1)(\frac{1}{3}v_1)^2}{\frac{1}{2}m_1v_1^2} = \frac{1}{3}$$

7. **C** Total linear momentum is conserved in a collision during which the net external force is zero. If kinetic energy is lost, then by definition, the collision is not elastic.

8. **D** Because the two carts are initially at rest, the initial momentum is zero. Therefore, the final total momentum must be zero.

9. **C** The linear momentum of the bullet must have the same magnitude as the linear momentum of the block in order for their combined momentum after impact to be zero. The block has momentum MV to the left, so the bullet must have momentum MV to the right. Since the bullet's mass is m, its speed must be $v = MV/m$.

10. **C** In a perfectly inelastic collision, kinetic energy is never conserved; some of the initial kinetic energy is always lost to heat and some is converted to potential energy in the deformed shapes of the objects as they lock together.

11. **B** Total linear momentum is conserved in the absence of external forces. If the final speed of both objects is 0, that means the total linear momentum after the collision is 0, which then implies that the total linear momentum before the collision is also 0. As Object 2 has half the mass of Object 1, Object 2 must have twice the initial speed of Object 1 (and must be traveling in the opposite direction) in order for the total linear momentum to equal 0.

Section II: Free Response

1. (A) First, draw a free-body diagram:

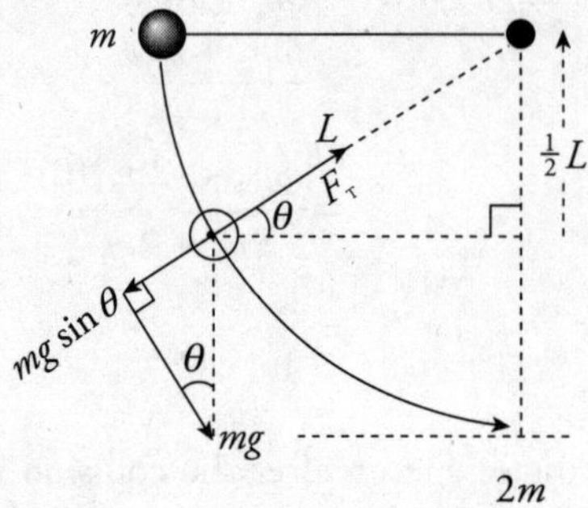

The net force toward the center of the steel ball's circular path provides the centripetal force. From the geometry of the diagram, you get

$$F_T - mg\sin\theta = \frac{mv^2}{L} \qquad (*)$$

In order to determine the value of mv^2, use Conservation of Mechanical Energy:

$$K_i + U_i = K_f + U_f$$
$$0 + mgL = \frac{1}{2}mv^2 + mg(\frac{1}{2}L)$$
$$\frac{1}{2}mgL = \frac{1}{2}mv^2$$
$$mgL = mv^2$$

Substituting this result into Equation (*), you get

$$F_T - mg\sin\theta = \frac{mgL}{L}$$
$$F_T = mg(1 + \sin\theta)$$

Now, from the free-body diagram, $\sin\theta = \frac{1}{2}L/L = \frac{1}{2}$, so

$$F_T = mg(1 + \frac{1}{2}) = \frac{3}{2}mg$$

(B) Apply Conservation of Energy to find the speed of the ball just before impact:

$$K_i + U_i = K_f + U_f$$
$$0 + mgL = \frac{1}{2}mv^2 + 0$$
$$v = \sqrt{2gL}$$

Now use Conservation of Linear Momentum and Conservation of Kinetic Energy for the elastic collision to derive the expressions for the speeds of the ball, v_1, and the block, v_2, immediately after the collision. Applying Conservation of Linear Momentum yields

$$m\sqrt{2gL} = mv_1 + 4mv_2$$
$$\sqrt{2gL} = v_1 + 4v_2$$
$$v_1 = \sqrt{2gL} - 4v_2$$

Applying the Conservation of Kinetic Energy for the elastic collision yields

$$\frac{1}{2}m(\sqrt{2gL})^2 = \frac{1}{2}mv_1^2 + \frac{1}{2}(4m)v_2^2$$
$$2gL = v_1^2 + (4m)v_2^2$$

Plugging in the expression for v_1 from the Conservation of Linear Momentum yields

$$2gL = (\sqrt{2gL} - 4v_2)^2 + 4v_2^2$$
$$2gL = 2gL - 8\sqrt{2gL}v_2 + 16v_2^2 + 4v_2^2$$
$$0 = 20v_2^2 - 8\sqrt{2gL}v_2$$
$$0 = 2v_2(10v_2 - 4\sqrt{2gL})$$
$$0 = 10v_2 - 4\sqrt{2gL}$$
$$10v_2 = 4\sqrt{2gL}$$
$$v_2 = \frac{2\sqrt{2}}{5}\sqrt{gL}$$

(C) Using the velocity of the block immediately after the collision, v_2, solve for the velocity of the ball immediately after the collision:

$$v_1 = \sqrt{2gL} - 4v_2$$

$$v_1 = \sqrt{2gL} - 4\left(\frac{2\sqrt{2}}{5}\sqrt{gL}\right)$$

$$v_1 = \sqrt{2gL} - \frac{4 \cdot 2}{5}\sqrt{2gL}$$

Factor out $\sqrt{2gL}$ to get

$$v_1 = \sqrt{2gL}\left(1 - \frac{8}{5}\right)$$

$$v_1 = -\frac{3}{5}\sqrt{2gL}$$

Now, apply Conservation of Mechanical Energy to find

$$K_i + U_i = K_f + U_f$$

$$\frac{1}{2}mv_1'^2 + 0 = 0 + mgh$$

$$h = \frac{v_1'^2}{2g} = \frac{\left(\frac{3}{5}\sqrt{2gL}\right)^2}{2g} = \frac{9}{25}L$$

2. (A) By Conservation of Linear Momentum, $mv = (m + M)v'$, so $v' = \frac{mv}{m+M}$.

Now, by Conservation of Mechanical Energy,

$$K_i + U_i = K_f + U_f$$

$$\frac{1}{2}(m+M)v'^2 + 0 = 0 + (m+M)gy$$

$$\frac{1}{2}v'^2 = gy$$

$$\frac{1}{2}\left(\frac{mv}{m+M}\right)^2 = gy$$

$$v = \frac{m+M}{m}\sqrt{2gy}$$

(B) Use the result derived in part (A) to compute the kinetic energy of the block and bullet immediately after the collision:

$$K' = \frac{1}{2}(m+M)v'^2 = \frac{1}{2}(m+M)\left(\frac{mv}{m+M}\right)^2 = \frac{1}{2}\frac{m^2v^2}{m+M}$$

Since $K = \frac{1}{2}mv^2$, the difference is

$$\begin{aligned}\Delta K = K' - K &= \frac{1}{2}\frac{m^2v^2}{m+M} - \frac{1}{2}mv^2 \\ &= \frac{1}{2}mv^2\left(\frac{m}{m+M} - 1\right) \\ &= K\left(\frac{-M}{m+M}\right)\end{aligned}$$

Therefore, the fraction of the bullet's original kinetic energy that was lost is $M/(m + M)$. This energy is manifested as heat (the bullet and block are warmer after the collision than before), and some was used to break the intermolecular bonds within the wooden block to allow the bullet to penetrate.

(C) From the geometry of the diagram,

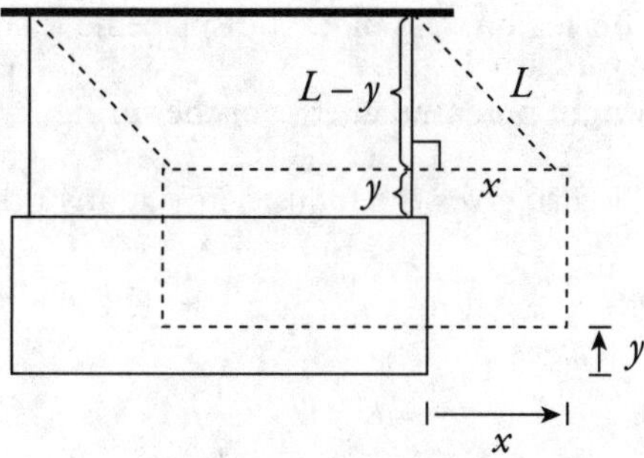

the Pythagorean Theorem implies that $(L - y)^2 + x^2 = L^2$. Therefore,

$$L^2 - 2Ly + y^2 + x^2 = L^2 \quad \Rightarrow \quad y = \frac{x^2}{2L}$$

(where you have used the fact that y^2 is small enough to be neglected). Substituting this into the result of part (A), derive the following equation for the speed of the bullet in terms of x and L instead of y:

$$v = \frac{m+M}{m}\sqrt{2gy} = \frac{m+M}{m}\sqrt{2g\frac{x^2}{2L}} = \frac{m+M}{m}x\sqrt{\frac{g}{L}}$$

(D) No; momentum is conserved only when the net external force on the system is zero (or at least negligible). In this case, the block and bullet feel a net nonzero force that causes it to slow down as it swings upward. Since its speed is decreasing as it swings upward, its linear momentum cannot remain constant.

CHAPTER 9 REVIEW QUESTIONS

Section I: Multiple Choice

1. **B, C** The acceleration of a simple harmonic oscillator is not constant, since the restoring force—and consequently, the acceleration—depends on the position. Eliminate (A). Choices (B) and (C) are defining characteristics of simple harmonic motion; therefore, (D) is also false. Period must also be independent of amplitude, as period and frequency are proportional to one another.

2. **B** The acceleration of the block has its maximum magnitude at the points where its displacement from equilibrium has the maximum magnitude (since $a = F/m = kx/m$). At the endpoints of the oscillation region, the potential energy is maximized and the kinetic energy (and hence the speed) is zero.

3. **D** By Conservation of Mechanical Energy, $K + U_S$ is a constant for the motion of the block. At the endpoints of the oscillation region, the block's displacement, x, is equal to $\pm A$. Since $K = 0$ here, all the energy is in the form of potential energy of the spring, $\frac{1}{2}kA^2$. Because $\frac{1}{2}kA^2$ gives the total energy at these positions, it also gives the total energy at any other position.

   Using the equation $U_S(x) = \frac{1}{2}kx^2$, find that, at $x = \frac{1}{2}A$.

$$K + U_S = \frac{1}{2}kA^2$$

$$K + \frac{1}{2}k(\frac{1}{2}A)^2 = \frac{1}{2}kA^2$$

$$K = \frac{3}{8}kA^2$$

   Therefore,

$$K/E = \frac{3}{8}kA^2 \Big/ \frac{1}{2}kA^2 = \frac{3}{4}$$

4. **C** The maximum speed of the block is given by the equation $v_{max} = A\sqrt{k/m}$. Therefore, v_{max} is inversely proportional to $\sqrt{m}$. If m is increased by a factor of 2, then v_{max} will decrease by a factor of $\sqrt{2}$.

5. **D** The period of a spring-block simple harmonic oscillator is independent of the value of g. (Recall that $T = 2\pi\sqrt{m/k}$.) Therefore, the period will remain the same.

6. **D** The frequency of a spring-block simple harmonic oscillator is given by the equation $f = (1/2\pi)\sqrt{k/m}$. Squaring both sides of this equation, you get $f^2 = (k/4\pi^2)(1/m)$. Therefore, if f^2 is plotted versus $(1/m)$, then the graph will be a straight line with slope $k/4\pi^2$. (Note: The slope of the line whose equation is $y = ax$ is a.)

7. **C** For small angular displacements, the period of a simple pendulum is essentially independent of amplitude.

8. **D** Combining Hooke's Law with Newton's Second Law, you get

$$F = kx = ma \Rightarrow a = \frac{kx}{m} = \frac{50\text{ N/m} \cdot 4\text{ m}}{20\text{ kg}} = 10\text{ m/s}^2$$

9. **B** By Conservation of Mechanical Energy, the energy of the block is the same throughout the motion. At the amplitude, the block has potential energy $U = \frac{1}{2}kA^2$ and zero kinetic energy. At the equilibrium position, the block has kinetic energy $K = \frac{1}{2}mv^2$ and zero potential energy. Applying the Conservation of Mechanical Energy to these two points in the motion yields,

$$\frac{1}{2}kA^2 + 0 = 0 + \frac{1}{2}mv^2$$

$$kA^2 = mv^2$$

$$k = \frac{mv^2}{A^2} = \frac{10\text{ kg} \cdot (4\text{ m/s})^2}{(2\text{ m})^2} = 40\text{ kg/s}^2$$

The period of the block can then be calculated using the following equation:

$$T = 2\pi\sqrt{\frac{m}{k}} = 2\pi\sqrt{\frac{10\text{ kg}}{40\text{ kg/s}^2}} = \pi\text{ s} \approx 3\text{ s}$$

10. **B** The frequency of a spring-block simple harmonic oscillator is independent of the amplitude. The equation for the frequency of a spring-block simple harmonic oscillator is $f = \frac{1}{2\pi}\sqrt{\frac{k}{m}}$. The frequency is inversely proportional to the square root of the mass, so decreasing the mass of the block by a factor of 4 would increase the frequency by a factor of 2.

Section II: Free Response

1. (A) Since the spring is compressed to 3/4 of its natural length, the block's position relative to equilibrium is $x = -\frac{1}{4}L$. Therefore, from $F_S = -kx$, find

$$a = \frac{F_S}{m} = \frac{-k(-\frac{1}{4}L)}{m} = \frac{kL}{4m}$$

(B) Let v_1 denote the velocity of Block 1 just before impact, and let v_1' and v_2' denote, respectively, the velocities of Block 1 and Block 2 immediately after impact. By Conservation of Linear Momentum, write $mv_1 = mv_1' + mv_2'$, or

$$v_1 = v_1' + v_2' \qquad (1)$$

The initial kinetic energy of Block 1 is $\frac{1}{2}mv_1^2$. If half is lost to heat, then $\frac{1}{4}mv_1^2$ is left to be shared by Block 1 and Block 2 after impact: $\frac{1}{4}mv_1^2 = \frac{1}{2}mv_1'^2 + \frac{1}{2}mv_2'^2$, or

$$v_1^2 = 2v_1'^2 + 2v_2'^2 \qquad (2)$$

Square Equation (1) and multiply by 2 to give

$$2v_1^2 = 2v_1'^2 + 4v_1'v_2' + 2v_2'^2 \qquad (1')$$

Then subtract Equation (2) from Equation (1′):

$$v_1^2 = 4v_1'v_2' \qquad (3)$$

Square Equation (1) again,

$$v_1^2 = v_1'^2 + 2v_1'v_2' + v_2'^2$$

and substitute into this the result of Equation (3):

$$\begin{aligned} 4v_1'v_2' &= v_1'^2 + 2v_1'v_2' + v_2'^2 \\ 0 &= v_1'^2 - 2v_1'v_2' + v_2'^2 \\ 0 &= (v_1' - v_2')^2 \\ v_1' &= v_2' \qquad (4) \end{aligned}$$

Thus, combining Equations (1) and (4), find that

$$v_1' = v_2' = \frac{1}{2}v_1$$

(C) When Block 1 reaches its new amplitude position, A', all of its kinetic energy is converted to elastic potential energy of the spring. That is,

$$K_1' \rightarrow U_S' \quad \Rightarrow \quad \frac{1}{2}mv_1'^2 = \frac{1}{2}kA'^2$$

$$A'^2 = \frac{m}{k}v_1'^2$$

$$A'^2 = \frac{m}{k}(\frac{1}{2}v_1)^2$$

$$A'^2 = \frac{mv_1^2}{4k} \qquad (1)$$

But the original potential energy of the spring, $U_S = \frac{1}{2}k(-\frac{1}{4}L)^2$, gave K_1:

$$U_S \rightarrow K_1 \quad \Rightarrow \quad \frac{1}{2}k(-\frac{1}{4}L)^2 = \frac{1}{2}mv_1^2 \quad \Rightarrow \quad mv_1^2 = \frac{1}{16}kL^2 \qquad (2)$$

Substituting this result into Equation (1) gives

$$A'^2 = \frac{\frac{1}{16}kL^2}{4k} = \frac{L^2}{64} \quad \Rightarrow \quad A' = \frac{1}{8}L$$

(D) The period of a spring-block simple harmonic oscillator depends only on the spring constant k and the mass of the block. Since neither of these changes, the period will remain the same; that is, $T' = T_0$.

(E) As shown in part (B), Block 2's velocity as it slides off the table is $\frac{1}{2}v_1$ (horizontally). The time required to drop the vertical distance H is found as follows (calling *down* the positive direction):

$$\Delta y = v_{0y}t + \frac{1}{2}gt^2 \quad \Rightarrow \quad H = \frac{1}{2}gt^2 \quad \Rightarrow \quad t = \sqrt{\frac{2H}{g}}$$

Therefore,

$$R = (\frac{1}{2}v_1)t = \frac{1}{2}v_1\sqrt{\frac{2H}{g}}$$

Now, from Equation (2) of part (C), $v_1 = \sqrt{\frac{kL^2}{16m}}$, so

$$R = \frac{1}{2}\sqrt{\frac{kL^2}{16m}}\sqrt{\frac{2H}{g}} = \frac{L}{8}\sqrt{\frac{2kH}{mg}}$$

2. (A) By Conservation of Linear Momentum,

$$mv = (m+M)v' \quad \Rightarrow \quad v' = \frac{mv}{m+M}$$

(B) When the block is at its amplitude position (maximum compression of spring), the kinetic energy it (and the embedded bullet) had just after impact will become the potential energy of the spring:

$$K' \rightarrow U_S$$

$$\frac{1}{2}(m+M)\left(\frac{mv}{m+M}\right)^2 = \frac{1}{2}kA^2$$

$$A = \frac{mv}{\sqrt{k(m+M)}}$$

(C) Since the mass on the spring is $m + M$, $f = \frac{1}{2\pi}\sqrt{\frac{k}{m+M}}$

3. (A) By Conservation of Mechanical Energy, $K + U = E$, so

$$\frac{1}{2}Mv^2 + \frac{1}{2}k(\frac{1}{2}A)^2 = \frac{1}{2}kA^2$$

$$\frac{1}{2}Mv^2 = \frac{3}{8}kA^2$$

$$v = A\sqrt{\frac{3k}{4M}}$$

(B) Since the clay ball delivers no horizontal linear momentum to the block, horizontal linear momentum is conserved. Thus,

$$Mv = (M+m)v'$$

$$v' = \frac{Mv}{M+m} = \frac{MA}{M+m}\sqrt{\frac{3k}{4M}} = \frac{A}{M+m}\sqrt{\frac{3kM}{4}}$$

(C) Applying the general equation for the period of a spring-block simple harmonic oscillator,

$$T = 2\pi\sqrt{\frac{M+m}{k}}$$

(D) The total energy of the oscillator after the clay hits is $\frac{1}{2}kA'^2$, where A' is the new amplitude. Just after the clay hits the block, the total energy is

$$K' + U_S = \frac{1}{2}(M+m)v'^2 + \frac{1}{2}k(\frac{1}{2}A)^2$$

Substitute for v' from part (b), set the resulting sum equal to $\frac{1}{2}kA'^2$, and solve for A'.

$$\frac{1}{2}(M+m)\left(\frac{A}{M+m}\sqrt{\frac{3kM}{4}}\right)^2 + \frac{1}{2}k(\frac{1}{2}A)^2 = \frac{1}{2}kA'^2$$

$$\frac{A^2 \cdot 3kM}{8(M+m)} + \frac{1}{8}kA^2 = \frac{1}{2}kA'^2$$

$$A^2\left(\frac{3M}{M+m}+1\right) = 4A'^2$$

$$A' = \frac{1}{2}A\sqrt{\frac{3M}{M+m}+1}$$

(E) No, because the period depends only on the mass and the spring constant k.

(F) Yes. For example, if the clay had landed when the block was at $x = A$, the speed of the block would have been zero immediately before the collision and immediately after. No change in the block's speed would have meant no change in K, so no change in E, so no change in $A = \sqrt{2E/k}$.

CHAPTER 10 REVIEW QUESTIONS

Section I: Multiple Choice

1. **B** For a disc rotating at 8 rev/s, the time for 1 revolution is 1/8 s. Speed is distance divided by time, so for one rotation of a point on the rim,

$$v = \frac{2\pi r}{T} = \frac{2\pi(0.05\text{ m})}{(1/8\text{ s})} = 2.5\text{ m/s}$$

2. **B** Use Big 5 #3 for rotational motion:

$$\theta = \theta_0 + \omega_0 t + \frac{1}{2}\alpha t^2 \Rightarrow \alpha = \frac{2\theta}{t^2} = \frac{2(90\text{ rad})}{(15\text{ s})^2} = 0.80\text{ m/s}^2$$

3. **B** Apply the rotational definition for torque:

$$\tau = I\alpha = I\frac{\omega-\omega_0}{t} = \left(0.5\text{ kg}\cdot\text{m}^2\right)\frac{0-12\text{ rad/s}}{0.4\text{ s}} = 15\text{ N}\cdot\text{m}$$

4. **A** To maintain a constant angular velocity, the angular acceleration must equal zero. Therefore, the net torque on the grinding wheel also equals zero.

$$\Sigma\tau = I\alpha = 0$$

$$\tau_{motor} = \tau_{knife} = rF_f \sin 90° = r\mu_k F_N = (0.12 \text{ m})(0.28)(16 \text{ N}) = 0.54 \text{ N·m}$$

5. **A** Since the ladder isn't moving, it's in equilibrium, so $\Sigma F = 0$ and $\Sigma\tau = 0$. The ladder experiences a normal force and a frictional force where it touches the ground, both of which are unknown. Therefore, select that point as the pivot, and apply the torque condition for equilibrium,

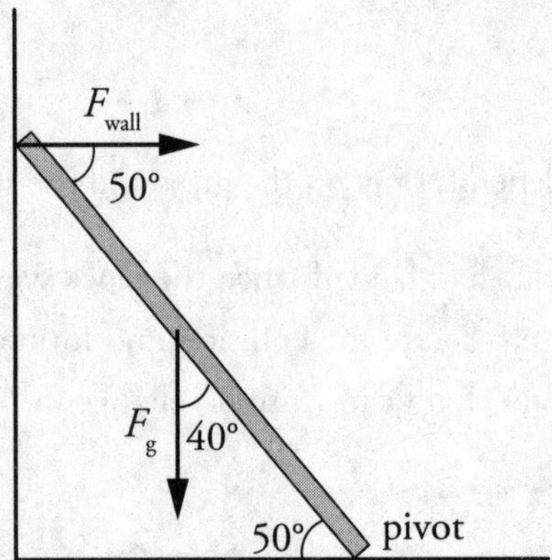

$$\Sigma\tau = 0 \Rightarrow \tau_g + \tau_{wall} = 0 \Rightarrow r_g F_g \sin 40° = r_{wall} F_{wall} \sin 50°$$

$$F_{wall} = \frac{r_g mg \sin 40°}{r_{wall} \sin 50°} = \frac{\left(\frac{L}{2}\right)(20 \text{ kg})(9.8 \text{ m/s}^2)\sin 40°}{L \sin 50°}$$

$$= 82 \text{ N}$$

By Newton's Third Law, if the wall exerts an 82 N force against the ladder, the ladder exerts an 82 N force against the wall.

6. **A** By Conservation of Momentum,

$$L_i = L_f \Rightarrow I_i\omega_i = I_f\omega_f \Rightarrow \omega_f = \frac{I_i\omega_i}{I_f} = \frac{I_i\omega}{4I_i} = \frac{\omega}{4}$$

7. **D** First, convert the initial angular velocity to units of rad/s:

$$20\frac{\text{rev}}{\text{min}}\times\frac{1\text{ min}}{60\text{ s}}\times\frac{2\pi\text{ rad}}{1\text{ rev}} = 2.09\text{ rad/s}$$

Since the child and merry-go-round are rotating without friction, angular momentum is conserved. The rotational inertia of the system is the sum of the rotational inertia of the child and the merry-go-round.

$$I_i\omega_i = I_f\omega_f$$

$$(I_{\text{child, i}} + I_m)\omega_i = (I_{\text{child, f}} + I_m)\omega_f$$

$$(mr_i^2 + I_m)\omega_i = (mr_f^2 + I_m)\omega_f$$

$$\omega_f = \frac{\left(mr_i^2 + I_m\right)\omega_i}{\left(mr_f^2 + I_m\right)}$$

$$\omega_f = \frac{\left((25\text{ kg})(2.0\text{ m})^2 + 1200\text{ kg}\cdot\text{m}^2\right)(2.1\text{ rad/s})}{\left((25\text{ kg})(0.5\text{ m})^2 + 1200\text{ kg}\cdot\text{m}^2\right)} = 2.25\text{ rad/s}$$

8. **C** The torque is $\tau = rF = (0.20\text{ m})(20\text{ N}) = 4\text{ N}\cdot\text{m}$.

9. **D** From the diagram,

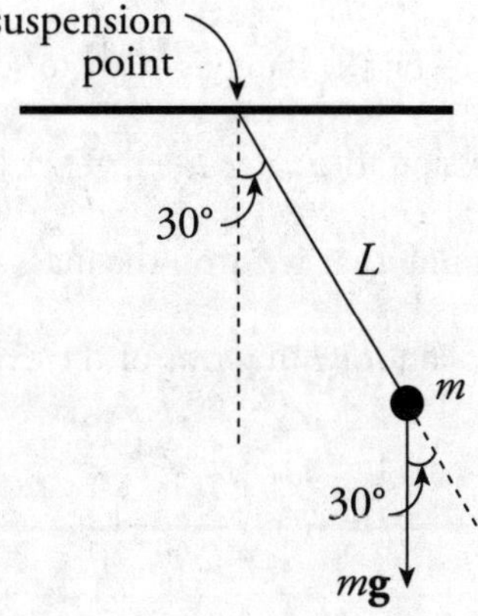

calculate that

$$\begin{aligned}\tau = rF\sin\theta &= Lmg\sin\theta \\ &= (0.80\text{ m})(0.50\text{ kg})(10\text{ N/kg})(\sin 30°) \\ &= 2.0\text{ N}\cdot\text{m}\end{aligned}$$

Section II: Free Response

1. (A) (i) The wheel will rotate according to the equation $\tau = I\alpha$. The torque is provided by the tension on the string that the mass hangs from, that acts at a distance to the rotation axis equal to the radius of the wheel. The tension depends on the mass of the hanging mass and the acceleration of the hanging mass. The acceleration of the hanging mass can be calculated from its starting height and the time it takes for the mass to hit the ground. The angular acceleration of the wheel is related to the tangential acceleration at its outer rim, which is the same as the acceleration of the hanging mass. Therefore, the quantities to be measured are the mass of the hanging object, *m*, that can be measured with any type of scale; the radius, *R*, of the wheel, that can be measured with a meter stick; the initial height, *h*, of the hanging mass, that can be measured with a meter stick; and the time it takes for the hanging mass to reach the ground, *t*, that can be measured with a stop watch.

(ii) Start with $\tau = I\alpha$. Torque is given by $\tau = r_{\perp}F = RT$, where *T* is the tension in the string, and angular acceleration is given by $\alpha = a/R$, where *a* is the acceleration of the hanging mass. Solving for the rotational inertia gives:

$$I = \frac{\tau}{\alpha} = \frac{RT}{a/R} = \frac{R^2T}{a}$$

Apply Newton's second law on the hanging mass to solve for *T*: $mg - T = ma \rightarrow T = m(g - a)$. Substitute into the expression for *I*: $I = R^2m\left(\frac{g}{a} - 1\right)$. Apply kinematics to express *a* as a function of *h* and *t* assuming that we drop the mass of rest: $h = \frac{1}{2}at^2 \rightarrow a = \frac{2h}{t^2}$. Finally, substitute into the expression for *I* in terms of the experimentally measured values:

$$I = R^2m\left(\frac{gt^2}{2h} - 1\right)$$

(B) Mount the wheel on a fixed horizontal axis. Cut a length of string long enough to wrap around the wheel and reach the ground. Fix one end of the string on the edge of the wheel, and wind the string around the wheel. Tie the free end if the string to the hanging mass. Turn the wheel to wind the string so that the string is taut and the mass is hanging from one side of the wheel. From rest, release the system, allowing the mass to drop and the wheel to spin.

(C) The rotational inertia of the wheel is independent of the hanging mass, so changing the value of the hanging mass should lead to approximately the same value of the rotational inertia:

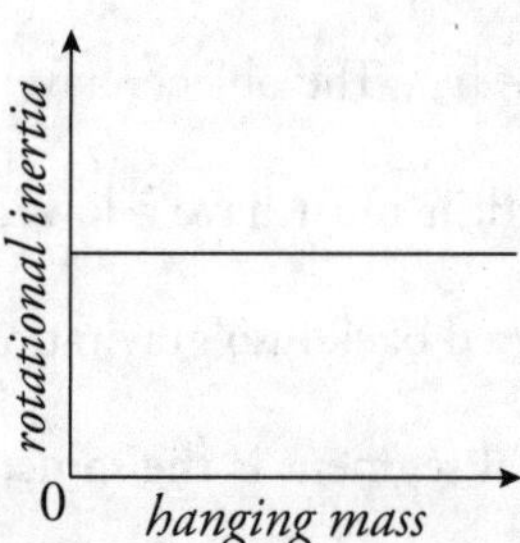

2. (A) Newton's second law for rotational motion states that $\tau_{\text{net}} = I\alpha$. In the absence of friction, the net torque is the torque applied by the cyclist, $\tau_{\text{app}} = I\alpha$. The angular acceleration is given by $\alpha = \frac{\omega - \omega_0}{t}$. The wheel starts from rest, so $\omega_0 = 0$. The final angular velocity is $\omega = \frac{v}{r} = \frac{2v}{D}$. Therefore, $\tau_{\text{app}} = I\left(\frac{\omega - 0}{t}\right) = \frac{I}{t}\left(\frac{2v}{D}\right) = \frac{2Iv}{tD}$.

(B) The kinetic friction force applies a frictional torque $\tau_f = r_{\perp}F = sf$. Apply Newton's second law for rotational motion to solve for α, $\tau_{\text{app}} - \tau_f = I\alpha \rightarrow \alpha = \frac{\tau_{\text{app}} - \tau_f}{I}$. The same rotational kinematic equations apply as before, with t' replacing t, $\alpha = \frac{2v / D}{t'}$. Set the two equations for α equal to each other, and solve for t', $\frac{\tau_{\text{app}} - \tau_f}{I} = \frac{2v / D}{t'} \rightarrow t' = \frac{2vI}{D(\tau_{\text{app}} - \tau_f)}$.

3. The balls roll without slipping, so no work is done by kinetic friction. Therefore, total mechanic energy of each ball is conserved, $\Delta K = -\Delta U$. The balls experience the same change in height, so as they roll off the first ramp, they've experienced the same change in gravitational potential energy in their respective ball-Earth systems. Therefore, they also experience the same change in kinetic energy, so their kinetic energies are the same at the bottom of the ramp, as the balls start from rest.

However, kinetic energy has both translational, $K_t = \frac{1}{2}mv^2$, and rotational, $K_r = \frac{1}{2}I\omega^2$, contributions. For spherical objects that roll without slipping, $\omega = \frac{v}{R}$, so the total kinetic energy is $K = \frac{1}{2}mv^2 + \frac{1}{2}I\omega^2 = \frac{1}{2}\left(m + \frac{I}{R^2}\right)v^2$. The objects have the same mass and radius, so for ball A to have a greater translational speed, it must have a lower rotational inertia. On the second incline, the total kinetic energy is converted back into gravitational potential energy. The total kinetic energies of the balls on the horizontal segment is the same, so the final gravitational potential energy, and therefore the final height, is the same.

Part VI
Additional Practice Tests

- Practice Test 2
- Practice Test 2: Diagnostic Answer Key and Explanations
- How to Score Practice Test 2
- Practice Test 3
- Practice Test 3: Answers and Explanations
- How to Score Practice Test 3
- Practice Test 4
- Practice Test 4: Answers and Explanations
- How to Score Practice Test 4

Practice Test 2

AP® Physics 1 Exam

SECTION I: Multiple-Choice Questions

DO NOT OPEN THIS BOOKLET UNTIL YOU ARE TOLD TO DO SO.

At a Glance

Total Time
90 minutes
Number of Questions
50
Percent of Total Grade
50%
Writing Instrument
Pen required

Instructions

Section I of this examination contains 50 multiple-choice questions. Fill in only the ovals for numbers 1 through 50 on your answer sheet.

CALCULATORS MAY BE USED ON BOTH SECTIONS OF THE AP PHYSICS 1 EXAM.

Indicate all of your answers to the multiple-choice questions on the answer sheet. No credit will be given for anything written in this exam booklet, but you may use the booklet for notes or scratch work. Please note that there are two types of multiple-choice questions: single-select and multi-select questions. After you have decided which of the suggested answers is best, completely fill in the corresponding oval(s) on the answer sheet. For single-select, you must give only one answer; for multi-select you must give BOTH answers in order to earn credit. If you change an answer, be sure that the previous mark is erased completely. Here is a sample question and answer.

Sample Question

Sample Answer

Chicago is a
(A) state
(B) city
(C) country
(D) continent

Use your time effectively, working as quickly as you can without losing accuracy. Do not spend too much time on any one question. Go on to other questions and come back to the ones you have not answered if you have time. It is not expected that everyone will know the answers to all the multiple-choice questions.

About Guessing

Many candidates wonder whether or not to guess the answers to questions about which they are not certain. Multiple-choice scores are based on the number of questions answered correctly. Points are not deducted for incorrect answers, and no points are awarded for unanswered questions. Because points are not deducted for incorrect answers, you are encouraged to answer all multiple-choice questions. On any questions you do not know the answer to, you should eliminate as many choices as you can, and then select the best answer among the remaining choices.

GO ON TO THE NEXT PAGE.

ADVANCED PLACEMENT PHYSICS 1 TABLE OF INFORMATION

CONSTANTS AND CONVERSION FACTORS			
Proton mass,	$m_p = 1.67 \times 10^{-27}$ kg	Electron charge magnitude,	$e = 1.60 \times 10^{-19}$ C
Neutron mass,	$m_n = 1.67 \times 10^{-27}$ kg	Coulomb's law constant,	$k = 1/4\pi\varepsilon_0 = 9.0 \times 10^9\ \text{N}\cdot\text{m}^2/\text{C}^2$
Electron mass,	$m_e = 9.11 \times 10^{-31}$ kg	Universal gravitational constant,	$G = 6.67 \times 10^{-11}\ \text{m}^3/\text{kg}\cdot\text{s}^2$
Speed of light,	$c = 3.00 \times 10^8$ m/s	Acceleration due to gravity at Earth's surface,	$g = 9.8\ \text{m/s}^2$

UNIT SYMBOLS									
	meter,	m	kelvin,	K	watt,	W	degree Celsius,	°C
	kilogram,	kg	hertz,	Hz	coulomb,	C		
	second,	s	newton,	N	volt,	V		
	ampere,	A	joule,	J	ohm,	Ω		

PREFIXES		
Factor	Prefix	Symbol
10^{12}	tera	T
10^{9}	giga	G
10^{6}	mega	M
10^{3}	kilo	k
10^{-2}	centi	c
10^{-3}	milli	m
10^{-6}	micro	μ
10^{-9}	nano	n
10^{-12}	pico	p

VALUES OF TRIGONOMETRIC FUNCTIONS FOR COMMON ANGLES

θ	$0°$	$30°$	$37°$	$45°$	$53°$	$60°$	$90°$
$\sin\theta$	0	$1/2$	$3/5$	$\sqrt{2}/2$	$4/5$	$\sqrt{3}/2$	1
$\cos\theta$	1	$\sqrt{3}/2$	$4/5$	$\sqrt{2}/2$	$3/5$	$1/2$	0
$\tan\theta$	0	$\sqrt{3}/3$	$3/4$	1	$4/3$	$\sqrt{3}$	∞

The following conventions are used in this exam.

I. The frame of reference of any problem is assumed to be inertial unless otherwise stated.
II. Assume air resistance is negligible unless otherwise stated.
III. In all situations, positive work is defined as work done <u>on</u> a system.
IV. The direction of current is conventional current: the direction in which positive charge would drift.
V. Assume all batteries and meters are ideal unless otherwise stated.

GO ON TO THE NEXT PAGE.

ADVANCED PLACEMENT PHYSICS 1 EQUATIONS

MECHANICS

$v_x = v_{x0} + a_x t$

$x = x_0 + v_{x0}t + \frac{1}{2}a_x t^2$

$v_x^2 = v_{x0}^2 + 2a_x(x - x_0)$

$\vec{a} = \frac{\sum \vec{F}}{m} = \frac{\vec{F}_{net}}{m}$

$|\vec{F}_f| \le \mu |\vec{F}_n|$

$a_c = \frac{v^2}{r}$

$\vec{p} = m\vec{v}$

$\Delta\vec{p} = \vec{F}\,\Delta t$

$K = \frac{1}{2}mv^2$

$\Delta E = W = F_{\parallel}d = Fd\cos\theta$

$P = \frac{\Delta E}{\Delta t}$

$\theta = \theta_0 + \omega_0 t + \frac{1}{2}\alpha t^2$

$\omega = \omega_0 + \alpha t$

$x = A\cos(2\pi ft)$

$\vec{\alpha} = \frac{\sum \vec{\tau}}{I} = \frac{\vec{\tau}_{net}}{I}$

$\tau = r_{\perp}F = rF\sin\theta$

$L = I\omega$

$\Delta L = \tau\,\Delta t$

$K = \frac{1}{2}I\omega^2$

$|\vec{F}_s| = k|\vec{x}|$

$U_s = \frac{1}{2}kx^2$

$\rho = \frac{m}{V}$

$\Delta U_g = mg\,\Delta y$

$T = \frac{2\pi}{\omega} = \frac{1}{f}$

$T_s = 2\pi\sqrt{\frac{m}{k}}$

$T_p = 2\pi\sqrt{\frac{\ell}{g}}$

$|\vec{F}_g| = G\frac{m_1 m_2}{r^2}$

$\vec{g} = \frac{\vec{F}_g}{m}$

$U_G = -\frac{Gm_1 m_2}{r}$

a = acceleration
A = amplitude
d = distance
E = energy
f = frequency
F = force
I = rotational inertia
K = kinetic energy
k = spring constant
L = angular momentum
ℓ = length
m = mass
P = power
p = momentum
r = radius or separation
T = period
t = time
U = potential energy
V = volume
v = speed
W = work done on a system
x = position
y = height
α = angular acceleration
μ = coefficient of friction
θ = angle
ρ = density
τ = torque
ω = angular speed

ELECTRICITY

$|\vec{F}_E| = k\left|\frac{q_1 q_2}{r^2}\right|$

$I = \frac{\Delta q}{\Delta t}$

$R = \frac{\rho\ell}{A}$

$I = \frac{\Delta V}{R}$

$P = I\,\Delta V$

$R_s = \sum_i R_i$

$\frac{1}{R_p} = \sum_i \frac{1}{R_i}$

A = area
F = force
I = current
ℓ = length
P = power
q = charge
R = resistance
r = separation
t = time
V = electric potential
ρ = resistivity

WAVES

$\lambda = \frac{v}{f}$

f = frequency
v = speed
λ = wavelength

GEOMETRY AND TRIGONOMETRY

Rectangle
$A = bh$

Triangle
$A = \frac{1}{2}bh$

Circle
$A = \pi r^2$
$C = 2\pi r$

Rectangular solid
$V = \ell wh$

Cylinder
$V = \pi r^2\ell$
$S = 2\pi r\ell + 2\pi r^2$

Sphere
$V = \frac{4}{3}\pi r^3$
$S = 4\pi r^2$

A = area
C = circumference
V = volume
S = surface area
b = base
h = height
ℓ = length
w = width
r = radius

Right triangle
$c^2 = a^2 + b^2$

$\sin\theta = \frac{a}{c}$

$\cos\theta = \frac{b}{c}$

$\tan\theta = \frac{a}{b}$

c a θ 90° b

GO ON TO THE NEXT PAGE.

THIS PAGE IS LEFT INTENTIONALLY BLANK.

GO ON TO THE NEXT PAGE.

AP PHYSICS 1

SECTION I

Note: To simplify calculations, you may use $g = 10 \text{ m/s}^2$ in all problems.

Directions: Each of the questions or incomplete statements is followed by four suggested answers or completions. Select the one that is best in each case and then fill in the corresponding circle on the answer sheet.

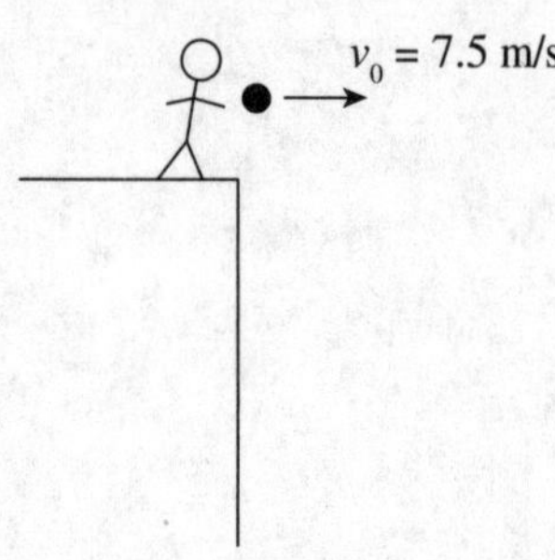

1. An object is thrown horizontally to the right off a high cliff with an initial speed of 7.5 m/s. Which arrow best represents the direction of the object's velocity after 2 seconds? (Assume air resistance is negligible.)

(A)

(B)

(C) 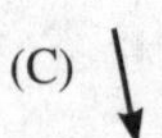

(D)

Questions 2–4 refer to the following figure:

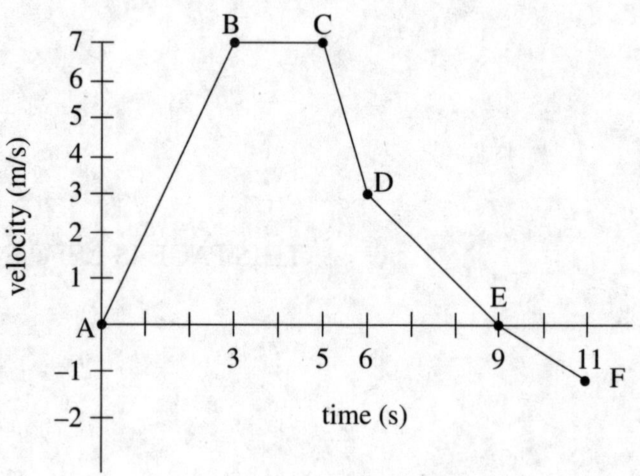

2. Rank the change in momentum for each segment from least (most negative) to greatest (most positive). Assume that the mass of the object remains constant.

(A) BC, EF, DE, AB, CD
(B) AB, CD, DE, EF, BC
(C) BC, EF, DE, CD, AB
(D) CD, DE, EF, BC, AB

GO ON TO THE NEXT PAGE.

3. During which segment is the magnitude of average acceleration greatest?
   (A) AB
   (B) BC
   (C) CD
   (D) DE

4. What is the total distance traveled by the object?
   (A) 32 m
   (B) 34 m
   (C) 35 m
   (D) 38 m

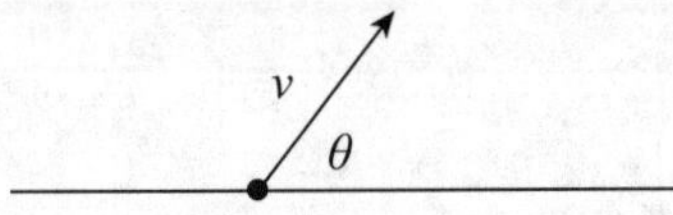

5. A ball is thrown with an initial velocity v at an angle θ above the ground. If the acceleration due to gravity is $-g$, which of the following is the correct expression of the time it takes for the ball to reach its highest point, y, from the ground?
   (A) $v^2 \sin \theta/g$
   (B) $-v \cos \theta/g$
   (C) $v \sin \theta/g$
   (D) $v^2 \cos \theta/g$

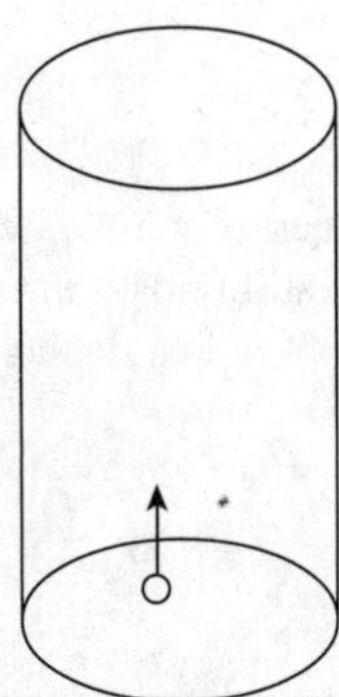

6. A bubble in a glass of water releases from rest at the bottom of the glass and rises at acceleration a to the surface in t seconds. How much farther does the bubble travel in its last second than in its first second? Assume that the journey takes longer than 2 seconds.
   (A) $a(t + 1\text{ s})^2$
   (B) $a(t - 1\text{ s})(1\text{ s})$
   (C) at^2
   (D) $a(t + 1\text{ s})(1\text{ s})$

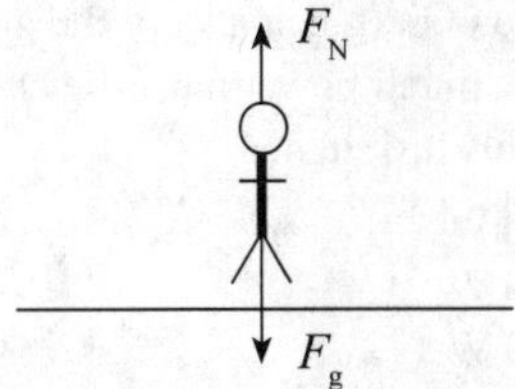

7. A person standing on a horizontal floor is acted upon by two forces: the downward pull of gravity and the upward normal force of the floor. These two forces
   (A) have equal magnitudes and form an action/reaction pair
   (B) have equal magnitudes and do not form an action/reaction pair
   (C) have unequal magnitudes and form an action/reaction pair
   (D) have unequal magnitudes and do not form an action/reaction pair

8. Which of the following graphs best represents the force of friction on an object that starts at rest and eventually starts sliding across a level surface because of a linearly increasing horizontal force?

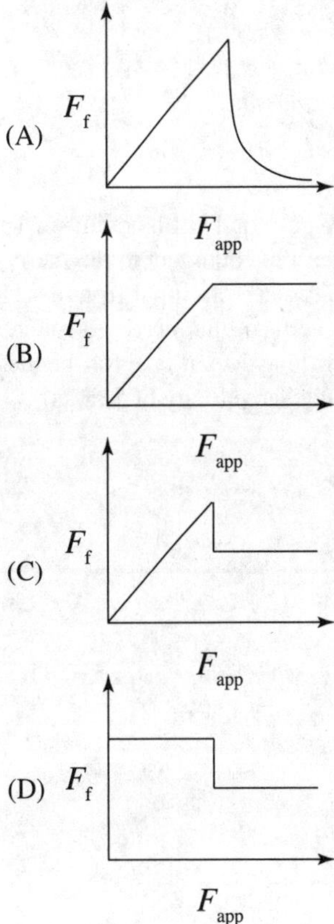

GO ON TO THE NEXT PAGE.

9. A crane lifts a 600 kg beam from the ground to a height of 20 m. How much power must the crane provide to accomplish this task in 4 s?

(A) 1,470 W
(B) 3,000 W
(C) 29,400 W
(D) 117,600 W

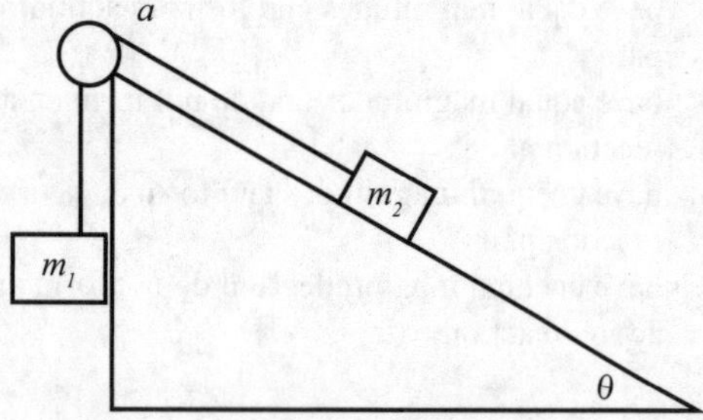

10. Consider the above configuration of two blocks of masses attached via a massless rope and pulley over a frictionless inclined plane. What is the acceleration of the blocks?

(A) $(m_2 - m_1)\, g/(m_1 + m_2)$
(B) $(m_2 \sin\theta - m_1)\, g/(m_1 + m_2)$
(C) $(m_2 \cos\theta - m_1)\, g/(m_1 + m_2)$
(D) g

11. A child sits on a merry-go-round with a radius of 1.5 m. Her father exerts a force 120 N tangent to the merry-go-round at its outer edge to start rotating it from rest. If the combined moment of inertia of the merry-go-round and child is 250 kg·m^2, how long does it take for the merry-go-round to reach an angular velocity of 5 rad/s?

(A) 0.14 s
(B) 6.94 s
(C) 10.4 s
(D) 36.0 s

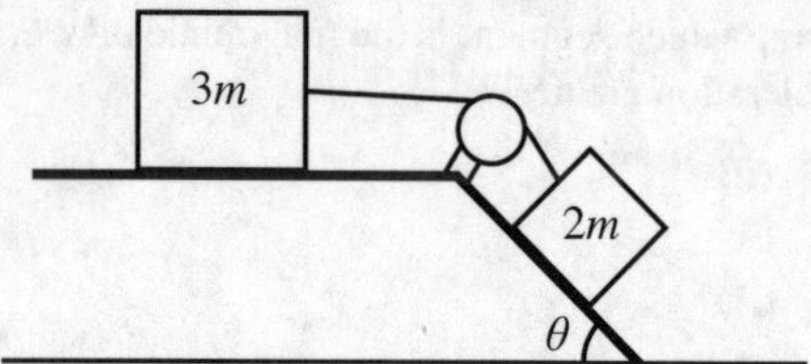

12. In the figure above, two blocks of mass $3m$ and $2m$ are attached together. The plane is frictionless and the pulley is frictionless and massless. The inclined portion of the plane creates an angle θ with the horizontal floor. What is the acceleration of the block $2m$ if both blocks are released from rest (gravity = g)?

(A) $2mg$
(B) $\left(\frac{2}{5}\right) g \sin\theta$
(C) $\left(\frac{2}{3}\right) g \sin\theta$
(D) $\left(\frac{3}{5}\right) g \sin\theta$

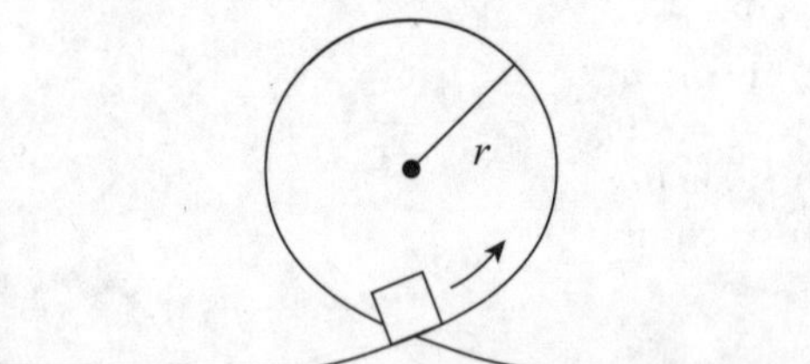

13. If a roller coaster cart of mass m was not attached to the track, it would still remain in contact with a track throughout a loop of radius r as long as

(A) $v \le \sqrt{rg}$
(B) $v \ge \sqrt{rg}$
(C) $v \le \sqrt{(rg/m)}$
(D) $v \ge \sqrt{(rg/m)}$

GO ON TO THE NEXT PAGE.

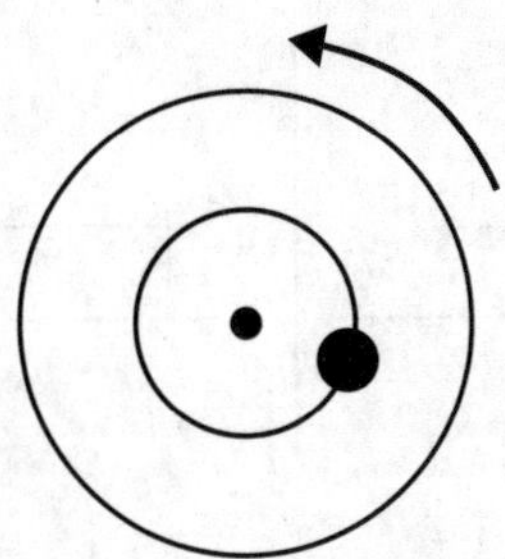

14. The diagram above shows a top view of an object of mass M on a circular platform of mass $2M$ that is rotating counterclockwise. Assume the platform rotates without friction. Which of the following best describes an action the object can take that will increase the angular speed of the entire system?

(A) The object moves toward the center of the platform, increasing the total angular momentum of the system.
(B) The object moves toward the center of the platform, decreasing the rotational inertia of the system.
(C) The object moves away from the center of the platform, increasing the total angular momentum of the system.
(D) The object moves away from the center of the platform, decreasing the rotational inertia of the system.

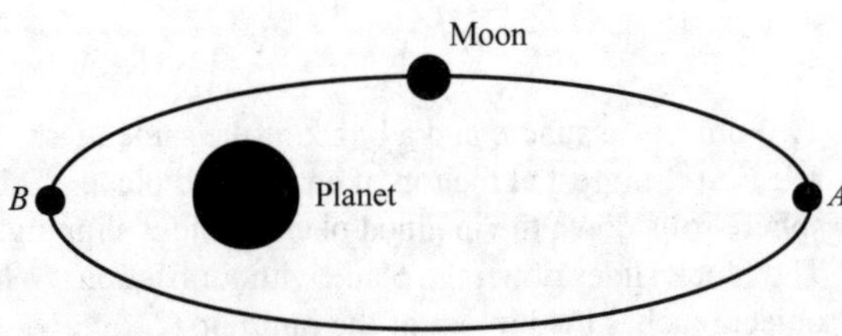

15. A moon has an elliptical orbit about the planet as shown above. The moon's mass is much smaller than the planet's. Which of the following quantities is/are (approximately) conserved as the moon moves from point A to point B?

I. The moon's mechanical energy

II. The moon's angular momentum

III. The moon's linear momentum

(A) I and II only
(B) I and III only
(C) II and III only
(D) I, II, and III

16. A sphere starts from rest atop a hill with a constant angle of inclination and is allowed to roll down the hill, without slipping. What force provides the torque that causes the sphere to rotate?

(A) Static friction
(B) Kinetic friction
(C) The normal force of the hill on the sphere
(D) Gravity

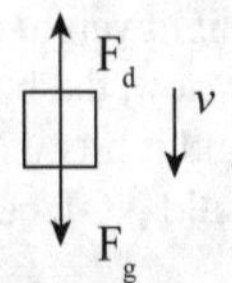

17. Which of the following correctly describes the motion of a real object in free fall? Assume that the object experiences drag force proportional to speed and that it strikes the ground before reaching terminal speed.

(A) It will fall with increasing speed and increasing acceleration.
(B) It will fall with increasing speed and decreasing acceleration.
(C) It will fall with decreasing speed and increasing acceleration.
(D) It will fall with decreasing speed and decreasing acceleration.

18. Which of the following concerning uniform circular motion is true?

(A) The centrifugal force is the action/reaction pair of the centripetal force.
(B) The centripetal acceleration and velocity point in the same direction.
(C) The velocity of the object in motion changes, whereas the acceleration of the object is constant.
(D) A satellite undergoing uniform circular motion is falling toward the center even though its path is circular.

GO ON TO THE NEXT PAGE.

19. A girl of mass m and a boy of mass $2m$ are sitting on opposite sides of a lightweight seesaw with its fulcrum in the center. Right now, the boy and girl are equally far from the fulcrum, and it tilts in favor of the boy. Which of the following would NOT be a possible method to balance the seesaw?
    (A) Move the boy to half his original distance from the fulcrum.
    (B) Move the girl to double her original distance from the fulcrum.
    (C) Allow a second girl of mass m to join the first.
    (D) Move the fulcrum to half its original distance from the boy.

20. A satellite of mass m is in orbit at an altitude of one-third of Earth's radius, R_e. How much work must its thrusters do to put it in an orbit of one-half R_e? (G is the gravitational constant, and M is Earth's mass.)
    (A) $\dfrac{GMm}{24R_e}$
    (B) $\dfrac{GMm}{12R_e}$
    (C) $\dfrac{GMm}{8R_e}$
    (D) $\dfrac{GMm}{2R_e}$

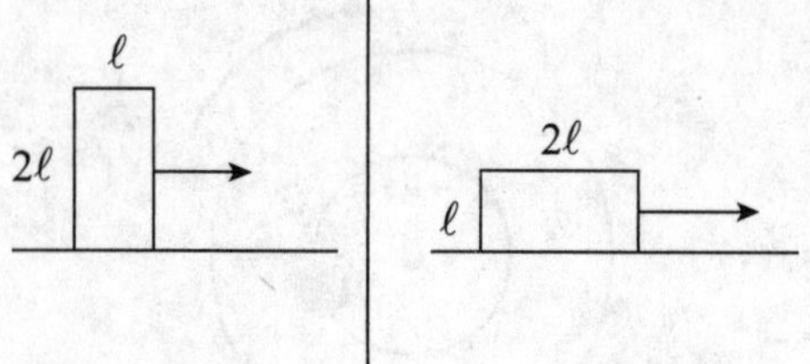

21. A wooden block experiences a frictional force, $\mathbf{F}_f$, as it slides across a table. If a block of the same material with half the height and twice the length was to slide across the table in the same direction, what would be the frictional force it experienced?
    (A) $(1/2)\mathbf{F}_f$
    (B) $\mathbf{F}_f$
    (C) $2\mathbf{F}_f$
    (D) $4\mathbf{F}_f$

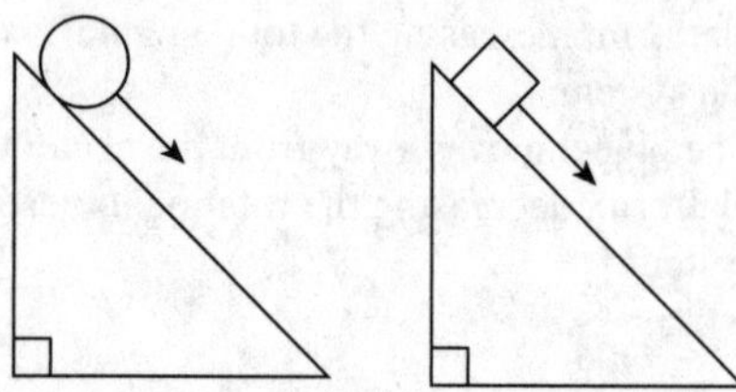

22. Two objects, a sphere and a block of the same mass, are released from rest at the top of an inclined plane. The sphere rolls down the inclined plane without slipping. The block slides down the plane without friction. Which object reaches the bottom of the ramp first?
    (A) The sphere, because it gains rotational kinetic energy, but the block does not
    (B) The sphere, because it gains mechanical energy due to the torque exerted on it, but the block does not
    (C) The block, because it does not lose mechanical energy due to friction, but the sphere does
    (D) The block, because it does not gain rotational kinetic energy, but the sphere does

GO ON TO THE NEXT PAGE.

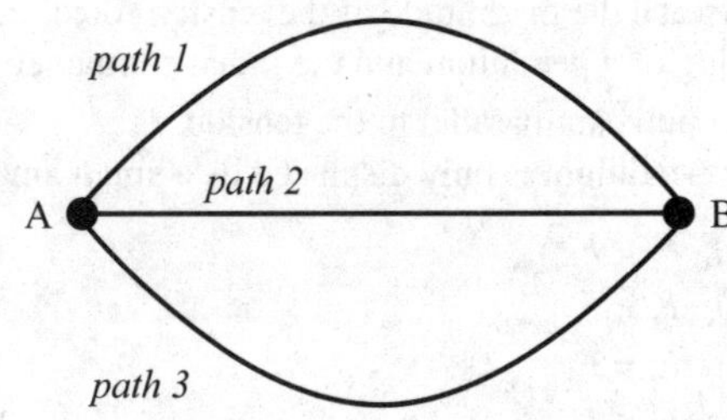

23. In the diagram above, each of the three frictionless tracks shown here contains an object of mass *m.* Each object starts at point A with the same initial horizontal velocity v_0, which in each case is sufficient to allow the object to reach the end of the track at point B. The masses remain in contact with the tracks throughout their motions, and the displacement A to B is the same in each case. Path 1 is directed up, path 2 horizontally, and path 3 down. The total length of paths1 and 3 are equal. If t_1, t_2, and t_3 are the total travel times between A and B for paths 1, 2, and 3, respectively, what is the relation among these times?
(A) $t_3 < t_2 < t_1$
(B) $t_2 < t_3 < t_1$
(C) $t_2 < t_1 = t_3$
(D) $t_2 = t_3 < t_1$

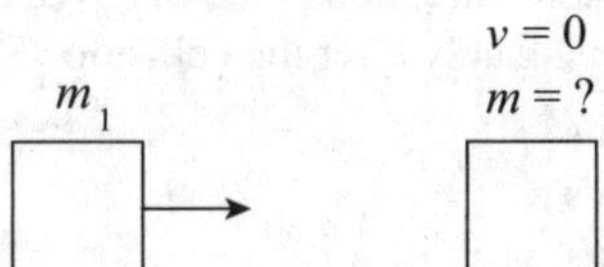

24. An object of mass m_1 experiences a linear, elastic collision with a stationary object of unknown mass. In addition to m_1, what is the minimum necessary information that would allow you to determine the mass of the second object?
(A) The final velocity of object 1
(B) The initial speed of object 1
(C) The final velocity of object 2
(D) Any 2 of the above values

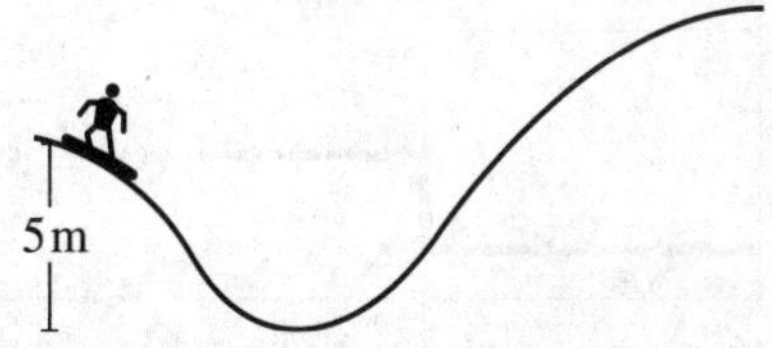

25. A 60-kg snowboarder is riding down a hill at 15 m/s when he realizes that he's about to hit a 5-m dip in the terrain with a hill behind it. Friction from the snow does –6000 J of work on him before he slows to a stop. How high up the next hill, relative to the bottom of the dip, does he get before he stops?
(A) 1.3 m
(B) 6.3 m
(C) 16.5 m
(D) 26.7 m

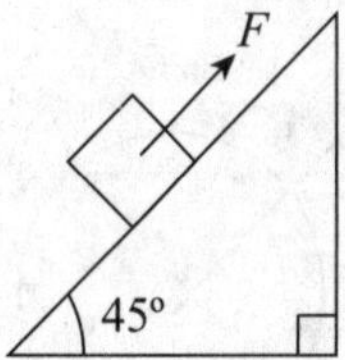

26. A box of mass *m* is sitting on an incline of 45° and it requires an applied force *F* up the incline to get the box to begin to move. What is the maximum coefficient of static friction?

(A) $\left(\dfrac{\sqrt{2}F}{mg}\right) - 1$
(B) $\left(\dfrac{\sqrt{2}F}{mg}\right)$
(C) $\left(\dfrac{\sqrt{2}F}{mg}\right) + 1$
(D) $\left(\dfrac{2F}{mg}\right) - 1$

GO ON TO THE NEXT PAGE.

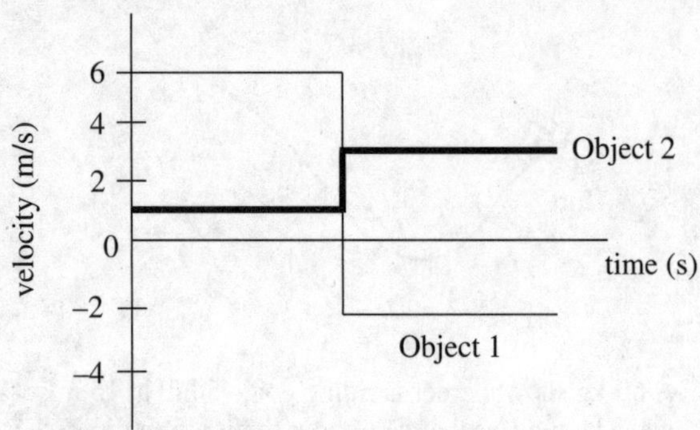

27. The graph above shows the velocities of two objects undergoing a head-on collision. Given that Object 2 has 4 times the mass of Object 1, which type of collision is it?

(A) Perfectly elastic
(B) Perfectly inelastic
(C) Inelastic
(D) Cannot be determined

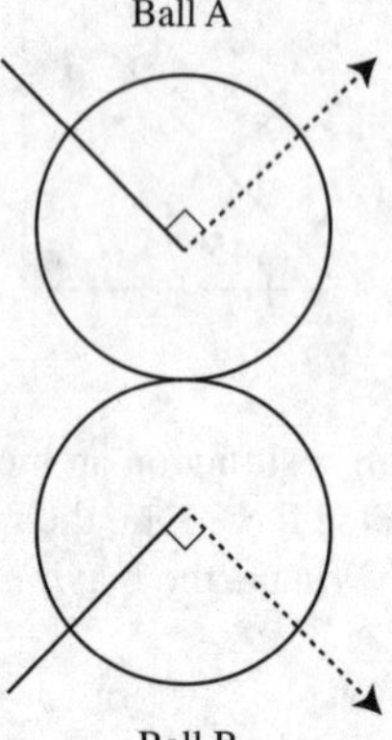

28. The picture above depicts the collision of two balls of equal mass. Which arrow best indicates the direction of the impulse on Ball A from Ball B during the collision?

(A) 

(B) 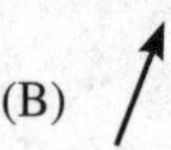

(C) 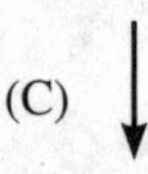

(D)

29. Which of the following best describes the relationship between the magnitude of the tension force, F_T, in the string of a pendulum and the radial component of gravity that pulls antiparallel to the tension, $F_{g,\ radial}$? Assume that the pendulum is only displaced by a small amount.

(A) $F_T > F_{g,\ radial}$
(B) $F_T \geq F_{g,\ radial}$
(C) $F_T = F_{g,\ radial}$
(D) $F_T \leq F_{g,\ radial}$

30. A ball bounces straight up off the ground with an initial speed v_0. When the ball is at two-thirds of the maximum height in its trajectory, what is its speed?

(A) $\frac{1}{9}v_0$
(B) $\frac{1}{3}v_0$
(C) $\frac{\sqrt{3}}{3}v_0$
(D) $\frac{\sqrt{6}}{3}v_0$

31. A 5.0-kg bowling ball traveling at 7.0 m/s strikes a 1.6-kg bowling pin. After the collision, the bowling ball continues in the same direction at 5.2 m/s. What is the speed of the pin immediately after the collision?

(A) 1.8 m/s
(B) 5.3 m/s
(C) 5.6 m/s
(D) 8.3 m/s

32. A young girl at a waterpark rides down a waterslide with a height of h and slides from rest to speed v at the bottom of the slide. Then, she rides down a slide with a height of $2h$. If there is negligible friction on the slides, how fast is she going at the bottom of the second slide?

(A) $\frac{\sqrt{2}}{2}v$
(B) $\sqrt{2}\,v$
(C) $2v$
(D) $4v$

GO ON TO THE NEXT PAGE.

33. A projectile at the peak of its flight has $\frac{1}{4}$ of its initial kinetic energy at launch. What was its launch angle?

Assume negligible air resistance.

(A) 30°
(B) 60°
(C) 76°
(D) 86°

34. The Gravitron is a carnival ride that looks like a large cylinder. People stand inside the cylinder against the wall as it begins to spin. Eventually, it is rotating fast enough that the floor can be removed without anyone falling. Given that the coefficient of friction between a person's clothing and the wall is μ, the tangential speed is v, and the radius of the ride is r, what is greatest mass that a person can be to safely go on this ride?

(A) $\mu v^2/(rg)$
(B) $r^2v^2/(\mu g)$
(C) $rg/(\mu v^2)$
(D) None of the above

35. The spring in the launcher of a pinball machine has a spring constant of 150 N/m. If the spring is compressed 3 cm, how fast can it launch a pinball with a mass of 0.08 kg?

(A) 1.30 m
(B) 1.69 m
(C) 1.84 m
(D) 13.0 m

36. A student is experimenting with a simple spring-block oscillator that has spring constant k and amplitude A. The block attached to the spring has a mass of M. If the student places a small block of mass m on top of the original block, which of the following is true?

(A) The small block is most likely to slide off when the original block is at maximum displacement from the equilibrium position but will not slide off as long as the coefficient of static friction between the blocks is greater than $kA/[(M + m)g]$.
(B) The small block is most likely to slide off when the original block is at the equilibrium position but will not slide off as long as the coefficient of static friction between the blocks is greater than $kA/[(M + m)g]$.
(C) The small block is most likely to slide off when the original block is at maximum displacement from the equilibrium position but will not slide off as long as the coefficient of static friction between the blocks is greater than $(M + m)g/(kA)$.
(D) The small block is most likely to slide off when the original block is at the equilibrium position but will not slide off as long as the coefficient of static friction between the blocks is greater than $(M + m)g/(kA)$.

37. A horse accelerates at a constant rate from rest to 15 m/s in 2.4 s. How far does the horse travel in that time?

(A) 6.3 m
(B) 18 m
(C) 36 m
(D) 43 m

38. A 62,000-kg maglev train traveling at 110 m/s brakes over a distance of 2,800 m on a horizontal track. What magnitude of force do its brakes exert?

(A) 122,000 N
(B) 134,000 N
(C) 244,000 N
(D) 268,000 N

GO ON TO THE NEXT PAGE.

Questions 39–41 refer to the following scenario:

In a theater, a winch is used to lower a 75-kg performer in a harness down onto the stage below. The cable is wound around the drum of the winch that has a diameter of 15 cm.

39. What is the torque applied by the winch to lower the performer at a constant speed?
 (A) 0 N·m
 (B) 5.6 N·m
 (C) 55 N·m
 (D) 110 N·m

40. To lower the performer 12 m, how many revolutions does the drum rotate through?
 (A) 12.7 revolutions
 (B) 25.5 revolutions
 (C) 80 revolutions
 (D) 160 revolutions

41. During the last portion of the performer's descent, the drum undergoes 2.5 revolutions as the performer slows to a stop from a speed of 0.8 m/s. What is the magnitude of the angular acceleration of the drum during this period?
 (A) 3.6 rad/s^2
 (B) 5.7 rad/s^2
 (C) 14 rad/s^2
 (D) 23 rad/s^2

42. How much power does the engine of a 1300-kg car need to provide to accelerate the car from 0 to 25 m/s in 7 s at constant rate?
 (A) 46,000 W
 (B) 58,000 W
 (C) 116,000 W
 (D) 406,000 W

43. A box of mass m is pulled a distance d across a horizontal surface with a force of mg at an angle of 30° above the horizontal. If the coefficient of kinetic friction between the box and the surface is $\mu_k = \frac{1}{3}$, what is the total work done on the box?
 (A) $\frac{mgd}{6}\left(\sqrt{3}+1\right)$
 (B) $\frac{mgd}{6}\left(3\sqrt{3}-2\right)$
 (C) $\frac{mgd}{6}\left(3\sqrt{3}-1\right)$
 (D) $\frac{mgd}{6}\left(3\sqrt{3}+5\right)$

44. A tennis player hits a 60-g tennis ball that is approaching her at 32 m/s. Her racket is in contact with the ball for 5 ms, after which the ball leaves her racket traveling opposite to its initial direction at 36 m/s. If the ball's initial velocity was in the positive direction, what is the average force exerted on the ball by her racket?
 (A) –816 N
 (B) –48 N
 (C) 48 N
 (D) 816 N

45. An 85-g apple drops 9 m to the ground below. It hits the ground with a speed of 13 m/s. How much work is done by air resistance on the apple during its descent?
 (A) –0.315 J
 (B) –0.282 J
 (C) 0.282 J
 (D) 0.315 J

GO ON TO THE NEXT PAGE.

Directions: For each of the questions 46–50, two of the suggested answers will be correct. Select the two answers that are best in each case, and then fill in both of the corresponding circles on the answer sheet.

46. An object traveling at x m/s can stop at a distance d m with a maximum negative acceleration. If the car is traveling at $2x$ m/s, which of the following statements are true? Select two answers.
    (A) The stopping time is doubled.
    (B) The stopping time is quadrupled.
    (C) The stopping distance is doubled.
    (D) The stopping distance is quadrupled.

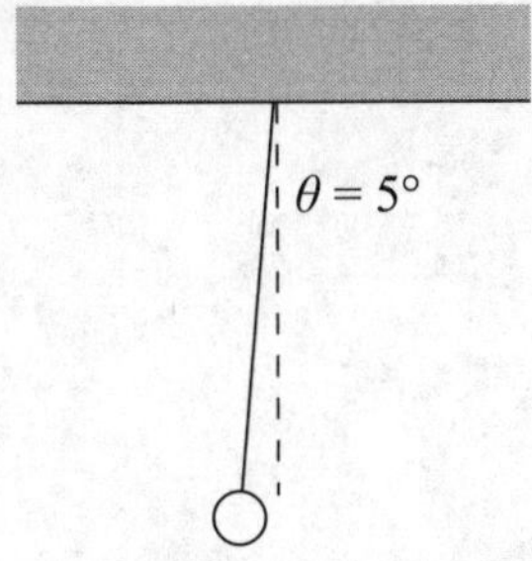

47. A ball with a 2 kg mass is attached to a massless 0.5 m string and becomes a simple pendulum when the string is pulled to an angle of $\theta = 5°$, allowing it to oscillate. Which of the following will change the period of the pendulum? Select two answers.
    (A) Replacing the mass with a 1 kg mass
    (B) Changing the initial extension of the pendulum to a 10° angle
    (C) Replacing the string with a 0.25 m string
    (D) Moving the pendulum to the surface of the Moon

48. In a billiards game, the 1 and 2 balls are at rest and in contact with one another. The cue ball then strikes the 1 ball in line with the 1 and 2 balls, resulting in the cue ball coming to rest, the 1 ball remaining motionless, and the 2 ball being shot off the end. Which of the following statements is true? Select two answers.
    (A) The cue ball exerts a force on the 2 ball.
    (B) The reaction force to the cue ball striking the 1 ball causes the motion of the 2 ball.
    (C) The force that acts on the 2 ball is equal in magnitude to the force the cue ball exerts on the 1 ball.
    (D) The reaction force to the cue ball striking the 1 ball stops the cue ball.

49. A warehouse worker uses a forklift to move a crate from the ground to an elevated shelf. Which of the following must be true of this process? Select two answers.
    (A) The mechanical energy of the crate is conserved.
    (B) Gravity does positive work on the crate.
    (C) The forklift transfers mechanical energy to the crate.
    (D) The total work done on the crate is zero.

50. A paintball hits a wall and sticks to it. Which of the following would decrease the impulse delivered to the wall by the paintball? Select two answers.
    (A) Adding padding to the wall
    (B) Reducing the speed of the paintball
    (C) Changing the material of the paintball so that it takes longer to deform and break open
    (D) Using paintballs that are the same size but less dense

GO ON TO THE NEXT PAGE.

AP PHYSICS 1

SECTION II

Free-Response Questions

Time—90 minutes

Percent of total grade—50

General Instructions

Use a separate piece of paper to answer these questions. Show your work. Be sure to write CLEARLY and LEGIBLY. If you make an error, you may save time by crossing it out rather than trying to erase it.

GO ON TO THE NEXT PAGE.

AP PHYSICS 1

SECTION II

Directions: Questions 1–5 here are as follows: one experimental design question (worth 12 points), one quantitative/qualitative translation question (worth 12 points), one paragraph argument short-answer question (worth 7 points), and two additional short-answer questions (worth 7 points each). You have a total of 90 minutes to complete this section. Show your work for each part in the space provided after that part.

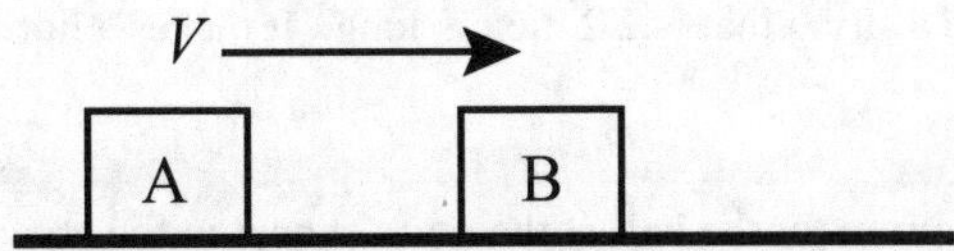

1. An experiment is conducted in which Block A with a mass of m_A is slid to the right across a frictionless table. Block A collides with Block B, which is initially at rest, of an unknown mass and sticks to it.

    (A) Describe an experimental procedure that determines the velocities of the blocks before and after a collision. Include all the additional equipment you need. You may include a labeled diagram of your setup to help in your description. Indicate what measurements you would take and how you would take them. Include enough detail so that the experiment could be repeated with the procedure you provide.

    (B) If Block A has a mass of 0.5 kg and starts off with a speed of 1.5 m/s and the experiment is repeated, the velocity of the blocks after they collide are recorded to be 0.25 m/s. What is the mass of Block B?

    (C) How much kinetic energy was lost in this collision from part (B)?

GO ON TO THE NEXT PAGE.

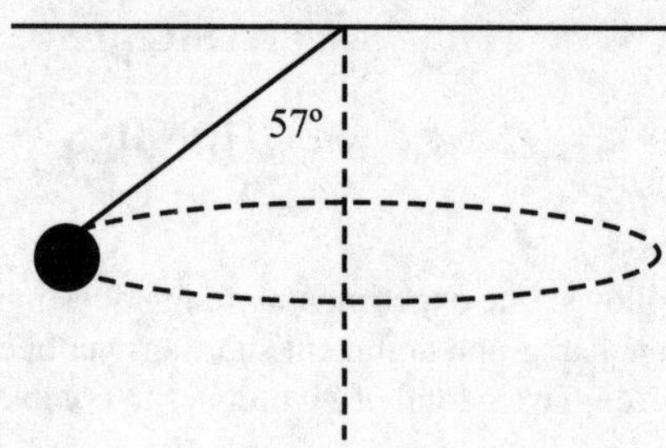

2. A conical pendulum is hanging from a string that is 2.2 meters long. It makes a horizontal circle. The mass of the ball at the end of the string is 0.5 kg.

(A) Below, make a free-body diagram for the ball at the point shown in the above illustration. Label each force with an appropriate letter.

(B) Write out Newton's Second Law in both the *x*- and *y*-direction in terms used in your free-body diagram.

(C) Calculate the centripetal acceleration from your free-body diagram.

(D) What is the radius of the circle that the ball is traveling in?

(E) What is the speed of the ball?

GO ON TO THE NEXT PAGE.

3. A fire-fighting airplane flies toward a forest fire at a constant velocity and elevation and drops water out of its tanks to douse the fire that is burning on flat terrain. Ignore the effects of air resistance.

(A) Explain why the pilot must open the water tanks before the plane is directly over the fire.

(B) It takes 8 s for all the water to drain out of the tanks. If the goal is to distribute the water over a distance of 600 m. How fast should the plane fly as it releases the water?

(C) The plane flies directly toward the fire at the speed determined in (B) and at an elevation of 150 m above the ground. At what horizontal distance from the fire should the pilot open his tanks?

GO ON TO THE NEXT PAGE.

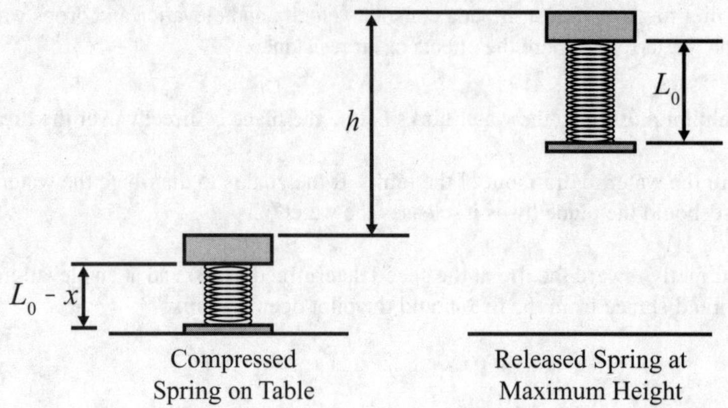

4. An experiment is designed to calculate the spring constant k of a vertical spring for a jumping toy. The toy is compressed a distance of x from its natural length of L_0, as shown on the left in the diagram, and then released. When the toy is released, the top of the toy reaches a height of h in comparison to its previous height and the spring reaches its maximum extension. The experiment is repeated multiple times and replaced with different masses m attached to the spring. The spring itself has negligible mass.

(A) Derive an expression for the height h in terms of m, x, k, and any other constants provided in the formula sheet.

(B) To standardize the experiment, the compressed distance x is set to 0.020 m. The following table shows the data for different values of m.

	***m* (kg)**	***h* (m)**
	0.020	0.49
	0.030	0.34
	0.040	0.28
	0.050	0.19
	0.060	0.18

i. What quantities should be graphed so that the slope of a best-fit straight line through the data points can help us calculate the spring constant k?

ii. Fill in the blank column in the table above with calculated values. Also include a header with units.

GO ON TO THE NEXT PAGE.

(C) On the axes below, plot the data and draw the best-fit straight line. Label the axes and indicate scale.

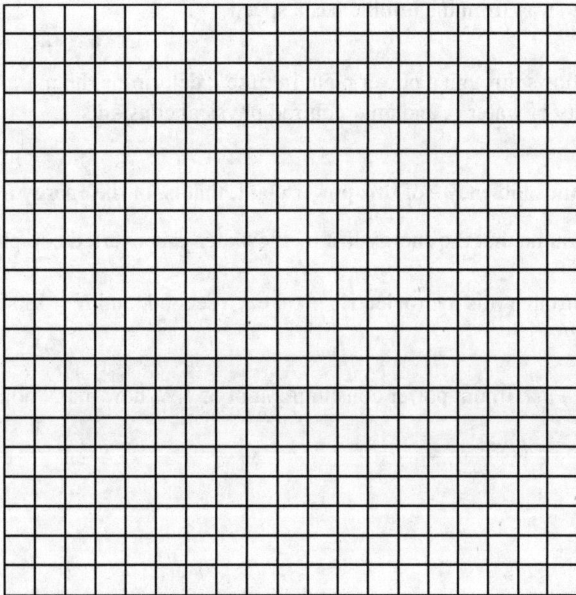

(D) Using your best-fit line, calculate the numerical value of the spring constant.

(E) Describe an experimental procedure that determines the height h in the experiment, given that the toy is only momentarily at that maximum height. You may include a labeled diagram of your setup to help in your description.

GO ON TO THE NEXT PAGE.

5. Hydroelectric plants use flowing water to generate electricity. Water that is approximately motionless in a storage reservoir flows down through a height H to have speed of v_{in} as it encounters a turbine. The flowing water turns the turbine, which generates electricity. Water flowing away from the turbine has a speed v_{out}.

(A) If a volume V of water flows through a power plant in time t, determine the maximum power the plant could produce in terms of H, the density of water ρ, and any required physical constants.

(B) If $v_{out} = \frac{1}{2}v_{in}$, what is the efficiency e of this power plant, that is, of the energy available, what fraction is converted to electrical energy? Assume that the energy lost by the water is converted completely to electrical energy.

(C) If 3 m^3 of water flows through this hydroelectric plant each second, and H = 180 m, how much electrical power is produced? (ρ = 1000 kg/m^3)

(D) If a home typically has a maximum power consumption of 55 kW, how many homes could his hydroelectric plant power?

STOP

END OF EXAM

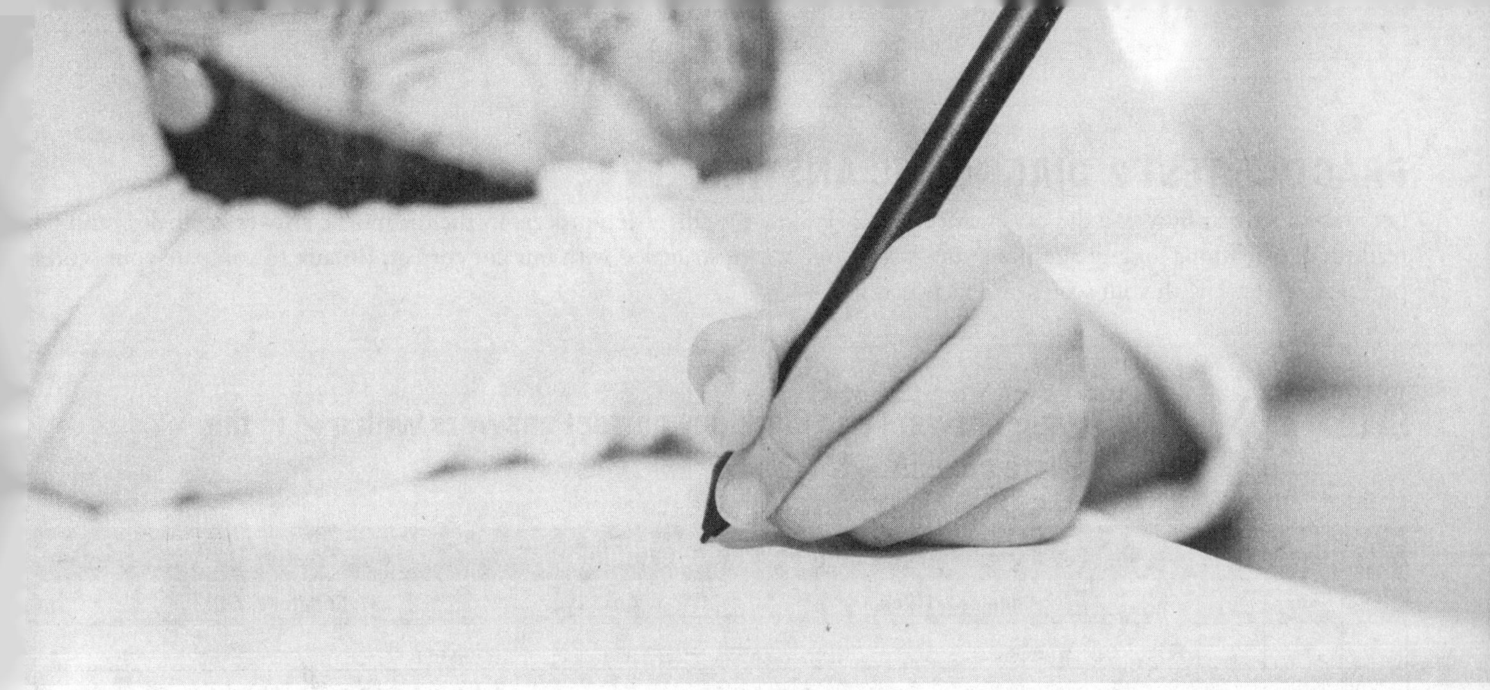

Practice Test 2: Diagnostic Answer Key and Explanations

PRACTICE TEST 2: DIAGNOSTIC ANSWER KEY

Let's take a look at how you did on Practice Test 2. Follow the three-step process in the diagnostic answer key below and go read the explanations for any questions you got wrong, or you struggled with but got correct. Be sure to compare your scores on Practice Test 1 with your scores on Practice Test 2.

Check your answers and mark any correct answers with a ✔ in the appropriate column.

Section I: Multiple Choice							
Q #	**Ans.**	**✔**	**Chapter #, Title**	**Q #**	**Ans.**	**✔**	**Chapter #, Title**
1	C		**4,** Projectile Motion	21	B		**5,** Friction
2	D		**4,** Momentum	22	D		**7,** Conservation of Mechanical Energy **10,** Rotational Kinetic Energy
3	C		**4,** Acceleration	23	A		**7,** Conservation of Mechanical Energy
4	C		**4,** Acceleration	24	D		**8,** Collisions
5	C		**4,** Projectile Motion	25	B		**7,** Conservation of Energy with Nonconservative Forces
6	B		**4,** Uniformly Accelerated Motion and the Big Five	26	A		**5,** Friction **5,** Inclined Planes
7	B		**5,** Newton's 1st Law **5,** Newton's 3rd Law	27	A		**8,** Collisions
8	C		**5,** Friction	28	D		**8,** Impulse
9	C		**7,** Power	29	B		**9,** Pendulums
10	B		**5,** Pulleys **5,** Inclined Planes	30	C		**7,** Conservation of Mechanical Energy
11	B		**10,** Rotational Motion	31	C		**8,** Collisions
12	B		**5,** Pulleys **5,** Inclined Planes	32	B		**7,** Conservation of Mechanical Energy
13	B		**6,** Uniform Circular Motion	33	B		**4,** Projectile Motion **7,** Kinetic Energy
14	B		**10,** Angular Momentum	34	D		**5,** Friction **6,** Uniform Circular Motion
15	A		**7,** Conservation of Mechanical Energy **8,** Momentum **10,** Angular Momentum	35	A		**7,** Conservation of Mechanical Energy **9,** Simple Harmonic Motion
16	A		**10,** Torque	36	A		**5,** Friction **9,** Simple Harmonic Motion
17	B		**5,** Newton's 2nd Law	37	B		**4,** Uniformly Accelerated Motion and the Big Five
18	D		**6,** Uniform Circular Motion	38	B		**7,** The Work-Energy Theorem
19	D		**10,** Equilibrium	39	C		**10,** Torque
20	A		**7,** Conservation of Mechanical Energy	40	B		**10,** Rotational Motion

Section I: Multiple Choice (Continued)							
Q #	Ans.	✔	Chapter #, Title	Q #	Ans.	✔	Chapter #, Title
41	A		**10,** Rotational Motion	46	A, D		**4,** Uniformly Accelerated Motion and the Big Five
42	B		**7,** The Work-Energy Theorem **7,** Power	47	C, D		**9,** Pendulums
43	C		**7,** Work	48	C, D		**5,** Newton's 3rd Law
44	A		**8,** Impulse	49	C, D		**7,** Energy
45	A		**7,** Conservation of Energy with Nonconservative Forces	50	B, D		**8,** Impulse

Section II: Free Response			
Q #	Ans.	✔	Chapter #, Chapter Title, Section Title
1(A)	See Explanation		**8,** Collisions
1(B)	See Explanation		**8,** Collisions
1(C)	See Explanation		**7,** Kinetic Energy
2(A)	See Explanation		**5,** Dynamics: An Overall Strategy
2(B)	See Explanation		**5,** Newton's 2nd Law
2(C)	See Explanation		**5,** Newton's 2nd Law **6,** Uniform Circular Motion
2(D)	See Explanation		**6,** Uniform Circular Motion
2(E)	See Explanation		**6,** Uniform Circular Motion
3(A)	See Explanation		**4,** Projectile Motion
3(B)	See Explanation		**4,** Projectile Motion
3(C)	See Explanation		**4,** Projectile Motion
4(A)	See Explanation		**7,** Conservation of Mechanical Energy **9,** Simple Harmonic Motion
4(B)	See Explanation		**9,** Simple Harmonic Motion
4(C)	See Explanation		**9,** Simple Harmonic Motion
4(D)	See Explanation		**9,** Simple Harmonic Motion
4(E)	See Explanation		**4,** Uniformly Accelerated Motion and the Big Five
5(A)	See Explanation		**7,** Potential Energy **7,** Power
5(B)	See Explanation		**7,** Conservation of Energy with Nonconservative Forces
5(C)	See Explanation		**7,** Power
5(D)	See Explanation		**7,** Power

Tally your correct answers from Step 1 by chapter. For each chapter, write the number of correct answers in the appropriate box. Then, divide your correct answers by the number of total questions (which we've provided) to get your percent correct.

CHAPTER 4 TEST SELF-EVALUATION

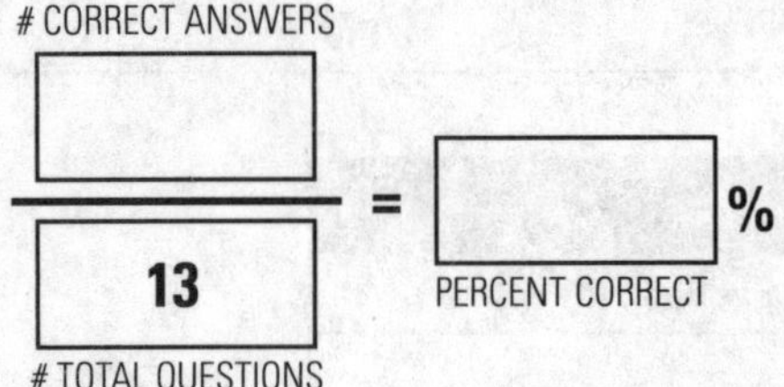

CHAPTER 5 TEST SELF-EVALUATION

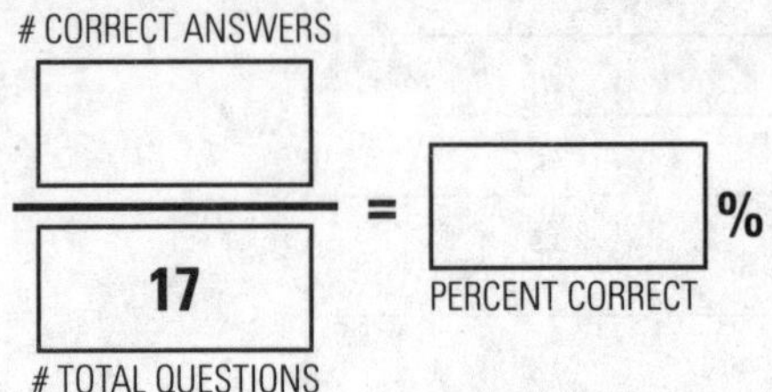

CHAPTER 6 TEST SELF-EVALUATION

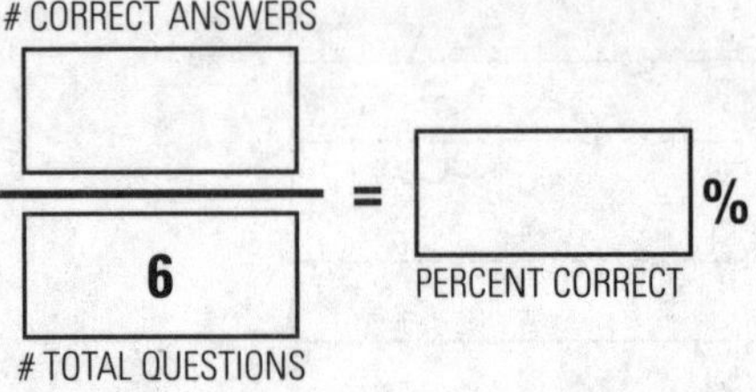

CHAPTER 7 TEST SELF-EVALUATION

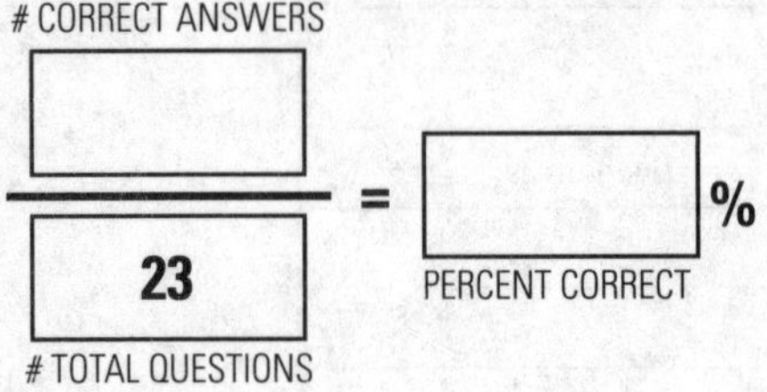

CHAPTER 8 TEST SELF-EVALUATION

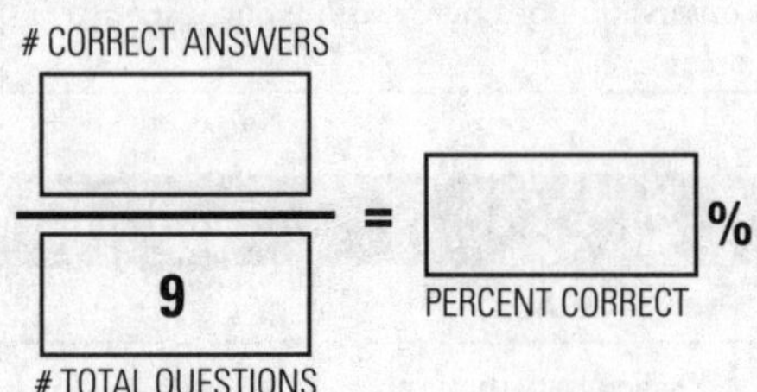

CHAPTER 9 TEST SELF-EVALUATION

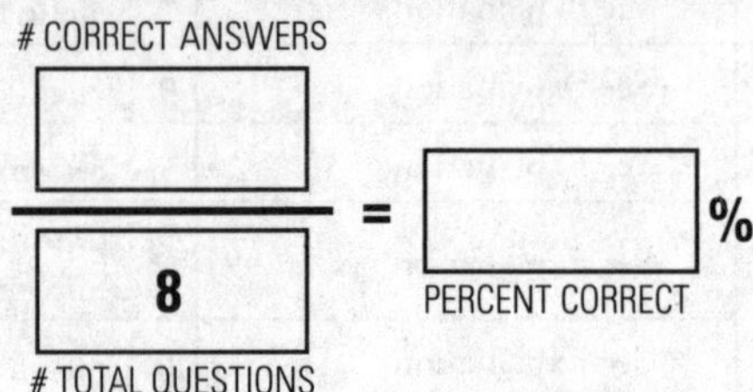

CHAPTER 10 TEST SELF-EVALUATION

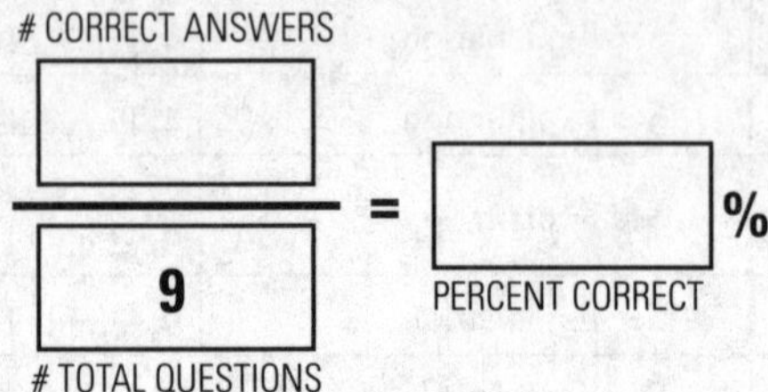

Use the results above to customize your study plan. You may want to start with, or give more attention to, the chapters with the lowest percents correct.

PRACTICE TEST 2: ANSWERS AND EXPLANATIONS

Section I: Multiple-Choice Questions

1. **C** Since acceleration due to gravity is 10 m/s^2, the vertical speed of the object after 2 seconds will be 20 m/s. Because the horizontal speed will not be affected, the direction will be mostly down but slightly to the right.

2. **D** Mass is not changing, so you need to consider only the changes in velocity. For each segment in order, the change in velocity is +7 m/s, 0 m/s, –4 m/s, –3 m/s, and –2 m/s. The question does not specify magnitude only, so all negatives will come before the others.

3. **C** Acceleration is the change in velocity divided by the change in time. You found the changes in velocity for each segment in the previous question, and the times for each segment are, in order, 3 s, 2 s, 1 s, 3 s, and 2 s. Divide the results in the previous question by these values, and the highest result (ignoring sign because the problem specifies magnitude only) is for segment CD.

4. **C** In a velocity-versus-time graph, displacement is the area beneath the curve. If the question had asked for displacement, then anything beneath the *x*-axis would be added as a negative displacement. However, distance cannot be negative, so there is no distinction between positive and negative velocity for this question.

   Splitting the area into triangles and rectangles like this should help:

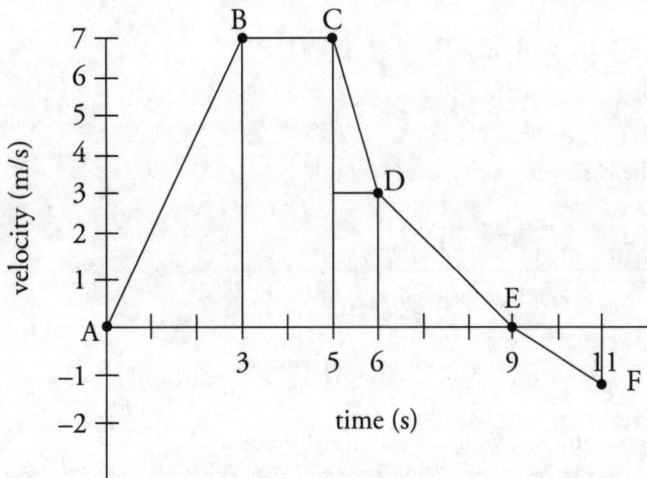

From t = 0 to 3 s, the area is $\frac{1}{2}$(3 s)(7 m/s) = 10.5 m. From t = 3 to 5 s, the area is (2 s)(7 m/s) = 14 m. From t = 5 to 6 s, the area is $\frac{1}{2}$(1 s)(7 m/s – 3 m/s) + (1 s)(3 m/s) = 5 m. From t = 6 to 7 s, the area is $\frac{1}{2}$(3 s)(3 m/s) = 4.5 m. From t = 9 to 11 s, the positive area is $\frac{1}{2}$(2 s)(1 m/s) = 1 m. Therefore, the total distance is 10.5 m + 14 m + 5 m + 4.5 m + 1 m = 35 m.

5. **C** At the highest point from the ground, the ball has a vertical velocity of zero. Therefore, applying the formula $v_y = v_{0y} + a_y t$ and rearranging it for t, it becomes $t = -v_{0y}/-g$. Substituting $v_{0y} = v \sin \theta$ into the equation, $t = v \sin \theta / g$.

6. **B** A trick for solving problems of this nature is to make up numbers for a and t. Pretend that total time = 10 seconds and acceleration = 2 m/s². Then make a table:

Time, s	Speed, m/s
0	0
1	2
9	18
10	20

So in the first second, v_{avg} = 1 m/s and d = 1 m. In the last second (the tenth), v_{avg} = 19 m/s and d = 19 m. The answer to the question is 19 m – 1 m = 18 m. Which answer choice gives you 18 m?

(A) $a(t + 1\text{ s})^2 = (2\text{ m/s}^2)(10\text{ s} + 1\text{ s})^2 = (2\text{ m/s}^2)(11\text{ s})^2 = (2\text{ m/s}^2)(121\text{ s}^2) = 242\text{ m}$

(B) $a(t - 1\text{ s})(1\text{ s}) = (2\text{ m/s}^2)(10\text{ s} - 1\text{ s})(1\text{ s}) = (2\text{ m/s}^2)(9\text{ s})(1\text{ s}) = 18\text{ m}$

(C) $at^2 = (2\text{ m/s}^2)(10\text{ s})^2 = (2\text{ m/s}^2)(100\text{ s}^2) = 200\text{ m}$

(D) $a(t + 1\text{ s})(1\text{ s}) = (2\text{ m/s}^2)(10\text{ s} + 1\text{ s})(1\text{s}) = (2\text{ m/s}^2)(11\text{ s})(1\text{ s}) = 22\text{ m}$

Therefore, (B) is the correct answer.

If you really want to do it algebraically, here's one way. Solve for the distance traveled in the first second.

$$d = v_0 \Delta t + \frac{1}{2} a \Delta t^2$$

$$= (0\text{ m/s})(1\text{ s}) + (1/2)(a)(1\text{ s})^2$$

$$= (1/2)(1\text{ s}^2)a$$

Next, solve for the distance in the final second.

$$d = v_0 \Delta t + \frac{1}{2} a \Delta t^2$$

$$= a(t - 1\text{ s})(1\text{ s}) + \frac{1}{2}(a)(1\text{ s})^2$$

$$= a(t - 1\text{ s})(1\text{ s}) + \frac{1}{2}(1\text{ s}^2)(a)$$

The difference between the two is $a(t - 1\text{ s})(1\text{ s})$ meters.

7. **B** Because the floor is horizontal, the weight and normal force balance each other out, so these two forces do have equal magnitudes. Eliminate (C) and (D). These two forces, however, do not form an action/reaction pair. In order to form an action/reaction pair, two forces must come from two objects acting on each other. One action/reaction pair in this situation is Earth's gravitational pull on the person (the person's weight) and the person's pull on Earth. The other pair is the normal

force of the floor on the person (pushing up) and the normal force of the person on the floor (pushing down).

8. **C** The force of friction on a stationary object will exactly oppose the force applied to it until it reaches a maximum. This eliminates (B) and (D). Once enough force has been applied, the friction will change from static friction to kinetic friction. Kinetic friction is a constant value, so this eliminates (A).

9. **C** Apply the definition of power: $P_{crane} = W_{crane} / t$. The two forces doing work on the beam are the force from the crane and gravity, so $W_{crane} = -W_g = \Delta U_g = mg\Delta h$. Therefore,

$$P_{crane} = \frac{mg\Delta h}{t} = \frac{(600 \text{ kg})(9.8 \text{ m/s}^2)(20 \text{ m})}{4 \text{ s}} = 29{,}400 \text{ W}$$

10. **B** Use Process of Elimination: you need sine in the equation for the downward pull of the second mass along the incline of the plane, so you can conclude that the answer is (B). Apply Newton's Second Law to the system of the two masses connected by the rope. Choose clockwise rotation of the pulley as the positive direction. Therefore, the forces acting on the system are $m_2 g \sin\theta$ in the positive direction and $m_1 g$ in the negative direction. So, according to Newton's Second Law,

$$\text{a} = \frac{\Sigma \text{F}}{m_{system}} = \frac{m_2 g \sin\theta - m_1 g}{m_1 + m_2} = \frac{(m_2 \sin\theta - m_1)g}{m_1 + m_2}$$

The choice of identifying the direction of the pulley as positive is arbitrary.

11. **B** The father's force exerts a torque on the merry-go-round, $\tau = r_{\perp}F = (1.5 \text{ m})(120 \text{ N}) = 180 \text{ N} \cdot \text{m}$. This torque causes an angular acceleration on the merry-go-round, $\alpha = \frac{\tau}{I} = \frac{180 \text{ N} \cdot \text{m}}{250 \text{ kg} \cdot \text{m}^2} = 0.72 \text{ rad/s}^2$. Apply rotational kinematics to solve for the time:

$$\omega = \omega_0 + \alpha t \rightarrow t = \frac{\omega}{\alpha} = \frac{5 \text{ rad/s}}{0.72 \text{ rad/s}^2} = 6.94 \text{ s}$$

12. **B** Because all the masses are attached and moving as a single unit, the acceleration of any block in the system is the same as any other. The net force on the blocks is $F_{net} = (2M)g \sin\theta$. Therefore,

$$F_{net} = ma$$

$$(2M)g\sin\theta = (5M)a$$

$$a = \left(\frac{2}{5}\right)g\sin\theta$$

13. **B** At the top of the loop, the cart will remain in contact with the track as long as the centripetal force is at least as much as the force of gravity on the cart.

$$F_c \geq F_g$$
$$\frac{mv^2}{r} \geq mg$$
$$\frac{v^2}{r} \geq g$$
$$v \geq \sqrt{rg}$$

14. **B** The rotating platform and the object resting on it constitute an isolated system. Angular momentum must remain constant, so eliminate (A) and (C). To decide between the remaining choices, remember that a system's rotational inertia increases as the objects in that system get farther from the axis of rotation. For a point particle, $I = mr^2$. To decrease rotational inertia, move the object toward the center of the platform.

15. **A** If the moon and planet are treated as a single system, then the system's total mechanical energy (kinetic energy + gravitational potential energy), linear momentum, and angular momentum are all conserved. However, that's not necessarily true for each object individually. Let's start with momentum. The moon is moving toward the top of the page when traveling to A but toward the bottom of the page when traveling to B. Clearly, the moon's velocity changes a lot, so its momentum changes as well. Statement III is false. Eliminate (B), (C), and (D). The answer must be (A). (Any change in the moon's momentum is coupled with an opposite change in the planet's momentum, which is how total momentum stays constant. In this situation, the lighter object contains nearly all of the system's kinetic energy and angular momentum; the planet's K and L are negligibly small. So not only are the system's energy and angular momentum conserved, so is the planet's.)

16. **A** Gravity points straight down and does affect the object, but it does not provide the torque. Eliminate (D). The normal force points perpendicular to the plane and does no work on the object, nor does it provide the torque. Eliminate (C). In this case, friction provides the torque. Since the ball is rolling down the hill and not sliding down it, it is not kinetic friction. Eliminate (B). This is similar to when you drive a car and the tires grip the road; without static friction, the tires would not grip and you would not be able to go anywhere.

17. **B** An object's terminal speed is the highest speed it can reach in free fall. If it has not yet reached that speed, its speed must still be increasing. This eliminates (C) and (D). The problem says that the object will experience a drag force that increases as speed increases. Therefore, the opposing force will increase over time, which means the net force will decrease (since the force from gravity will be constant). If net force decreases, so will acceleration.

18. **D** Centripetal force is a name given to the net force of an object undergoing uniform circular motion. Therefore, it is not a separate force and does not have an action/reaction pair. This eliminates (A). Since the speed is constant but direction changes, acceleration must be perpendicular to velocity. Eliminate (B). Not only does velocity change with time, but the acceleration also changes because the direction of the centripetal acceleration always points to the center of the circle. This eliminates (C). A satellite undergoing uniform circular motion is in fact *falling* (accelerating due to gravity) toward the center but doesn't *get closer* due to its tangential velocity. Its velocity is always tangent to its circular path.

19. **D** To balance the seesaw, you need to balance the torques. Since $\tau = Fr \sin \theta$, the boy currently provides double the torque. Choices (B) and (C) would double the torque on the girl's side, and (A) would cut the boy's torque in half. Choice (D) would cut the boy's torque in half, but it would also increase the girl's torque, creating a new imbalance.

20. **A** The work done by the thrusters must change the energy of the system to put the satellite in the new orbit. First, determine the kinetic energy of each orbit by equating the centripetal force acting on the satellite to the force of Earth's gravitational pull,

$$F_c = F_g \rightarrow m\frac{v^2}{r} = G\frac{Mm}{r^2} \rightarrow \frac{1}{2}mv^2 = \frac{GMm}{2r} \rightarrow K = \frac{GMm}{2r}$$

Then, apply the energy equation, $K_1 + U_1 + W_{thrusters} = K_2 + U_2$, with $r_1 = \frac{4}{3}R_e$ and $r_2 = \frac{3}{2}R_e$, as r is defined as the distance from the center of the Earth:

$$\frac{GMm}{2r_1} - \frac{GMm}{r_1} + W_{thrusters} = \frac{GMm}{2r_2} - \frac{GMm}{r_2}$$

$$GMm\left(-\frac{1}{2r_1}\right) + W_{thrusters} = GMm\left(-\frac{1}{2r_2}\right)$$

$$W_{thrusters} = GMm\left(\frac{1}{2r_1} - \frac{1}{2r_2}\right) = GMm\left(\frac{1}{2\left(\frac{4}{3}R_e\right)} - \frac{1}{2\left(\frac{3}{2}R_e\right)}\right) = \frac{GMm}{R_e}\left(\frac{3}{8} - \frac{1}{3}\right) = \frac{GMm}{24R_e}$$

21. **B** The same two materials would be in contact, so the coefficient of friction would remain the same. The object's mass would also be the same, so the normal force would remain constant. Since $F_f = \mu F_N$, and neither of those terms changed, the magnitude of the frictional force will also be unchanged.

22. **D** Both objects begin with no kinetic energy, and they undergo the same decrease in potential energy. Therefore, they finish with the same *total* kinetic energy. The block doesn't have any rotational K, so all of its K is translational. But the sphere is rotating, so some of its K is translational, and some is rotational:

$$K_{\text{total, block}} = K_{\text{total, sphere}}$$

$$K_{\text{trans, block}} + 0 = K_{\text{trans, sphere}} + K_{\text{rot, sphere}}$$

Thus, once they start moving, the block has more K_{trans} than the sphere. Its center moves faster, so the block finishes first.

23. **A** The paths all have the same horizontal displacement, x. So focus on the horizontal component of velocity, v_x; don't worry about total speed or Conservation of Energy. Instead, look at the normal force. On path 2, the normal force points straight up, with no x-component. Thus, v_x is constant the whole time. On path 1 (up first), the normal force points up and to the left at the beginning, giving it a component in the negative x direction. Thus, v_x decreases at the beginning, and then increases back to v_{0x}. On path 3 (down first), the normal force points down and to the right at the beginning, giving it a component in the positive x direction. Thus, v_x increases at the beginning, and then decreases back to v_{0x}. In short, path 3 not only has a higher average *speed* than path 2, but path 3 also has a higher average *forward velocity* than path 2. That's why $t_3 < t_2$.

24. **D** For an elastic collision, both kinetic energy and momentum will be conserved. Writing these statements as equations gives

$$\frac{1}{2}m_1v_{1,0}^2 + \frac{1}{2}m_2v_{2,0}^2 = \frac{1}{2}m_1v_{1,f}^2 + \frac{1}{2}m_2v_{2,f}^2$$

$$m_1v_{1,0} + m_2v_{2,0} = m_1v_{1,f} + m_2v_{2,f}$$

You're told m_1 is a given in the question, and $v_{2,0}$ is also known to be 0 m/s since the object started at rest. This leaves m_2, $v_{1,0}$, $v_{1,f}$, and $v_{2,f}$ as unknowns. With two equations, you can have two unknowns. Therefore, (D) is correct.

25. **B** Friction from the snow is a nonconservative force that removes mechanical energy from the snowboarder.

$$K_i + U_i + W_{\text{friction}} = K_f + U_f$$

Let the bottom of the dip be where $h = 0$.

$$\frac{1}{2}mv_i^2 + mgh_i + W_{\text{friction}} = 0 + mgh_f$$

$$\frac{1}{2}(60\text{ kg})(15\text{ m/s})^2 + (60\text{ kg})(10\text{ m/s}^2)(5\text{ m}) - 6000\text{ J} = (60\text{ kg})(10\text{ m/s}^2)h_f \rightarrow h_f = 6.3\text{ m}$$

26. **A** The force of static friction and the force of gravity are acting down the incline in this situation. When the box just begins to move upward, the forces in both directions are equal and the force of static friction is at its maximum. Therefore, you have the equation

$$F = \mu_s mg \cos 45° + mg \sin 45° \longrightarrow \mu_s = \left(\frac{\sqrt{2}\,F}{mg}\right) - 1$$

27. **A** First, perfectly inelastic can be immediately eliminated because the objects have different velocities after the collision. Next, recall that a collision is perfectly elastic if kinetic energy is conserved.

$$K_0 = K_f$$

$$\frac{1}{2}m_1 v_{1,0}^2 + \frac{1}{2}m_2 v_{2,0}^2 = \frac{1}{2}m_1 v_{1,f}^2 + \frac{1}{2}m_2 v_{2,f}^2$$

$$\frac{1}{2}m_1(6\text{ m/s})^2 + \frac{1}{2}(4\,m_1)(1\text{ m/s})^2 = \frac{1}{2}m_1(2\text{ m/s})^2 + \frac{1}{2}(4\,m_1)(3\text{ m/s})^2$$

$$20\,m_1\text{ J} = 20\,m_1\text{ J}$$

Because kinetic energy was conserved, the collision must be perfectly elastic.

28. **D** In the picture, neither ball experiences any change in its horizontal motion. Ball A is originally moving down, but moves up after the collision. Therefore, the impulse it received is entirely up.

29. **B** Think of the pendulum as undergoing circular motion. The centripetal force would be equal to the difference between tension and the radial component of gravity.

$$F_c = F_t - F_{g,\text{ radial}}$$

$$\frac{mv^2}{r} = F_t - F_{g,\text{ radial}}$$

So long as the pendulum is in motion (as long as $v > 0$), tension must have a larger magnitude than the radial component of gravity. However, when $v = 0$ (at the maximum displacement), the two will be equal. Therefore, (B) is the correct answer.

30. **C** Apply Conservation of Energy to derive an expression for the maximum height H, where the instantaneous velocity of the ball is momentarily zero:

$$K_1 + U_1 = K_2 + U_2 \rightarrow \frac{1}{2}mv_0^2 + 0 = 0 + mgH \rightarrow H = \frac{v_0^2}{2g}$$

Apply Conservation of Energy again to determine the ball's speed v at $\frac{2}{3}H$:

$$K_1 + U_1 = K_2 + U_2$$

$$\frac{1}{2}mv_0^2 + 0 = \frac{1}{2}mv^2 + mg\left(\frac{2}{3}H\right)$$

$$\frac{1}{2}mv_0^2 + 0 = \frac{1}{2}mv^2 + \frac{2}{3}mg\left(\frac{v_0^2}{2g}\right)$$

$$v_0^2 = v^2 + \frac{2}{3}v_0^2$$

$$v^2 = \frac{1}{3}v_0^2$$

$$v = \frac{\sqrt{3}}{3}v_0$$

31. **C** Apply Conservation of Momentum: $m_b\mathbf{v}_b + m_p\mathbf{v}_p = m_b\mathbf{v}_b' + m_p\mathbf{v}_p'$. Define the positive direction as the original direction of the bowling ball, and substitute the given values to solve for v_p'.

$$(5.0\text{ kg})(7.0\text{ m/s}) + 0 = (5.0\text{ kg})(5.2\text{ m/s}) + (1.6\text{ kg})v_p'$$

$$v_p' = 5.6\text{ m/s}$$

32. **B** There is no friction acting on the waterslides, so apply the Conservation of Energy. Define $h = 0$ at the bottom of the slides, so that the initial potential energy is completely converted to kinetic energy at the bottom.

$$mgh_i = \frac{1}{2}mv_f^2 \rightarrow gh_i = \frac{1}{2}v_f^2 \rightarrow v_f = \sqrt{2gh_i}$$

For the waterslide of height h,

$$v = \sqrt{2gh}$$

For the waterslide of height $2h$, the girl's final speed, v', is

$$v' = \sqrt{2g(2h)} = \sqrt{2}\sqrt{2gh} = \sqrt{2}v$$

33. **B** At the top of its trajectory, the y-component of the projectile's velocity is 0, and the x-component is the same as it was at launch, $v_0 \cos\theta$. If the kinetic energy at the peak is $\frac{1}{4}$ of the initial kinetic energy, then

$$\frac{1}{2}m(v_0 \cos\theta)^2 = \frac{1}{4}\left(\frac{1}{2}mv_0^2\right) \rightarrow v_0^2 \cos^2\theta = \frac{1}{4}v_0^2 \rightarrow \cos^2\theta = \frac{1}{4} \rightarrow \cos\theta = \frac{1}{2} \rightarrow \theta = \cos^{-1}\frac{1}{2} = 60°$$

34. **D** First, draw a free-body diagram:

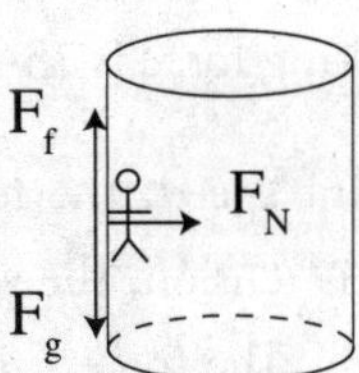

In order to stay in place, the force of friction needs to have a magnitude equal to the force of gravity.

$$F_f = F_g$$
$$\mu F_N = mg$$

Next, solve for the normal force. This is possible because the person is undergoing uniform circular motion, so set the normal force equal to centripetal force.

$$F_c = F_N$$
$$\frac{mv^2}{r} = F_N$$

Finally, substitute this value into the previous result.

$$\mu\left(\frac{mv^2}{r}\right) = mg$$

Notice that the mass actually drops out of the equation. The only variables that matter are the coefficient of friction, speed, and the radius.

35. **A** The compressed spring of the launcher stores elastic potential energy, which it transfers to the pinball as potential energy.

$$U_s = K_{ball} \rightarrow \frac{1}{2}kx^2 = \frac{1}{2}mv^2 \rightarrow v^2 = \frac{kx^2}{m} \rightarrow v = \sqrt{\frac{kx^2}{m}} = \sqrt{\frac{(150 \text{ N/m})(0.03 \text{ cm})^2}{0.08 \text{ kg}}} = 1.30 \text{ m}$$

36. **A** First, in order for the small block to slide off, there would have to be a big force affecting the larger block. Hooke's Law ($F = -kx$) states that this will happen at maximum displacement from the equilibrium position. Eliminate (B) and (D).

Second, the equation given in (C) indicates that the heavier the top block is, the more likely it is to slide off. But this is obviously not true, so (A) must be correct.

37. **B** Since the acceleration is not known, apply the kinematic equation that does not include acceleration, $\Delta x = \frac{1}{2}(v_0 + v)t = \frac{1}{2}(0 + 15 \text{ m/s})(2.4 \text{ s}) = 18 \text{ m}$.

38. **B** The train travels on a horizontal track, so the only force doing work on the train is the force from the brakes. Applying the Work–Energy Theorem,

$$W = \Delta K \rightarrow Fd = 0 - \frac{1}{2}mv_0^2 \rightarrow F = -\frac{mv_0^2}{2d} = -\frac{(62{,}000 \text{ kg})(110 \text{ m/s})^2}{2(2{,}800 \text{ m})} = -134{,}000 \text{ N}$$

Therefore, the magnitude of the braking force is 134,000 N.

39. **C** To lower the performer at a constant speed, the tension in the cable must equal the weight of the performer, $T = mg$. To create this tension, the winch applies a torque $\tau = r_{\perp}F$ in the direction opposite to the rotation of the drum. The force F is equal in magnitude to the tension, and the lever arm is the radius of the drum as F is tangential to the drum. Therefore,

$$\tau = \left(\frac{1}{2}d\right)T = \frac{1}{2}dmg = \frac{1}{2}(0.15 \text{ m})(75 \text{ kg})(9.8 \text{ m/s}^2) = 55 \text{ N}\cdot\text{m}$$

40. **B** Relate angular displacement $\Delta\theta$ to the distance Δs on the surface of the drum where the cable is wound, remembering that there are 2π radians in a revolution.

$$\Delta\theta = \frac{\Delta s}{r} = \frac{12 \text{ m}}{\frac{1}{2}(0.15 \text{ m})} = 160 \text{ rad} \times \frac{1 \text{ rev}}{2\pi \text{ rad}} = 25.5 \text{ rev}$$

41. **A** The initial angular speed of drum is $\omega_0 = \frac{v_0}{r} = \frac{2v_0}{d}$. Apply the equation of rotational kinematics without time:

$$\omega^2 = \omega_0^2 + 2\alpha\Delta\theta \rightarrow 0 = \left(\frac{2v_0}{d}\right)^2 + 2\alpha\Delta\theta$$

$$|\alpha| = \frac{2v_0^2}{d^2\Delta\theta} = \frac{2(0.8 \text{ m/s})^2}{(0.15 \text{ m})^2\left(2.5 \text{ rev} \times \frac{2\pi \text{ rad}}{1 \text{ rev}}\right)} = 3.6 \text{ rad/s}^2$$

42. **B** The power provided by the engine is the work it does on the car per unit time. By the Work–Energy Theorem, the work done on the car is its change in kinetic energy.

$$P = \frac{W}{\Delta t} = \frac{\Delta K}{\Delta t} = \frac{\frac{1}{2}mv^2 - 0}{\Delta t} = \frac{\frac{1}{2}(1300 \text{ kg})(25 \text{ m/s})^2}{7 \text{ s}} = 58{,}000 \text{ W}$$

43. **C** The total work done on the box is the sum of the work done by each force acting on the box: the applied force, gravity, and friction. The friction force depends on the normal force. Apply Newton's Second Law in the y-direction to solve for the normal force:

$$\Sigma F_y = F_n + F_{app,y} - mg = 0$$

$$F_n = mg - F_{app,y} = mg - mg\sin 30^\circ = mg\left(1 - \frac{1}{2}\right) = \frac{mg}{2}$$

Then, apply the equation for work done by a force $W = Fd\cos\theta$ to solve for the total work:

$$W_{total} = W_{app} + W_{gravity} + W_{friction}$$

$$W_{total} = F_{app}d\cos 30^\circ + mgd\cos 90^\circ + \mu_k F_n d\cos 180^\circ = mgd\left(\frac{\sqrt{3}}{2}\right) + 0 + \frac{1}{3}\left(\frac{mg}{2}\right)d(-1)$$

$$= mgd\left(\frac{\sqrt{3}}{2} - \frac{1}{6}\right) = \frac{mgd}{6}\left(3\sqrt{3} - 1\right)$$

44. **A** The racket exerts a force on the ball over a time that changes the ball's momentum.

$$\Delta\vec{p} = \vec{F}\Delta t \rightarrow \vec{F} = \frac{\Delta\vec{p}}{\Delta t} = \frac{m(\vec{v} - \vec{v}_0)}{\Delta t} = \frac{(0.06\text{ kg})\left[(-36\text{ m/s}) - 32\text{ m/s}\right]}{0.005\text{ s}} = -816\text{ N}$$

45. **A** Consider the energy of the apple under the influences of the nonconservative force of air resistance: $K_1 + U_1 + W_{ar} = K_2 + U_2 \rightarrow 0 + mgh_1 + W_{ar} = \frac{1}{2}mv_2^2 + 0 \rightarrow W_{ar} = \frac{1}{2}mv_2^2 - mgh_1$. Substitute the given values in SI units.

$$W_{ar} = \frac{1}{2}(0.085\text{ kg})(13\text{ m/s})^2 - (0.085\text{ kg})\left(9.8\frac{\text{m}}{\text{s}^2}\right)(9\text{ m}) = -0.315\text{ J}$$

46. **A, D** According to the formula $v^2 = v_0^2 + 2ad$, the initial velocity, v_0, can be related to the stopping distance, d. If v is zero, and the equation is rearranged for d, it becomes $d = v_0^2/2a$. Since the car is decelerating, a is negative. Therefore, if v_0 is doubled, the stopping distance, d, is quadrupled. To determine the relationship between v_0 and t, it becomes $t = -v_0/a$. Therefore, if v_0 is doubled, the stopping time, t, is doubled.

47. **C, D** The period for a pendulum is $T = 2\pi\sqrt{\frac{L}{g}}$. Making changes to the length of the string or changing the acceleration of gravity (by changing the location from the Earth to the Moon) will cause changes in the period.

48. **C, D** The cue ball does not contact the 2 ball, so the cue ball cannot exert a force on the 2 ball, making (A) incorrect. The reaction force to the cue ball striking the 1 ball is the force the 1 ball exerts on the cue ball, so that (B) is incorrect. Furthermore, this reaction force is the net force that acts on the cue ball to stop its motion, so (D) is correct. The 1 ball remains motionless, so it experiences no net force. Therefore, the force the cue ball exerts on the 1 ball is equal in magnitude to the force the 2 ball exerts on the 1 ball. The force the 2 ball exerts on the 1 ball is the reaction force to the energy the 1 ball exerts on it; the magnitude of the force on the 2 ball is equal to the force the cue ball exerts on the 1 ball, making (C) correct.

49. **C, D** The forklift exerts an upward force to move the crate upward, so it does positive work on the crate. Positive work done on a system transfers energy to the system. However, the crate starts and ends at rest, so it doesn't experience an overall change in kinetic energy. Therefore, the total work done on the crate is zero. The kinetic energy of the crate does not change, but it ends up with more gravitational potential energy than it started with, so the total mechanical energy of the crate is not conserved, which eliminates (A). Gravity acts downward in the opposite direction to the crate's displacement, doing negative work on the crate, eliminating (B).

50. **B, D** The impulse delivered to the wall by the paintball is equal in magnitude to the change in momentum of the paintball. Since $p = mv$, the final momentum of the paintball is zero, and reducing the impulse can be achieved by reducing the mass or the initial velocity, making (B) correct. Mass is density times volume, so reducing the density of the paintball reduces its mass, making (D) correct. Both (A) and (C) would reduce the average force experienced by the wall by increasing the time it takes for the paintball to stop but would not change the impulse, which is the product of average force and time.

Section II: Free-Response Questions

1. (A) One experimental procedure would be to mark off two distances on the table—one for Block A before the collision, and one for the combined blocks after the collision. Push Block A to give it an initial speed. Use a stopwatch to measure the time it takes for the blocks to cross the marked distances. The speeds are the distances divided by the time.

   (B)

$$m_A v_i = (m_A + m_B) v_f$$

$$m_B = \left(\frac{m_A v_i}{v_f}\right) - m_A = \left[\frac{(0.5\,\text{kg})(1.5\,\text{m/s})}{0.25\,\text{m/s}}\right] - 0.5\,\text{kg} = 2.5\,\text{kg}$$

(C)

$$K_{\text{initial}} = \frac{1}{2}(0.5\,\text{kg})(1.5\,\text{m/s})^2 = 0.56\,\text{J}$$

$$K_{\text{final}} = \frac{1}{2}(3\,\text{kg})(0.25\,\text{m/s})^2 = 0.09\,\text{J}$$

The total kinetic energy lost is equal to

$$K_{\text{initial}} - K_{\text{final}} = 0.56\,\text{J} - 0.09\,\text{J} = 0.47\,\text{J}$$

2. (A) As shown below, a free-body diagram would include only the tensions in the string and the force of gravity. Because the pendulum makes a horizontal circle, take some care to draw the direction of the force represented by the tension along the path of the string.

(B) Where θ is the angle between F_T and the normal,

$$\Sigma F_x = ma_x \qquad \Sigma F_y = ma_y$$
$$F_T \sin\theta = ma_x \quad \text{And} \quad F_T \cos\theta - F_g = ma_y$$

(C) The centripetal force is the net force in the *x*-direction. However, you need to use some information from the *y*-direction. Because the conical pendulum travels in a horizontal circle, there is no acceleration in the *y*-direction and so Newton's Second Law in the *y*-direction becomes

$$F_T \cos\theta - F_g = 0$$

or

$$F_T \cos\theta = F_g$$

$$F_T = \frac{F_g}{\cos\theta}$$

$$F_T = \frac{0.5\text{ kg}(10\text{ m/s}^2)}{\cos 57°}$$

$$F_T = 9.2\text{ N}$$

Knowing this, you can solve for the centripetal acceleration. It is the same as the acceleration in the *x*-direction.

$$F_T \sin\theta = ma_c$$
$$\frac{F_T \sin\theta}{m} = a_c$$
$$\frac{9.2 \text{ N} \sin 57°}{0.5 \text{ kg}} = a_c$$
$$15.4 \text{ m/s}^2 = a_c$$

(D) The radius the ball travels can be found using some geometry. The length of the string is 2.2 m at an angle of 57 degrees. This means

$$\sin\theta = \frac{opp}{hyp}$$
$$\sin 57° = \frac{r}{2.2 \text{ m}}$$

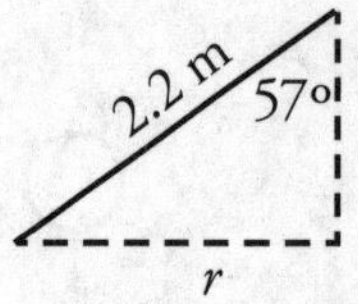

2.2 m
57°
r

$$r = 1.84 \text{ m}$$

(E) The ball will travel at a speed of

$$a_c = \frac{v^2}{r}$$
$$v^2 = a_c r$$
$$v = \sqrt{a_c r}$$
$$v = \sqrt{15.4 \text{ m/s}^2 \times 1.84 \text{ m}}$$
$$v = 5.32 \text{ m/s}^2$$

3. (A) When the pilot opens the tanks to drop the water, the initial velocity of the water is equal to that of the plane, so the water has an initial horizontal velocity. It takes time for the water to fall to the ground and, in that time, the water will travel a horizontal distance determined by its horizontal velocity and time. Therefore, for the water to douse the fire, the water must be released before the plane is directly above the fire.

(B) Apply the definition of velocity, $v = \frac{d}{t} = \frac{600 \text{ m}}{8 \text{ s}} = 75 \text{ m/s}$.

(C) Apply the kinematic equations in the y-direction to determine the time it takes for the water to hit the ground, $y = y_0 + v_{0y}t - \frac{1}{2}gt^2 \rightarrow 0 = y_0 + (0)t - \frac{1}{2}gt^2 \rightarrow t = \sqrt{\frac{2y_0}{g}}$. Apply the kinematic equations in the x-direction to determine the horizontal distance the water travels between its release and hitting the ground,

$$x = x_0 + v_{ox}t = 0 + v_{ox}\sqrt{\frac{2y_0}{g}} = (75 \text{ m/s})\sqrt{\frac{2(150 \text{ m})}{9.8 \text{ m/s}^2}} = 415 \text{ m}$$

4. (A) The spring observes Conservation of Mechanical Energy,

$$mgh = \frac{1}{2}kx^2$$

$$h = \frac{kx^2}{2mg}$$

(B) (i) The value of $1/m$ would help in determining the spring constant when you take the slope of the line.

(ii)

$1/m$ (kg^{-1})	**m (kg)**	**h (m)**
50	0.020	0.49
30	0.030	0.34
25	0.040	0.28
20	0.050	0.19
17	0.060	0.18

(C)

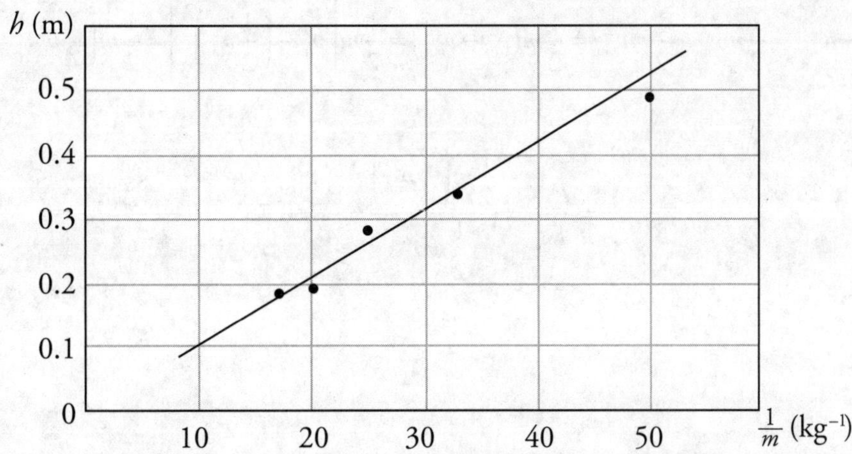

(D) To calculate the slope, select two points on our best-fit line. Here is one example:

$$\text{slope} = \frac{(0.42 - 0.10)\ \text{m}}{(40 - 10)\ \text{kg}^{-1}} = \frac{0.32\ \text{m}}{30\ \text{kg}^{-1}} = 1.07 \times 10^{-2}\ \text{m} \cdot \text{kg}$$

From part (A),

$$h = \frac{kx^2}{2mg} \rightarrow \text{slope} = \frac{kx^2}{2g}$$

Note: Values between 450 N/m and 550 N/m are acceptable.

$$k = \frac{2g(\text{slope})}{x^2} = \frac{2(9.8\ \text{m/s}^2)(1.07 \times 10^{-2}\ \text{m} \cdot \text{kg})}{(0.02\ \text{m})^2} = 524\,\text{N/m}$$

(E) One example would be to use a meter stick and video camera. Hold the meter stick next to the toy and allow it to jump up. Record the toy with a video camera. With this you can watch the video in slow motion later to determine its height. You can also find the height by recording the time t it takes to fall from its highest point to the lowest. Then, using kinematics equations, you can determine the height.

5. (A) As water flows down through a height H, the potential energy is converted to kinetic energy that turns the turbine. Therefore, the total available energy from the water is $E = U = mgH$. A volume V of water has mass $m = \rho V$. The available power is

$$P = \frac{E}{t} = \frac{mgH}{t} = \frac{\rho VgH}{t}$$

(B) The energy converted to electrical energy is equal to the energy lost by the water $E_{\text{elec}} = K_{\text{in}} - K_{\text{out}} = \frac{1}{2}mv_{\text{in}}^2 - \frac{1}{2}mv_{\text{out}}^2 = \frac{1}{2}m(v_{\text{in}}^2 - v_{\text{out}}^2) = \frac{1}{2}m\left[v_{\text{in}}^2 - \left(\frac{1}{2}v_{\text{in}}\right)^2\right] = \frac{1}{2}m\left(\frac{3}{4}v_{\text{in}}^2\right)$. The efficiency is

$$e = \frac{E_{\text{elec}}}{K_{in}} = \frac{\frac{1}{2}m\left(\frac{3}{4}v_{\text{in}}^2\right)}{\frac{1}{2}mv_{\text{in}}^2} = \frac{3}{4} = 0.75$$

(C) The electrical power produced is the efficiency of the plant times the available power.

$$P_{elec} = eP = \frac{e\rho VgH}{t} = \frac{(0.75)(1000\text{ kg/m}^3)(3\text{ m}^3)(9.8\text{ m/s}^2)(180\text{ m})}{1\text{ s}} = 3.97\times10^6\text{ W}$$

(D) The number of homes this hydroelectric plant can power is

$$n = \frac{P_{elec}}{P_{max}\text{ per house}} = \frac{3.97\times10^6\text{ W}}{55\times10^3\text{ W}} \approx 72\text{ homes}$$

HOW TO SCORE PRACTICE TEST 2

Section I: Multiple Choice

________ × 1.5 = ________

Number Correct (out of 50) | Weighted Section I Score (Do not round)

Section II: Free Response

Question 1: ________ (out of 12) × 1.25 = ________ (Do not round)

Question 2: ________ (out of 12) × 1.25 = ________ (Do not round)

Question 3: ________ (out of 7) × 2.143 = ________ (Do not round)

Question 4: ________ (out of 7) × 2.143 = ________ (Do not round)

Question 5: ________ (out of 7) × 2.143 = ________ (Do not round)

Sum = ________

Weighted Section II Score (Do not round)

AP Score Conversion Chart Physics 1	
Composite Score Range	AP Score
107–150	5
90–106	4
73–89	3
56–72	2
0–55	1

Composite Score

________ + ________ = ________

Weighted Section I Score | Weighted Section II Score | Composite Score (Round to nearest whole number)

Practice Test 3

AP® Physics 1 Exam

SECTION I: Multiple-Choice Questions

DO NOT OPEN THIS BOOKLET UNTIL YOU ARE TOLD TO DO SO.

At a Glance

Total Time
90 minutes
Number of Questions
50
Percent of Total Grade
50%
Writing Instrument
Pen required

Instructions

Section I of this examination contains 50 multiple-choice questions. Fill in only the ovals for numbers 1 through 50 on your answer sheet.

CALCULATORS MAY BE USED ON BOTH SECTIONS OF THE AP PHYSICS 1 EXAM.

Indicate all of your answers to the multiple-choice questions on the answer sheet. No credit will be given for anything written in this exam booklet, but you may use the booklet for notes or scratch work. Please note that there are two types of multiple-choice questions: single-select and multi-select questions. After you have decided which of the suggested answers is best, completely fill in the corresponding oval(s) on the answer sheet. For single-select, you must give only one answer; for multi-select you must give BOTH answers in order to earn credit. If you change an answer, be sure that the previous mark is erased completely. Here is a sample question and answer.

Sample Question — Sample Answer

Chicago is a

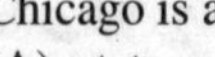

(A) state
(B) city
(C) country
(D) continent

Use your time effectively, working as quickly as you can without losing accuracy. Do not spend too much time on any one question. Go on to other questions and come back to the ones you have not answered if you have time. It is not expected that everyone will know the answers to all the multiple-choice questions.

About Guessing

Many candidates wonder whether or not to guess the answers to questions about which they are not certain. Multiple-choice scores are based on the number of questions answered correctly. Points are not deducted for incorrect answers, and no points are awarded for unanswered questions. Because points are not deducted for incorrect answers, you are encouraged to answer all multiple-choice questions. On any questions you do not know the answer to, you should eliminate as many choices as you can, and then select the best answer among the remaining choices.

GO ON TO THE NEXT PAGE.

ADVANCED PLACEMENT PHYSICS 1 TABLE OF INFORMATION

CONSTANTS AND CONVERSION FACTORS	
Proton mass, $m_p = 1.67 \times 10^{-27}$ kg	Electron charge magnitude, $e = 1.60 \times 10^{-19}$ C
Neutron mass, $m_n = 1.67 \times 10^{-27}$ kg	Coulomb's law constant, $k = 1/4\pi\varepsilon_0 = 9.0 \times 10^9\ \text{N}\cdot\text{m}^2/\text{C}^2$
Electron mass, $m_e = 9.11 \times 10^{-31}$ kg	Universal gravitational constant, $G = 6.67 \times 10^{-11}\ \text{m}^3/\text{kg}\cdot\text{s}^2$
Speed of light, $c = 3.00 \times 10^8$ m/s	Acceleration due to gravity at Earth's surface, $g = 9.8\ \text{m/s}^2$

UNIT SYMBOLS								
	meter,	m	kelvin,	K	watt,	W	degree Celsius,	°C
	kilogram,	kg	hertz,	Hz	coulomb,	C		
	second,	s	newton,	N	volt,	V		
	ampere,	A	joule,	J	ohm,	Ω		

PREFIXES		
Factor	Prefix	Symbol
10^{12}	tera	T
10^{9}	giga	G
10^{6}	mega	M
10^{3}	kilo	k
10^{-2}	centi	c
10^{-3}	milli	m
10^{-6}	micro	μ
10^{-9}	nano	n
10^{-12}	pico	p

VALUES OF TRIGONOMETRIC FUNCTIONS FOR COMMON ANGLES

θ	$0°$	$30°$	$37°$	$45°$	$53°$	$60°$	$90°$
$\sin\theta$	0	1/2	3/5	$\sqrt{2}/2$	4/5	$\sqrt{3}/2$	1
$\cos\theta$	1	$\sqrt{3}/2$	4/5	$\sqrt{2}/2$	3/5	1/2	0
$\tan\theta$	0	$\sqrt{3}/3$	3/4	1	4/3	$\sqrt{3}$	∞

The following conventions are used in this exam.

I. The frame of reference of any problem is assumed to be inertial unless otherwise stated.
II. Assume air resistance is negligible unless otherwise stated.
III. In all situations, positive work is defined as work done on a system.
IV. The direction of current is conventional current: the direction in which positive charge would drift.
V. Assume all batteries and meters are ideal unless otherwise stated.

GO ON TO THE NEXT PAGE.

ADVANCED PLACEMENT PHYSICS 1 EQUATIONS

MECHANICS

$v_x = v_{x0} + a_x t$

$x = x_0 + v_{x0}t + \frac{1}{2}a_x t^2$

$v_x^2 = v_{x0}^2 + 2a_x(x - x_0)$

$\vec{a} = \frac{\sum \vec{F}}{m} = \frac{\vec{F}_{net}}{m}$

$|\vec{F}_f| \le \mu |\vec{F}_n|$

$a_c = \frac{v^2}{r}$

$\vec{p} = m\vec{v}$

$\Delta\vec{p} = \vec{F}\,\Delta t$

$K = \frac{1}{2}mv^2$

$\Delta E = W = F_{\parallel}d = Fd\cos\theta$

$P = \frac{\Delta E}{\Delta t}$

$\theta = \theta_0 + \omega_0 t + \frac{1}{2}\alpha t^2$

$\omega = \omega_0 + \alpha t$

$x = A\cos(2\pi ft)$

$\vec{\alpha} = \frac{\sum \vec{\tau}}{I} = \frac{\vec{\tau}_{net}}{I}$

$\tau = r_{\perp}F = rF\sin\theta$

$L = I\omega$

$\Delta L = \tau\,\Delta t$

$K = \frac{1}{2}I\omega^2$

$|\vec{F}_s| = k|\vec{x}|$

$U_s = \frac{1}{2}kx^2$

$\rho = \frac{m}{V}$

$\Delta U_g = mg\,\Delta y$

$T = \frac{2\pi}{\omega} = \frac{1}{f}$

$T_s = 2\pi\sqrt{\frac{m}{k}}$

$T_p = 2\pi\sqrt{\frac{\ell}{g}}$

$|\vec{F}_g| = G\frac{m_1 m_2}{r^2}$

$\vec{g} = \frac{\vec{F}_g}{m}$

$U_G = -\frac{Gm_1 m_2}{r}$

a = acceleration
A = amplitude
d = distance
E = energy
f = frequency
F = force
I = rotational inertia
K = kinetic energy
k = spring constant
L = angular momentum
ℓ = length
m = mass
P = power
p = momentum
r = radius or separation
T = period
t = time
U = potential energy
V = volume
v = speed
W = work done on a system
x = position
y = height
α = angular acceleration
μ = coefficient of friction
θ = angle
ρ = density
τ = torque
ω = angular speed

ELECTRICITY

$|\vec{F}_E| = k\left|\frac{q_1 q_2}{r^2}\right|$

$I = \frac{\Delta q}{\Delta t}$

$R = \frac{\rho\ell}{A}$

$I = \frac{\Delta V}{R}$

$P = I\,\Delta V$

$R_s = \sum_i R_i$

$\frac{1}{R_p} = \sum_i \frac{1}{R_i}$

A = area
F = force
I = current
ℓ = length
P = power
q = charge
R = resistance
r = separation
t = time
V = electric potential
ρ = resistivity

WAVES

$\lambda = \frac{v}{f}$

f = frequency
v = speed
λ = wavelength

GEOMETRY AND TRIGONOMETRY

Rectangle
$A = bh$

Triangle
$A = \frac{1}{2}bh$

Circle
$A = \pi r^2$
$C = 2\pi r$

Rectangular solid
$V = \ell wh$

Cylinder
$V = \pi r^2 \ell$
$S = 2\pi r\ell + 2\pi r^2$

Sphere
$V = \frac{4}{3}\pi r^3$
$S = 4\pi r^2$

A = area
C = circumference
V = volume
S = surface area
b = base
h = height
ℓ = length
w = width
r = radius

Right triangle
$c^2 = a^2 + b^2$

$\sin\theta = \frac{a}{c}$

$\cos\theta = \frac{b}{c}$

$\tan\theta = \frac{a}{b}$

c a θ 90° b

GO ON TO THE NEXT PAGE.

THIS PAGE IS LEFT INTENTIONALLY BLANK.

GO ON TO THE NEXT PAGE.

AP PHYSICS 1

SECTION I

Note: To simplify calculations, you may use $g = 10 \text{ m/s}^2$ in all problems.

Directions: Each of the questions or incomplete statements is followed by four suggested answers or completions. Select the one that is best in each case and then fill in the corresponding circle on the answer sheet.

1. A section of a river flows with a velocity of 1 m/s due south. A kayaker who is able to propel her kayak at 1.5 m/s wishes to paddle directly east from one bank to the other. In what direction should she direct her kayak?

(A) 37° N of E
(B) 42° N of E
(C) 45° N of E
(D) 48° N of E

2. An object's initial velocity has components $v_{1x} = -1$ m/s and $v_{1y} = 4$ m/s. Its velocity 3 s later has components $v_{2x} = 4$ m/s and $v_{2y} = -2$ m/s. What is its average acceleration relative to the +x-axis during this time interval?

(A) 1.1 m/s² at −50.2°
(B) 1.1 m/s² at −39.8°
(C) 2.6 m/s² at −50.2°
(D) 2.6 m/s² at −39.8°

3. A student lives 1.2 km from school as the crow flies. On a particular day, it takes her 10 min to get from home to school. Which of the following must be true about this trip to school?

I. She traveled a distance of 1.2 km.
II. Her average speed was 2 m/s.
III. The magnitude of her average velocity was 2 m/s.

(A) I only
(B) II and III only
(C) III only
(D) I, II, and III

4. In which of the following position-versus-time graphs does the object in motion have a constant negative acceleration?

(A)

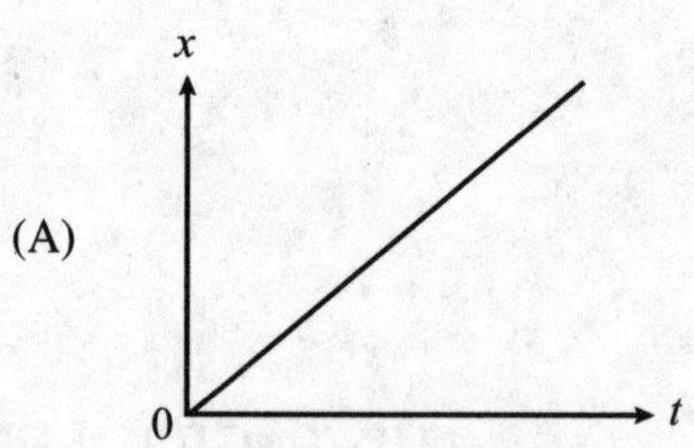

(B)

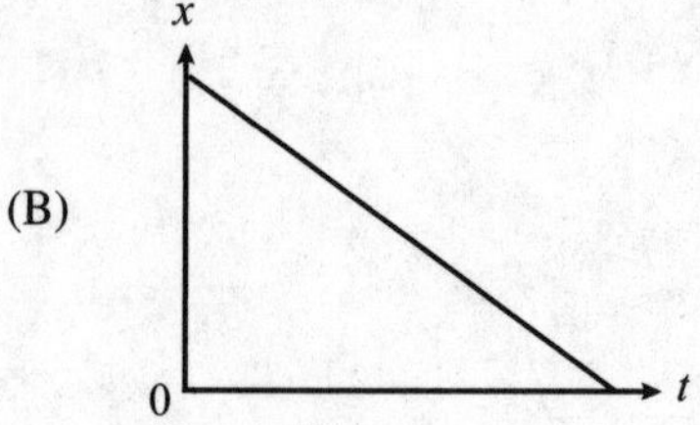

(C)

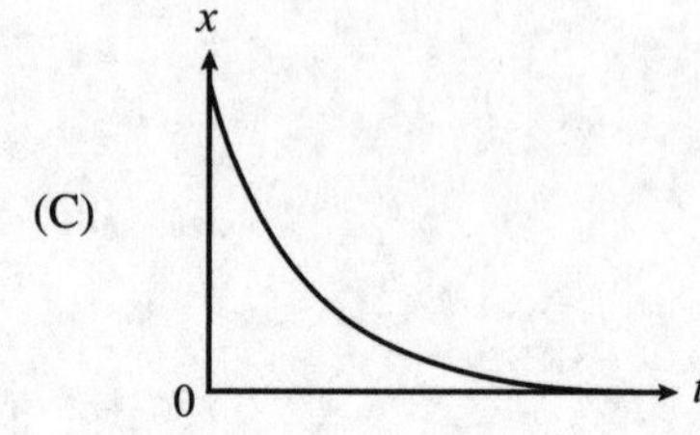

(D)

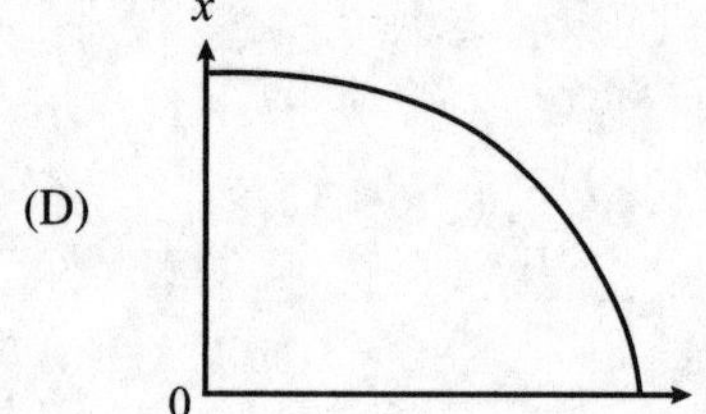

GO ON TO THE NEXT PAGE.

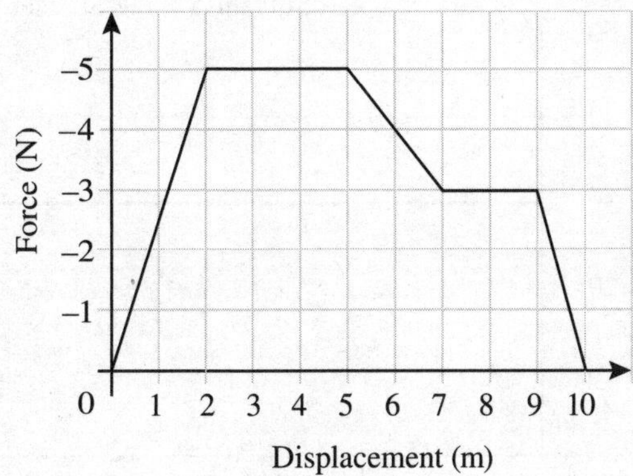

5. The graph above shows the magnitude of a variable force acting on an object moving in a straight line. The force acts in the same direction as the motion. How much work does the force do from $t = 0$ s to $t = 10$ s?
   (A) –3 J
   (B) 0 J
   (C) 29.5 J
   (D) 35.5 J

6. An airtanker is being used to fight a forest fire. It is flying with a ground speed of 85 m/s and maintaining an altitude of 300 m. If it is flying directly toward the fire, what approximate horizontal distance from the fire should its tanks be open in order for the water dropped to land on the fire? Assume that the fire is relatively localized.
   (A) 230 m
   (B) 250 m
   (C) 660 m
   (D) 1,700 m

7. Two cannons are fired from a cliff at a height of 50 m from the ground. Cannonball A is fired horizontally with an initial velocity of 40 m/s. Cannonball B is fired at a launch angle of 60° with an initial velocity of 80 m/s. Which cannonball will have a greater magnitude of displacement after two seconds?
   (A) Cannonball A
   (B) Cannonball B
   (C) Both cannonballs will have the same displacement.
   (D) Cannot be determined

8. A car initially at rest accelerates linearly at a constant rate for eight seconds. If the total displacement of the car was 600 m, what was the speed of the car after the eight seconds of acceleration?
   (A) 50 m/s
   (B) 100 m/s
   (C) 150 m/s
   (D) 200 m/s

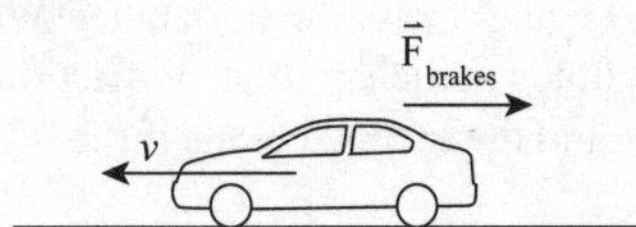

9. A car traveling at a speed of v_0 applies its brakes, skidding to a stop over a distance of x m. Assuming that the deceleration due to the brakes is constant, what would be the skidding distance of the same car if the braking were twice as effective (doubling the magnitude of deceleration)?
   (A) $0.25x$ m
   (B) $0.5x$ m
   (C) x m
   (D) $2x$ m

10. A student is pushing a 3-kg book across a table with a constant force of 30.0 N directed 10° below the horizontal. The coefficient of kinetic friction between the book and the table is 0.3. What is the magnitude of the force that the table exerts on the book during this motion?
    (A) 30.7 N
    (B) 37.0 N
    (C) 38.2 N
    (D) 45.0 N

GO ON TO THE NEXT PAGE.

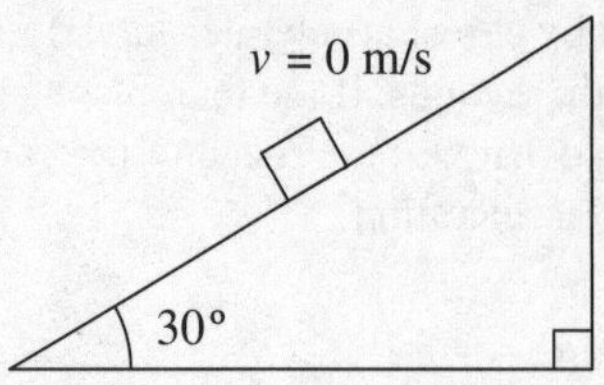

11. A box with a mass of 2 kg is placed on an inclined plane that makes a 30° angle with the horizontal. What is the minimum possible coefficient of static friction (μ_s) between the box and the inclined plane if the box remains at rest?
(A) 0.5
(B) 0.58
(C) 0.87
(D) 1

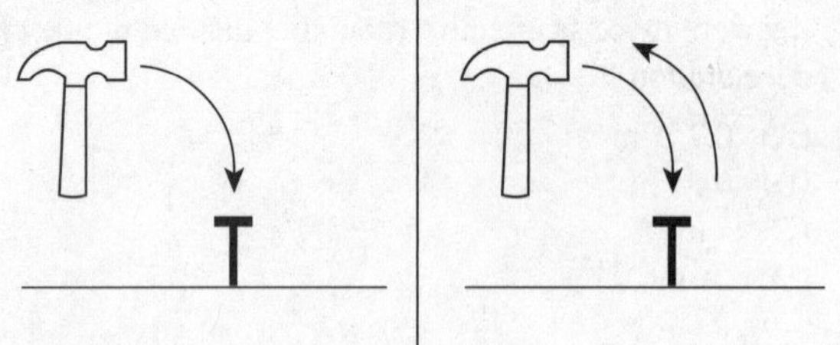

12. A construction worker strikes a nail with a hammer twice with the same initial velocity (v_0). The first time, the hammer comes to rest after hitting the nail. The second time, the hammer recoils after hitting the nail and bounces back toward the worker. Assuming the contact time in both strikes is the same, which strike imparts a greater impulse on the nail?
(A) The first strike
(B) The second strike
(C) The nail experiences the same impulse in both strikes.
(D) Cannot be determined

13. A box with a mass of 5 kg is sliding across a table at a speed of 2 m/s. The coefficient of kinetic friction between the box and table is $\mu = 0.25$. What is the minimum force that has to be applied on the box to maintain this speed?
(A) 0 N
(B) 10 N
(C) 12.5 N
(D) 25 N

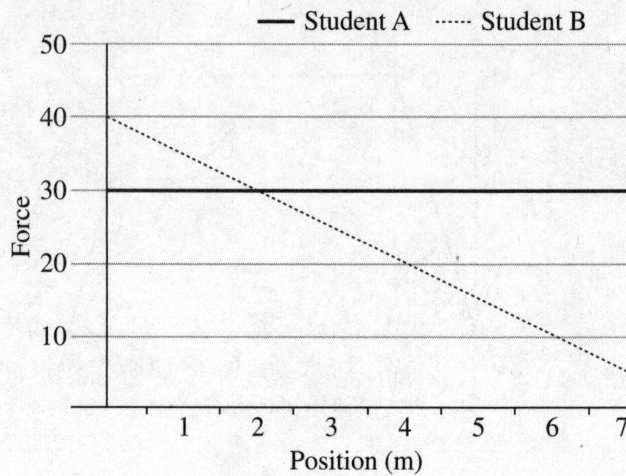

14. Two students each push a 30 kg box across the room. Student A applies a constant force of 30 N. Student B starts with a force of 40 N but gradually reduces the force due to fatigue. Given the force-versus-position graphs of the two students, at what position will both students have done the same amount of work?
(A) 2 m
(B) 3 m
(C) 4 m
(D) 5 m

15. A 2000 kg truck is initially traveling with a speed of 20 m/s. The driver applies the brakes and the truck slows to 10 m/s. How much work was done by the frictional force applied from the brakes?
(A) –300,000 J
(B) –100,000 J
(C) 100,000 J
(D) 300,000 J

GO ON TO THE NEXT PAGE.

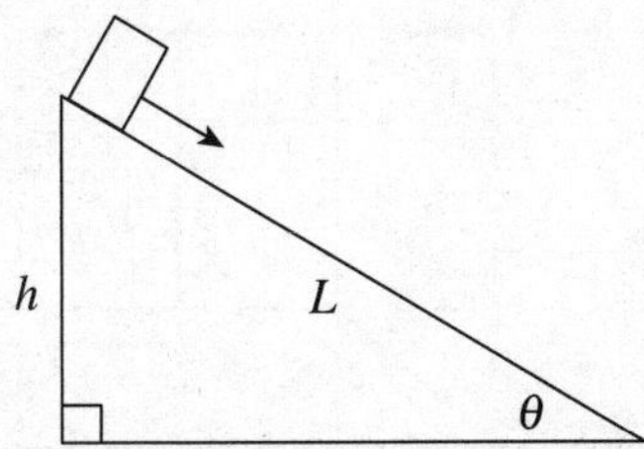

16. A box of mass m slides down a frictionless inclined plane of length L and height h. If the box is initially at rest, what is the speed of the box halfway down the inclined plane?

(A) $\sqrt{2gh}$
(B) $\sqrt{gh}$
(C) $\sqrt{\frac{gh}{2}}$
(D) $\sqrt{\frac{gh}{4}}$

17. A car with a mass of 1000 kg experiences a frictional force of 3500 N while driving at a constant speed of a 15 m/s. What is the power output of the car's engine?

(A) 3.5 kW
(B) 5.25 kW
(C) 35.0 kW
(D) 52.5 kW

18. A student launched a small rocket with a mass of 50 kg into the air with an initial velocity of 10 m/s in the positive vertical direction. The student then turns on the secondary engines of the rocket to apply a constant upward force that increases the velocity of the rocket to 15 m/s. If the force provided by the secondary engines did 4,500 J of work, how much work did the force of gravity do on the rocket?

(A) −625 J
(B) −750 J
(C) −1,250 J
(D) −1,375 J

19. A student drops a 1 kg rock off a cliff with a height of 20 m. The rock lands on the ground and comes to rest in 0.25 seconds. What was the magnitude of the average force that the rock experienced while coming to rest?

(A) 20 N
(B) 40 N
(C) 80 N
(D) 160 N

20. A plutonium atom with a mass of 244 Da (daltons) is initially at rest. The atom suddenly undergoes alpha decay, emitting an alpha particle with a mass of 4 Da with a velocity of 45 m/s east. What is the velocity of the decay product with a mass of 240 Da?

(A) 0.75 m/s west
(B) 0.75 m/s east
(C) 0.66 m/s west
(D) 0.66 m/s east

21. A 2000-kg truck traveling due north at 40.0 m/s collides with a 1500-kg car. The vehicles lock bumpers after the collision, and skid due north for 8 m before coming to a stop. If the coefficient of kinetic friction between the vehicles and the horizontal ground is 0.3, what was the car's velocity before the collision.

(A) 37.2 m/s north
(B) 37.2 m/s south
(C) 41.9 m/s north
(D) 41.9 m/s south

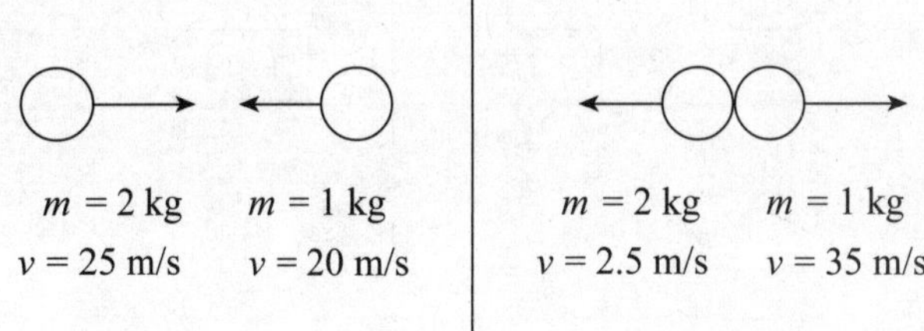

22. A 2 kg ball traveling at 25 m/s collides head on with a 1 kg ball traveling at 20 m/s. After impact, both objects reverse direction with the 2 kg ball traveling at 2.5 m/s and the 1 kg ball traveling at 35 m/s. What type of collision occurred?

(A) Inelastic
(B) Perfectly inelastic
(C) Elastic
(D) Cannot be determined

GO ON TO THE NEXT PAGE.

23. Which of the following statements is FALSE regarding perfectly inelastic collisions?

(A) The objects stick together after the collision.
(B) Momentum is conserved.
(C) If the two objects are of equal mass and collide head-on, they will exchange velocities.
(D) The maximum possible amount of kinetic energy is lost.

24. A 0.145 kg baseball is traveling at 40 m/s horizontally when it is struck by a baseball bat. The baseball leaves the bat at 50 m/s back in the direction it came from, but at an angle of 40° above the horizontal. What is the magnitude of the impulse imparted to the baseball?

(A) 1.45 N·s
(B) 4.66 N·s
(C) 12.3 N·s
(D) 13.1 N·s

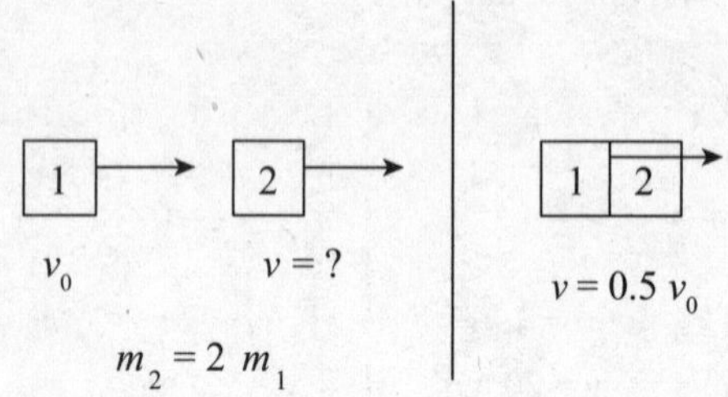

25. Object 1 travels with an initial speed of v_0 toward Object 2, which is traveling in the same direction as Object 1. Object 1 collides perfectly inelastically into Object 2, and the velocity after impact is $0.5v_0$. If Object 2 has twice the mass of Object 1, what must have been the initial speed of Object 2 (assuming no external forces)?

(A) $0.25v_0$
(B) $0.5v_0$
(C) $0.75v_0$
(D) v_0

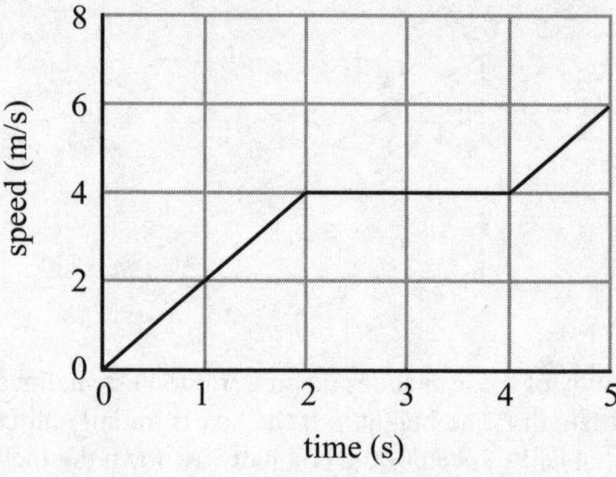

26. A 2 kg object is tied to a machine with a 0.5 m long string. The machine whirls the object in a horizontal circle at various speeds according to the speed-versus-time graph above. What is the magnitude of the centripetal acceleration of the 2 kg object at $t = 3$ s?

(A) 0 m/s²
(B) 8 m/s²
(C) 16 m/s²
(D) 32 m/s²

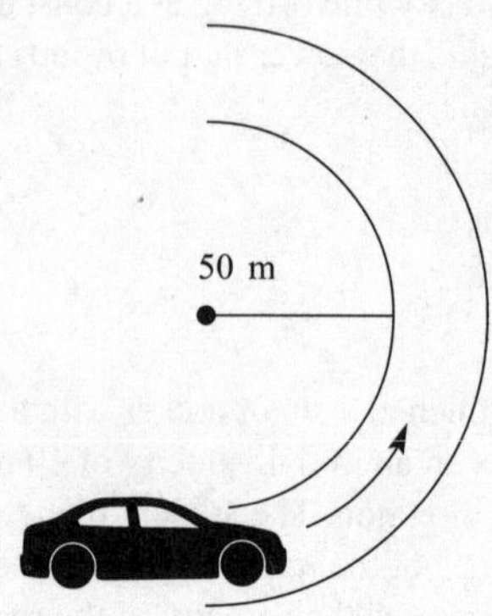

27. A car is driving on a flat curve with a radius of 50 m. If the coefficient of friction between the ground and the car's tires is 0.8, what is the maximum speed of the car in order to make the curve without sliding?

(A) 8 m/s²
(B) 12 m/s²
(C) 16 m/s²
(D) 20 m/s²

GO ON TO THE NEXT PAGE.

28. A planetoid orbits a planet in a circular path at constant speed. The planet has a mass of 4×10^{20} kg and a radius of 1.6×10^5 m. What is the speed of the planetoid if it is 3.2×10^5 m above the surface of the planet?

(A) 72 m/s
(B) 193 m/s
(C) 236 m/s
(D) 439 m/s

29. The magnitude of the gravitational force between two objects is F. If the distance between the two objects is tripled and the mass of one of the objects is doubled, what is the new magnitude of the gravitational force between the two objects?

(A) $\frac{1}{9}F$

(B) $\frac{2}{9}F$

(C) $\frac{1}{6}F$

(D) $\frac{1}{4}F$

30. A mechanical wheel initially at rest on the floor begins rolling forward with an angular acceleration of 2π rad/s^2. If the wheel has a radius of 2 m, what distance does the wheel travel in 3 seconds?

(A) 4π m
(B) 6π m
(C) 16π m
(D) 18π m

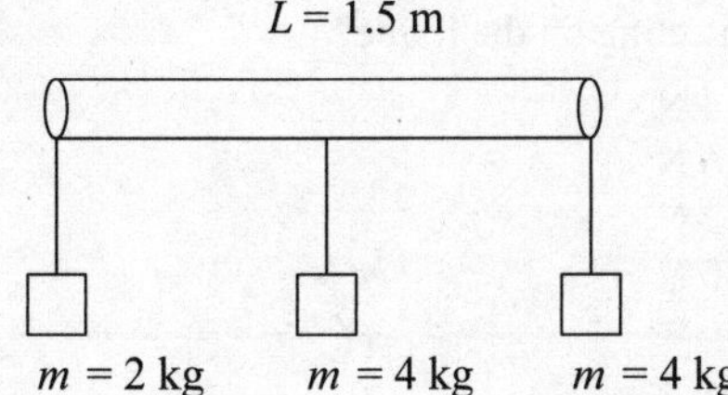

31. Three masses are attached to a 1.5 m long massless bar. Mass 1 is 2 kg and is attached to the far left side of the bar. Mass 2 is 4 kg and is attached to the far right side of the bar. Mass 3 is 4 kg and is attached to the middle of the bar. At what distance from the far left side of the bar can a string be attached to hold the bar up horizontally?

(A) 0.3 m
(B) 0.6 m
(C) 0.9 m
(D) 1.2 m

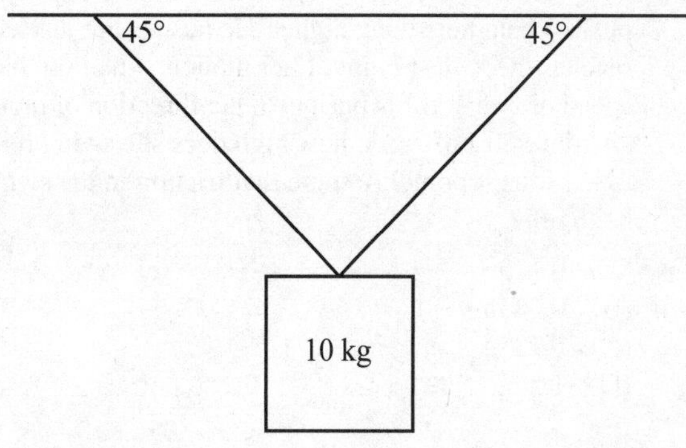

32. A 10 kg mass is attached to the ceiling with two strings (shown above). If the system is in static equilibrium, what is the tension in each of the strings?

(A) 50 N
(B) 70.7 N
(C) 100 N
(D) 141.4 N

33. A beginner skier is skiing slowly at 2 m/s and stops by gliding up a gentle slope. How high (vertically) above the bottom of the slope are they when they come to a stop? (Friction is negligible.)

(A) 0.10 m
(B) 0.20 m
(C) 0.41 m
(D) The answer cannot be determined without the angle of the slope.

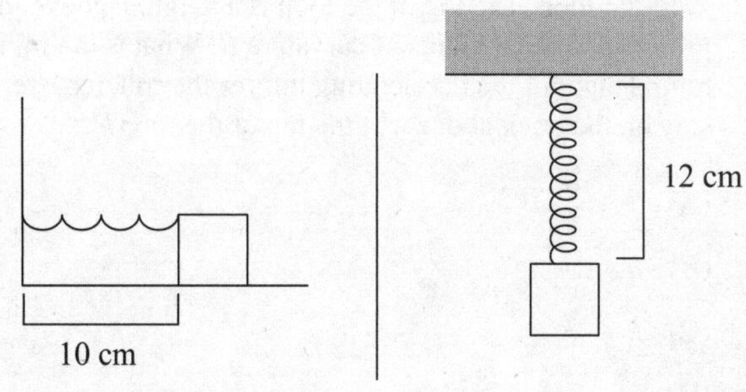

34. A horizontal spring with a block attached to the end has a natural length of 10 cm. When the spring is attached to the ceiling, the spring is stretched 2 cm beyond its natural length at equilibrium. If the spring constant of the block is 50 N/m, what is the mass of the block?

(A) 0.01 kg
(B) 0.05 kg
(C) 0.1 kg
(D) 0.5 kg

GO ON TO THE NEXT PAGE.

35. A boy gives his 18-kg younger sister on a swing a quick push to help her swing higher. He pushes her just as she reaches the lowest point of her motion, when she has a speed of 3 m/s. If his push is in her direction of motion and does 50 J of work, how high does she swing relative to her lowest point? Assume that friction in the swing is negligible.

(A) 0.17 m
(B) 0.28 m
(C) 0.73 m
(D) 1.5 m

36. A block attached to an ideal spring undergoes simple harmonic motion about its equilibrium position ($x = 0$) with amplitude A. What fraction of the total energy is in the form of potential energy when the block is at position $x = \frac{1}{4}A$?

(A) $\frac{1}{16}$

(B) $\frac{1}{4}$

(C) $\frac{3}{4}$

(D) $\frac{15}{16}$

37. A rollercoaster rolls from rest at the top of an acceleration hill a height H above the ground then rolls through a loop-the-loop. The top of the loop is a height h above the ground and has a radius of curvature R. What is the minimum height of the accelerating hill for the rollercoaster to stay on the track above it at the top of the loop?

(A) $h + \frac{R}{2}$

(B) $h + R$

(C) $h + 2R$

(D) $\frac{R}{2}$

38. A pottery wheel with a rotational inertia of 7 kg·m^2 rotates at a constant rate of 30 rad/s. An artist accidentally drops a 2.5-kg ball of clay onto the pottery wheel 0.3 m from its axis of rotation. If no net external torque acts on the system, and the rotational inertia of the clay is $I = mr^2$, what is the angular speed of the wheel with the clay?

(A) 22.3 rad/s
(B) 29.1 rad/s
(C) 29.5 rad/s
(D) 31.0 rad/s

39. A toy train engine is connected to toy train car A followed by toy train car B. The engine pulls with a force of 0.60 N. Car A has a mass of 500 g and car B has a mass of 300 g. What is the tension in the connector between cars A and B?

(A) 0.23 N
(B) 0.36 N
(C) 0.38 N
(D) 0.60 N

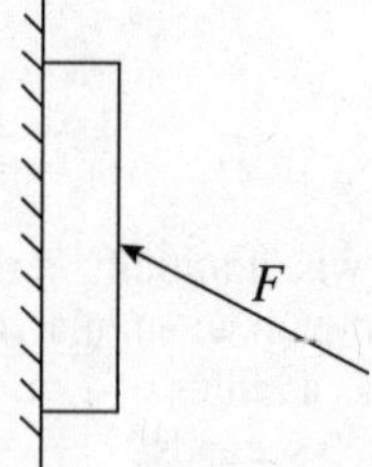

40. While deciding where to hang a picture frame, a man holds the 1.3-kg frame against the wall by applying a force of 15 N at an angle of 25° above the horizontal. The coefficient of static friction is 0.6. What is the force of friction acting on the frame?

(A) 6.4 N
(B) 7.6 N
(C) 8.2 N
(D) 9.0 N

GO ON TO THE NEXT PAGE.

41. A salad spinner has an internal 0.15-m radius spinning basket that spins at 26 rad/s to remove water from salad greens. The basket has a rotational inertia of 0.1 kg·m^2. To stop the basket, a piece of rubber is pressed against the outer edge of the basket, slowing it through friction. If rubber is pressed into the outer edge with a force of 5 N, and the coefficient of kinetic friction between the rubber and the basket is 0.35, how long does it take for the basket to stop?

(A) 1.5 s
(B) 3.5 s
(C) 5.2 s
(D) 9.9 s

42. When walking on a slackline or tight rope, why is it helpful for people to hold their arms out to the side?

(A) It decreases the angular acceleration resulting from a torque on the body.
(B) It raises the location of the body's center of mass, reducing the lever arm for forces exerted at the feet.
(C) Having an arm to either side of the body balances the torques on the body.
(D) It minimizes the impact of air currents.

43. Masses m_1 and m_2 are connected by a thin sting that passes over a massless pulley. The system is released from rest. What is the tension in the string?

(A) $(m_1 + m_2)g$

(B) $(m_2 - m_1)g$

(C) $\frac{(m_2 - m_1)^2}{m_1 + m_2}g$

(D) $\frac{2m_1m_2}{m_1 + m_2}g$

44. Two runners start at the same trailhead. Runner A takes a steep trail while runner B takes a trail with a more moderate slope, but they both meet again at the top of the hill to rest. Which of the following correctly describes the relationship between the work done by runner A, W_A, and the work done by runner B, W_B.

(A) $W_A > W_B$
(B) $W_A = W_B$
(C) $W_A < W_B$
(D) The relationship cannot be determined from the given information.

45. A 60-kg woman in an elevator descends 8 m at a constant speed. How much work does the normal force do during this motion?

(A) –4700 J
(B) 0 J
(C) 480 J
(D) 4700 J

Directions: For each of the questions 46–50, two of the suggested answers will be correct. Select the two answers that are best in each case, and then fill in both of the corresponding circles on the answer sheet.

46. Which of the following describes a particle that is slowing down? Select two answers.

(A) A particle has a positive velocity and a positive acceleration.
(B) A particle has a positive velocity and a negative acceleration.
(C) A particle has a negative velocity and a positive acceleration.
(D) A particle has a negative velocity and a negative acceleration.

47. A box slides from rest at a given height down and off of a ramp. Which of the follow statements are true? Select two answers.

(A) One of the forces acting on the box does no work during the motion.
(B) The work done by gravity on the box depends on the incline angle of the ramp.
(C) The magnitude of the work done by friction on the box is greater than the work done by gravity.
(D) The total work done on the box depends on the incline angle of the ramp.

GO ON TO THE NEXT PAGE.

48. For which combination of springs will the effective spring constant be equal to $2k$? Select two answers.

(A)

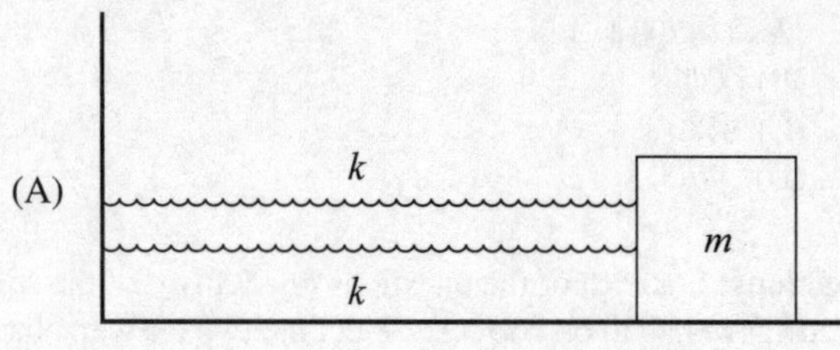

(B)

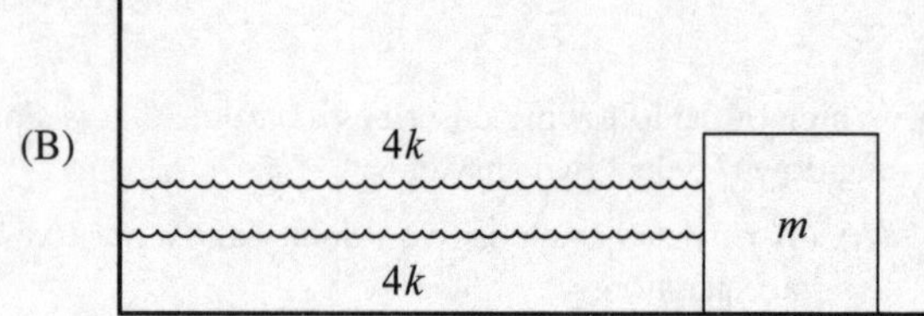

(C)

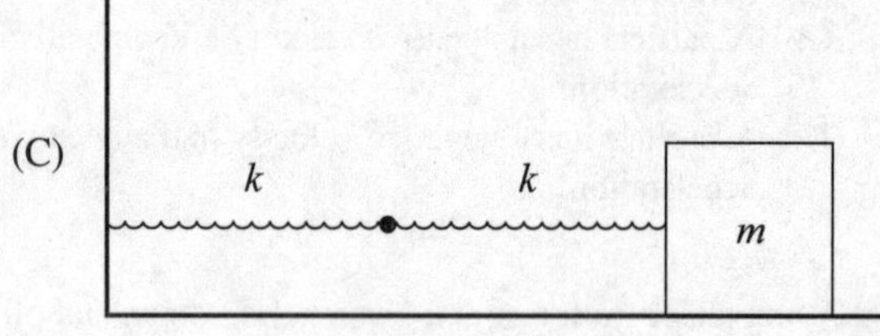

(D)

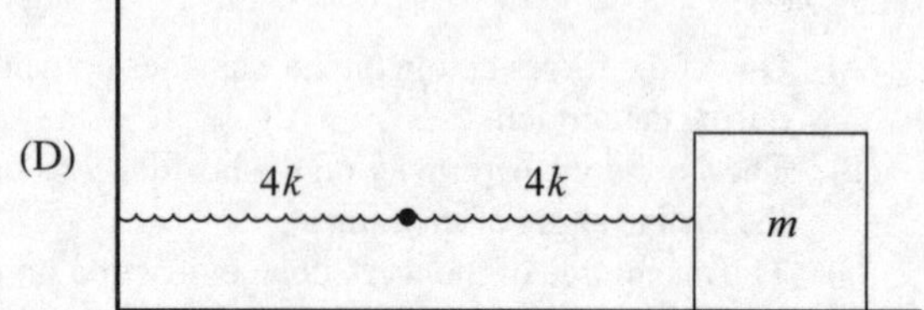

49. Which of the following must be true for an object at translational equilibrium? Select two answers.

(A) The kinetic energy of the object is 0.
(B) The net force on the object is 0.
(C) The acceleration of the object is 0.
(D) The net torque on the object is 0.

50. The biceps attach to the forearm close to the elbow joint. When lifting an object held in the hand without moving the elbow, the forearm has a constant angular speed. Which of the following is true? Select two answers.

(A) The forearm is in dynamic equilibrium.
(B) The biceps exert more force on the forearm than the object does.
(C) The net torque on the forearm does not change as the object is lifted.
(D) The object exerts an equal magnitude of torque on the forearm as the biceps do.

END OF SECTION I

DO NOT CONTINUE UNTIL INSTRUCTED TO DO SO.

AP PHYSICS 1

SECTION II

Free-Response Questions

Time—90 minutes

Percent of total grade—50

<u>General Instructions</u>

Use a separate piece of paper to answer these questions. Show your work. Be sure to write CLEARLY and LEGIBLY. If you make an error, you may save time by crossing it out rather than trying to erase it.

GO ON TO THE NEXT PAGE.

AP PHYSICS 1

SECTION II

Directions: Questions 1–5 here are as follows: one experimental design question (worth 12 points), one quantitative/qualitative translation question (worth 12 points), one paragraph argument short-answer question (worth 7 points), and two additional short-answer questions (worth 7 points each). You have a total of 90 minutes to complete this section. Show your work for each part in the space provided after that part.

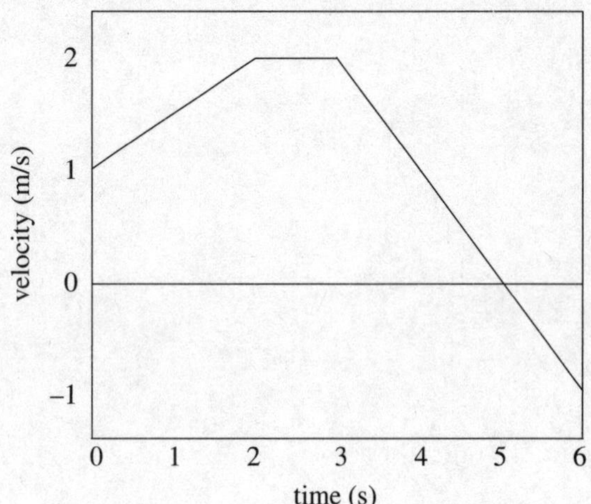

1. The motion of an object is given by the following velocity-versus-time graph.

(A) What is the displacement of the object from time $t = 0$ s to $t = 6$ s in the graph above?

(B) At what times is the speed of the object increasing?

(C) Make of a sketch of the object's position-versus-time graph during the time interval of $t = 0$ s to $t = 6$ s. Assume that the object begins at $x = 0$.

GO ON TO THE NEXT PAGE.

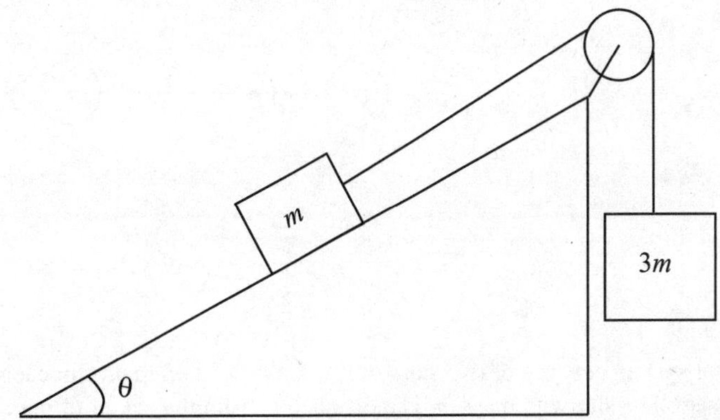

2. A small box of mass m is placed on an inclined plane with an angle of incline of θ. There is a coefficient of kinetic friction μ_k between the inclined plane and the small box. The small box is attached to a much heavier box of mass $3m$ by a pulley system, as shown above.

(A) Draw free-body diagrams of both masses, including all of the forces acting on each.

(B) Assuming a frictionless, massless pulley, determine the acceleration of the blocks once they are released from rest in terms of μ_k, g, and θ.

(C) If $\mu_k = 0.3$ and $\theta = 45°$, what distance is traveled by the blocks 3 s after being released from rest?

3. A 250-g volleyball is dropped from a height of 2 m. It bounces to a height of 1.8 m. Then, the volleyball is returned to a height 2 m, and a 60-g tennis ball is held centered, and slightly above the volleyball. Subsequently, the two are dropped simultaneously. The tennis ball collides with the volleyball just after the volleyball has bounced off of the ground. The volleyball continues up to a height of 0.3 m before returning to the ground again. (Neglect the effects of air resistance.)

(A) What percentage of its energy does the volleyball lose in its bounce? What happens to this lost energy?

(B) If it took 0.2 s for the volleyball to bounce, what was the average force on the ball during the bounce?

(C) How high does the tennis ball bounce?

GO ON TO THE NEXT PAGE.

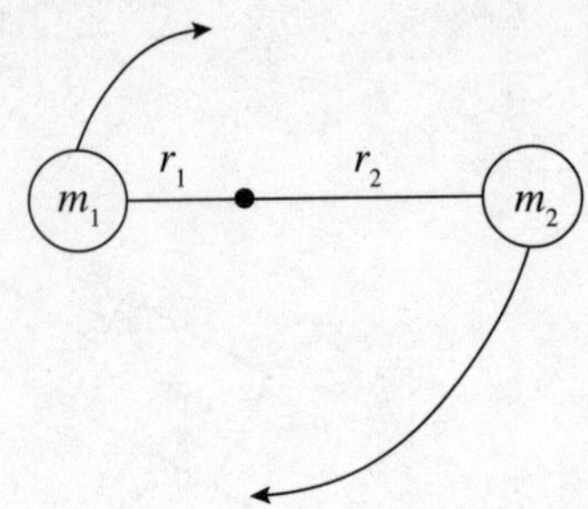

4. A particular binary star system consists of two stars of masses m_1 and m_2 that orbit each other in circular orbits about the center of mass of the system. The star with mass m_1 is a distance r_1 from the center of mass and orbits with a speed of v_1, and the star with mass m_2 is a distance r_2 from the center of mass and orbits with a speed of v_2.

(A) Derive an equation for the mass m_1 in terms of v_2, r_1, r_2, and the gravitational constant, G.

(B) Derive an equation for the ratio m_1/m_2 in terms of v_1 and v_2.

(C) If the star of mass m_2 is four times faster than the star of mass m_1, and the total mass of the system is 10^{31} kg, what are m_1 and m_2?

5. Two spheres (one hollow, one solid) are placed side by side on an inclined plane and released at the same time. Both spheres roll down the inclined plane without slipping.

(A) Using a free-body diagram, explain what force provides the torque that allows the spheres to roll down the inclined plane.

(B) Which sphere reaches the bottom of the inclined plane first and why?

(C) How do the kinetic energies of the two spheres compare at the bottom of the inclined plane?

STOP

END OF EXAM

Practice Test 3: Answers and Explanations

PRACTICE TEST 3: ANSWER KEY

1.	B	26.	D
2.	C	27.	D
3.	C	28.	C
4.	D	29.	B
5.	D	30.	D
6.	C	31.	C
7.	B	32.	B
8.	C	33.	B
9.	B	34.	C
10.	B	35.	C
11.	B	36.	A
12.	B	37.	A
13.	C	38.	B
14.	C	39.	A
15.	A	40.	A
16.	B	41.	D
17.	D	42.	A
18.	D	43.	D
19.	C	44.	B
20.	A	45.	A
21.	B	46.	B, C
22.	A	47.	A, D
23.	C	48.	A, D
24.	C	49.	B, C
25.	A	50.	B, C

PRACTICE TEST 3: ANSWERS AND EXPLANATIONS

Section I: Multiple-Choice Questions

1. **B** The kayaker's resultant velocity, **v**, will be the sum of the river's velocity, $\mathbf{v}_r$, and the velocity of her paddling, $\mathbf{v}_p$:

$$\mathbf{v} = \mathbf{v}_r + \mathbf{v}_p$$

Since the kayaker wishes to travel due east, the component of $\mathbf{v}_p$ in the N-S direction must cancel $\mathbf{v}_r$, so $v_{p,N\text{-}S}$ = 1° m/s due N. Since the magnitude of $\mathbf{v}_p$ is 1.5 m/s, her heading can be calculated using

1.5 m/s
1.0 m/s
θ

$$\sin\theta = \frac{1.0\ \text{m/s}}{1.5\ \text{m/s}}$$

$$\theta = \sin^{-1}\left(\frac{1}{1.5}\right) = 42°\ \text{N of E}$$

2. **C** The average acceleration is found using $\bar{\mathbf{a}} = \dfrac{\Delta\mathbf{v}}{\Delta t}$. Deal with each component separately.

$$a_x = \frac{\Delta v_x}{\Delta t} = \frac{v_{2x} - v_{1x}}{\Delta t} = \frac{4\ \text{m/s} - (-1\ \text{m/s})}{3\ \text{s}} = 1.667\ \text{m/s}^2$$

$$a_y = \frac{\Delta v_y}{\Delta t} = \frac{v_{2y} - v_{1y}}{\Delta t} = \frac{-2\ \text{m/s} - 4\ \text{m/s}}{3\ \text{s}} = -2\ \text{m/s}^2$$

It may help to draw the $\bar{\mathbf{a}}$ vector. Find the magnitude using the Pythagorean Theorem.

$$a = \sqrt{a_x^{\,2} + a_y^{\,2}} = \sqrt{\left(1.667\ \text{m/s}^2\right)^2 + \left(-2\ \text{m/s}^2\right)^2} = 2.6\ \text{m/s}^2$$

The angle this vector forms with the $+x$-axis can be evaluated using the equation

$$\theta = \tan^{-1}\left(\frac{a_y}{a_x}\right) = \tan^{-1}\left(\frac{-2\ \text{m/s}^2}{1.667\ \text{m/s}^2}\right) = -50.2°$$

3. **C** The student starts her journey at home and ends up at school 1.2 km away. It's unknown if she traveled in a straight line directly to school from home, so it is unknown if the distance she traveled was 1.2 km or greater than that if she took a detour. This makes I incorrect, eliminating (A) and (D). Not knowing the distance she travels also means her average speed is also unknown since average speed = distance/time. This makes II false, eliminating (B). Although the distance traveled is unknown, displacement is an overall change in position, so her displacement is known to have a magnitude of 1.2 km. So the magnitude of her average velocity can be calculated to be

$$\overline{v} = \frac{\Delta s}{\Delta t} = \frac{1200 \text{ m}}{600 \text{ s}} = 2 \text{ m/s}$$

Making III, and therefore (C), correct.

4. **D** The position-versus-time graphs depicted in (A) and (B) are linear, indicating constant velocity and zero acceleration. Eliminate (A) and (B). The position-versus-time graph in (C) has a negative velocity but has a positive acceleration. Eliminate (C). The position-versus-time graph in (D) has a negative velocity and a negative acceleration. Choice (D) is correct.

5. **D** The work done by a force that acts parallel to the displacement of the object is equal to the area under the graph of the magnitude of the force versus displacement. Split the area up into triangles, rectangles and trapezoids to find the total area.

$$W_{0\text{–}2\text{ s}} = \frac{1}{2}(2\text{ s})(5\text{ N}) = 5\text{ J}$$

$$W_{2\text{–}5\text{ s}} = (3\text{ s})(5\text{ N}) = 15\text{ J}$$

$$W_{5\text{–}7\text{ s}} = \frac{1}{2}(5\text{ N} + 3\text{ N})(2\text{ s}) = 8\text{ J}$$

$$W_{7\text{–}9\text{ s}} = (2\text{ s})(3\text{ N}) = 6\text{ J}$$

$$W_{0\text{–}2\text{ s}} = \frac{1}{2}(1\text{ s})(3\text{ N}) = 1.5\text{ J}$$

The total work done by the force is 5 J + 15 J + 8 J + 6 J + 1.5 J = 35.5 J.

6. **C** The horizontal distance that the water travels from its drop point is given by $x = v_{0x}t$. Since the water is dropped from the moving plane, its initial velocity will be the same as that of the airtanker. Therefore, the water's initial velocity is $\mathbf{v}_0 = (85 \text{ m/s}, 0 \text{ m/s})$. Thus, $v_{0x} = 85$ m/s. The time it will travel is equivalent to the water's flight time, which can be calculated using the y-direction:

$$\Delta y = v_{0y}t + \frac{1}{2}gt^2$$

$$-300 \text{ m} = 0 + \frac{1}{2}(-9.8 \text{ m/s}^2)t^2$$

$$t \approx 7.8 \text{ s}$$

Therefore, $x = v_{0x}t = (85 \text{ m/s})(7.8 \text{ s}) = 660 \text{ m}$.

7. **B** To compare the displacement of the two cannonballs, the x- and y-components of the displacement must be calculated separately. The x-component of displacement is given by $\Delta x = v_{0x}t$. Since Cannonball A is fired horizontally, $v_{0x,\text{A}} = 40$ m/s, and Cannonball B has an initial x-velocity of $v_{0x,\text{B}} = (80 \text{ m/s})\cos 60° = (80 \text{ m/s})\left(\frac{1}{2}\right) = 40 \text{ m/s}$. Because the x-component of the initial velocities and the times will be the same, $\Delta x_{\text{A}} = \Delta x_{\text{B}}$. Thus, only differences in the displacement in the y-direction will determine the difference in the magnitude of displacement. Applying Big Five #3 for both cannonballs yields,

$$\Delta y_{\text{A}} = v_{0y,\text{A}}t + \frac{1}{2}at^2 = (0 \text{ m/s})(2 \text{ s}) + \frac{1}{2}(-9.8 \text{ m/s}^2)(2 \text{ s})^2 = -19.6 \text{ m}$$

$$\Delta y_{\text{B}} = v_{0y,\text{B}}t + \frac{1}{2}at^2 = (80 \text{ m/s})\sin 60°(2 \text{ s}) + \frac{1}{2}(-9.8 \text{ m/s}^2)(2 \text{ s})^2$$

$$= 138.6 \text{ m} - 19.6 \text{ m} = 119 \text{ m}$$

Since $|\Delta y_{\text{A}}| < |\Delta y_{\text{B}}|$, Cannonball B will have the greater magnitude of displacement after 2 s.

8. **C** Use Big Five #1:

$$\Delta x = \frac{1}{2}(v_0 + v)t \Rightarrow 600 \text{ m} = \frac{1}{2}(0 \text{ m/s} + v) \cdot 8 \text{ s} \Rightarrow 600 \text{ m} = 4 \text{ s} \cdot v$$

$$\Rightarrow 150 \text{ m/s} = v$$

9. **B** The variables involved in this question are the initial velocity (given in both cases), the acceleration (constant), final velocity (0 in both cases), and displacement (the skidding distance). As the missing variable is time, use Big Five #5 to solve for the displacement:

$$v^2 = v_0^2 + 2a(x - x_0)$$

$$x - x_0 = \frac{v^2 - v_0^2}{2a}$$

As the initial position, x_0, and the final velocity, v^2, are equal to 0, this equation simplifies to

$$x = \frac{-v_0^2}{2a}$$

When the acceleration is doubled, the final position is halved. (Note that the acceleration in this problem is negative, as the brakes cause the car to decelerate.)

10. **B** There are four forces acting on the book: the push from the student, gravity, normal force, and friction. Of these four forces, normal force and friction are forces exerted by the table on the book. The normal force is calculated from knowing that the net force in the *y*-direction is zero:

$$F_{\text{net},y} = F_N - F_g - F_{\text{student},y} = 0$$

$$F_N = F_g + F_{\text{student},y}$$

$$F_N = (3\text{ kg})(10\text{ m/s}^2) + (30\text{ N})\sin 10° = 35.2\text{ N}$$

The force of friction is calculated using the normal force:

$$F_f = \mu F_N = (0.3)(35.2\text{ N}) = 10.6\text{ N}$$

The force on the book from the table will then be the vector sum of the normal and friction forces. Since the normal force is in the *y*-direction, and the friction force is in the *x*-direction, the magnitude of their sum is found using

$$|F_{\text{table}}| = \sqrt{F_N^2 + F_f^2} = \sqrt{(35.2\text{ N})^2 + (10.6\text{ N})^2} \approx 37.0\text{ N}$$

11. **B** There are free forces acting on the box: the weight of the box, the normal force of the inclined plane on the box, and the force of friction. The weight of the box can be separated into its parallel and perpendicular components.

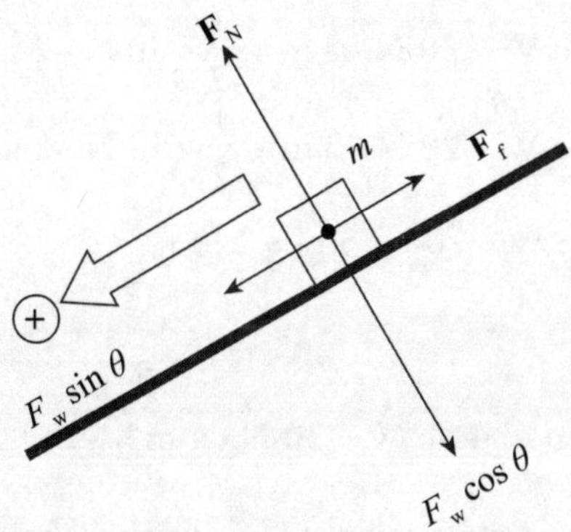

In order for the box to be at rest, the net force in both the parallel and perpendicular directions must be zero:

$$F_N - F_w \cos\theta = 0 \Rightarrow F_N = F_w \cos\theta \Rightarrow F_N = mg\cos\theta$$

$$F_w \sin\theta - F_f = 0 \Rightarrow F_w \sin\theta = F_f \Rightarrow mg\sin\theta = \mu_s F_N \Rightarrow$$

$$\mu_s = \frac{mg\sin\theta}{mg\cos\theta} = \frac{\sin\theta}{\cos\theta} = \tan\theta = \tan 30° = 0.58$$

12. **B** When the hammer hits the nail, it exerts an average force over a time interval on the nail, which imparts an impulse on the nail. According to Newton's Third Law, the hammer receives the same magnitude of average force in the opposite direction. Since the time of contact is the same, the hammer experiences the same magnitude of impulse as the nail. Impulse is also the change in momentum of an object. Therefore, the nail experiences a greater impulse when the hammer demonstrates a greater change in momentum. In the first strike, the change in momentum is $\Delta \mathbf{p}_1 = 0 - mv_0 = -mv_0$. In the second strike, the hammer recoils with some v in the direction opposite the initial velocity, so $\Delta \mathbf{p}_2 = -mv - mv_0$. $\Delta \mathbf{p}_2$ has a greater magnitude than $\Delta \mathbf{p}_1$, so the second strike imparts more momentum on the nail.

13. **C** The minimum force exerted to maintain the speed of 2 m/s will result in no acceleration. Thus, the force required will result in $F_{net} = 0$. The friction force acting against the motion is

$$F_f = \mu F_N = \mu mg = (0.25)(5\text{ kg})\left(10\text{ m/s}^2\right) = 12.5\text{ N}$$

Thus, the force required to maintain the speed is 12.5 N, just enough to counteract the friction force.

14. **C** On a force-versus-position graph, the work done is the area under the curve. At the 4 m position, the area under both student's curves is the same:

Student A:

$$0 \text{ m to } 2 \text{ m: } W = 30 \text{ N} \cdot 2 \text{ m} + \frac{1}{2} \cdot 10 \text{ N} \cdot 2 \text{ m} = 70 \text{ J}$$

$$2 \text{ m to } 4 \text{ m: } W = 20 \text{ N} \cdot 2 \text{ m} + \frac{1}{2} \cdot 10 \text{ N} \cdot 2 \text{ m} = 50 \text{ J}$$

$$0 \text{ m to } 4 \text{ m: } W = 70 \text{ J} + 50 \text{ J} = 120 \text{ J}$$

Student B:

$$0 \text{ m to } 4 \text{ m: } W = 30 \text{ N} \cdot 4 \text{ m} = 120 \text{ J}$$

15. **A** The work done by the brakes can be calculated using the Work–Energy Theorem:

$$W = \Delta K = K_{\text{final}} - K_{\text{initial}} = \frac{1}{2}mv^2 - \frac{1}{2}mv_o^{\,2} = \frac{1}{2}m(v^2 - v_o^{\,2}) =$$

$$\frac{1}{2}(2000 \text{ kg})(10 \text{ m/s})^2 - (20 \text{ m/s})^2 = \frac{1}{2}(2000 \text{ kg})(100 \text{ m}^2/\text{s}^2 - 400 \text{ m}^2/\text{s}^2)$$

$$= (1000 \text{ kg})(-300 \text{ m}^2/\text{s}^2) = -300{,}000 \text{ J}$$

Note that the work done is negative as the friction force of the brakes caused the truck to *lose kinetic energy.*

16. **B** Apply Conservation of Mechanical Energy:

$$K_i + U_i = K_f + U_f$$

$$0 + mgh = \frac{1}{2}mv_f^2 + \frac{mgh}{2}$$

$$\frac{mgh}{2} = \frac{1}{2}mv_f^2$$

$$gh = v_f^2$$

$$\sqrt{gh} = v_f$$

17. **D** In order to maintain a constant speed, the net force on the car must be zero. This means that the car's engine must be applying a force equal in magnitude but opposite in direction to the frictional force. The equation $P = Fv$ then yields,

$$P = Fv = 3500 \text{ N} \cdot 15 \text{ m/s} = 52{,}500 \text{ W} = 52.5 \text{ kW}$$

18. **D** The total work done on the rocket can be calculated using the Work–Energy Theorem:

$$W = \Delta K = K_{\text{final}} - K_{\text{initial}} = \frac{1}{2}mv^2 - \frac{1}{2}mv_o^2 = \frac{1}{2}m\left(v^2 - v_o^2\right) =$$

$$\frac{1}{2}(50 \text{ kg})(15 \text{ m/s})^2 - (10 \text{ m/s})^2 = \frac{1}{2}(50 \text{ kg})(225 \text{ m}^2/\text{s}^2 - 100 \text{ m}^2/\text{s}^2) = (25 \text{ kg})(125 \text{ m}^2/\text{s}^2) = 3{,}125 \text{ J}$$

This is the total work done on the rocket. As the rocket's secondary engines did 4,500 J of work, this means that the force of gravity must have done –1,375 J of work.

19. **C** The gravitational potential energy of the rock is converted into kinetic energy. This allows you to calculate the final speed of the rock when it lands:

$$mgh = \frac{1}{2}mv^2 \Rightarrow gh = \frac{1}{2}v^2 \Rightarrow v = \sqrt{2gh} = \sqrt{2\left(10 \text{ m/s}^2\right)\left(20 \text{ m}\right)} = 20 \text{ m/s}$$

Since the rock is moving downward, its velocity is –20 m/s. The average force experienced by the rock can be calculated using

$$\mathbf{F} = \frac{\Delta \mathbf{p}}{\Delta t} = \frac{\mathbf{p}_f - \mathbf{p}_i}{\Delta t} = \frac{0 - m\boldsymbol{v}}{\Delta t} = \frac{0 - (1 \text{ kg})(-20 \text{ m/s})}{0.25 \text{ s}} = 80 \text{ N}$$

20. **A** As the plutonium atom is initially at rest, it has an initial momentum of 0. By Conservation of Linear Momentum, the total final momentum must also be equal to 0.

$$m_{244}\mathbf{v}_{244,i} = m_4\mathbf{v}_{4,f} + m_{240}\mathbf{v}_{240,f}$$

$$0 = m_4\mathbf{v}_{4,f} + m_{240}\mathbf{v}_{240,f} \Rightarrow \mathbf{v}_{240,f} = -\frac{m_4\mathbf{v}_{4,f}}{m_{240}} = -\frac{4 \text{ Da} \cdot 45 \text{ m/s}}{240 \text{ Da}} = -0.75 \text{ m/s}$$

As the alpha particle was traveling east, the decay product must be traveling west, as its velocity is in the opposite direction.

21. **B** Friction does work to stop the vehicles after the collision, so apply the Work–Energy Theorem to determine the speed of the vehicles right after the collision,

$$W = \Delta K \rightarrow F_f d = 0 - \frac{1}{2}mv^2 \rightarrow \mu\left(m_t + m_c\right)gd = \frac{1}{2}\left(m_t + m_c\right)v^2 \rightarrow v = \sqrt{2\mu gd}$$

$$v = \sqrt{2(0.3)\left(10 \text{ m/s}^2\right)\left(8 \text{ m}\right)} = 6.93 \text{ m/s}$$

Then, apply Conservation of Momentum to the perfectly inelastic collision.

$$m_t\mathbf{v}_t + m_c\mathbf{v}_c = (m_t + m_c)\mathbf{v} \rightarrow \mathbf{v}_c = \frac{(m_t + m_c)\mathbf{v} - m_t\mathbf{v}_t}{m_c}$$

$$\mathbf{v}_c = \frac{(2000\text{ kg} + 1500\text{ kg})(7.02\text{ m/s}) - (2000\text{ kg})(40\text{ m/s})}{1500\text{ kg}} = -37.2\text{ m/s}$$

The negative sign indicates that the car was traveling south.

22. **A** The objects did not stick together after the collision, so (B) is wrong. To determine whether the collision is elastic or inelastic, calculate the total initial and final kinetic energies:

$$K_i = \frac{1}{2}m_{2\text{kg}}v_{2\text{kg,i}}^2 + \frac{1}{2}m_{1\text{kg}}v_{1\text{kg,i}}^2 = \frac{1}{2}(2\text{ kg})(25\text{ m/s})^2 + \frac{1}{2}(1\text{ kg})(20\text{ m/s})^2 = 825\text{ J}$$

$$K_f = \frac{1}{2}m_{2\text{kg}}v_{2\text{kg,f}}^2 + \frac{1}{2}m_{1\text{kg}}v_{1\text{kg,f}}^2 = \frac{1}{2}(2\text{ kg})(2.5\text{ m/s})^2 + \frac{1}{2}(1\text{ kg})(35\text{ m/s})^2 = 618.75\text{ J}$$

As kinetic energy was not conserved, the collision is inelastic.

23. **C** Choices (A), (B), and (D) are all true statements about perfectly inelastic collisions. Choice (C) is only true for elastic collisions.

24. **C** Impulse can be found using $\mathbf{J} = \Delta\mathbf{p}$. Remembering that impulse and momentum are vectors, the magnitude of the impulse can be calculated by finding the change in the momentum in both the *x*- and *y*-directions, and finding the resulting magnitude:

$$J = |\Delta\mathbf{p}| = \sqrt{\Delta p_x^{\ 2} + \Delta p_y^{\ 2}}$$

$$= m\sqrt{\Delta v_x^{\ 2} + \Delta v_y^{\ 2}}$$

$$= m\sqrt{(v_{x,\text{f}} - v_{x,\text{i}})^2 + (v_{y,\text{f}} - v_{y,\text{i}})^2}$$

$$= (0.145\text{ kg})\sqrt{((-50\text{ m/s})\cos 40° - 40\text{ m/s})^2 + ((-50\text{ m/s})\sin 40° - 0)^2}$$

$$\approx 12.3\text{ N·s}$$

25. **A** Total linear momentum is conserved in the absence of external forces:

$$\mathbf{p}_{1,i} + \mathbf{p}_{2,i} = \mathbf{p}_{1+2,f}$$

$$m_1v_{1,i} + m_2v_{2,i} = (m_1 + m_2)v_f$$

$$m_1v_0 + 2\,m_1v_{2,i} = (m_1 + 2\,m_1)(0.5\,v_0)$$

$$m_1v_0 + 2\,m_1v_{2,i} = 1.5\,m_1v_0$$

$$2\,m_1v_{2,i} = 0.5\,m_1v_0$$

$$v_{2,i} = 0.25\,v_0$$

26. **D** At time $t = 3$ s, the speed of the object is 4 m/s. The magnitude of the centripetal acceleration is given by

$$a_c = \frac{v^2}{r} = \frac{(4\text{ m/s})^2}{0.5\text{ m}} = 32\text{ m/s}^2$$

27. **D** When the car is making the curve, it is the force of friction providing the centripetal force:

$$F_c = F_f$$

$$\frac{mv^2}{r} = \mu F_N$$

$$\frac{mv^2}{r} = \mu mg$$

$$\frac{v^2}{r} = \mu g$$

$$v^2 = r\mu g$$

$$v = \sqrt{r\mu g} = \sqrt{50\text{ m}(0.8)(10\text{ m/s}^2)} = 20\text{ m/s}$$

28. **C** The gravitational pull of the planet provides the centripetal force on the planetoid, so

$$\mathbf{F}_g = \mathbf{F}_c$$

$$G\frac{Mm}{r^2} = \frac{mv^2}{r} \Rightarrow G\frac{M}{r} = v^2 \Rightarrow v = \sqrt{\frac{GM}{r}}$$

The radius between the planetoid and the planet is equal to the sum of the planet's radius and the height of the planetoid above the planet's surface: $r = 1.6 \times 10^5$ m + 3.2×10^5 m = 4.8×10^5 m. The speed of the planetoid can then be calculated as

$$v = \sqrt{\frac{GM}{r}} = \sqrt{\frac{(6.67 \times 10^{-11}\ \text{N} \cdot \text{m}^2/\text{kg}^2)(4 \times 10^{20}\ \text{kg})}{4.8 \times 10^5\ \text{m}}} = 236\ \text{m/s}$$

29. **B** The magnitude of the gravitational force is given by the equation:

$$F_G = G\frac{Mm}{r^2}$$

Doubling the mass of one object will double the magnitude of the gravitational force. Tripling the radius will decrease the magnitude of the gravitational force by a factor of nine. Combined, the new magnitude of the gravitational force is 2/9 of the original magnitude.

30. **D** Apply Big Five #3 for rotational motion:

$$\theta = \theta_0 + \omega_0 t + \frac{1}{2}\alpha t^2 = \frac{1}{2}(2\pi\ \text{rad/s}^2)(3\ \text{s})^2 = 9\pi\ \text{rad}$$

The circumference of the circle $d = 2\pi r = 2\pi(2\ \text{m}) = 4\pi$ m. 9π rad is equivalent to 4.5 rotations and a total distance traveled of $4.5 \cdot 4\pi\ \text{m} = 18\pi$ m.

31. **C** The position that the string can be attached to hold the bar up horizontally is the center of mass of the bar, which can be calculated using the center of mass equation with the far left side as the origin:

$$x(\text{cm}) = \frac{x_1 m_1 + x_2 m_2 + x_3 m_3}{m_1 + m_2 + m_3} = \frac{0\ \text{m}(2\ \text{kg}) + 1.5\ \text{m}(4\ \text{kg}) + 0.75\ \text{m}(4\ \text{kg})}{2\ \text{kg} + 4\ \text{kg} + 4\ \text{kg}}$$

$$= \frac{9}{10}\ \text{m} = 0.9\ \text{m}$$

32. **B** If the system is in equilibrium, then the net force acting on the mass is equal to 0. In the vertical direction, the forces acting on the mass are the weight of the mass and the vertical components of the two tension forces,

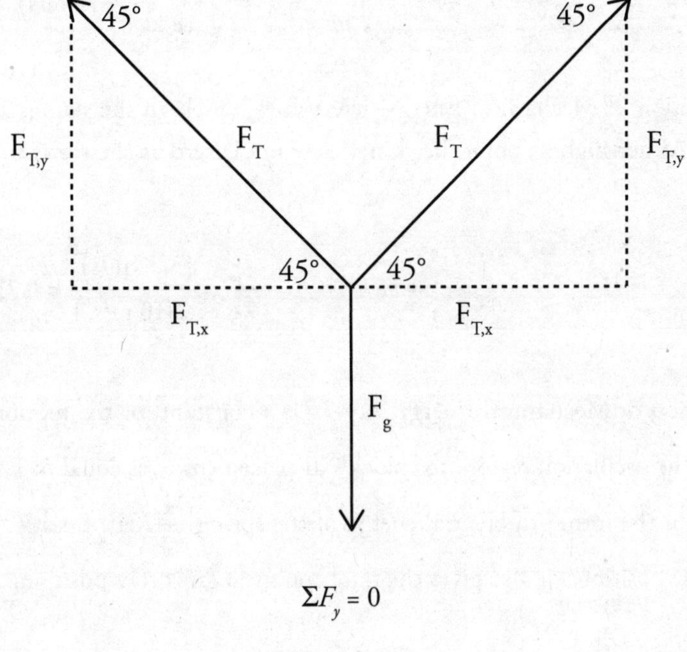

$$\Sigma F_y = 0$$

$$2F_{T,y} - F_g = 0$$

$$2F_{T,y} = F_g$$

$$2F_T \sin(45°) = mg$$

$$F_T = \frac{mg}{2\sin 45°} = \frac{10\text{ kg}\left(10\text{ m/s}^2\right)}{2\left(\frac{\sqrt{2}}{2}\right)} = \frac{100}{\sqrt{2}}\text{ N} = 70.7\text{ N}$$

33. **B** Apply Conservation of Energy to solve for the skier's height when they stop,

$$K_i + U_i = K_f + U_f \rightarrow \frac{1}{2}mv_i^2 + 0 = 0 + mgh_f \rightarrow h_f = \frac{v_i^2}{2g} = \frac{\left(2\text{ m/s}\right)^2}{2\left(9.8\text{ m/s}^2\right)} = 0.20\text{ m}$$

34. **C** When the spring is attached to the ceiling, the spring is at its natural length when the upward force of the spring balances the weight of the block. The increase in the natural length is the distance that the spring is stretched to exert the upward force:

$$F_{spring} = F_g$$

$$kx = mg$$

$$m = \frac{kx}{g} = \frac{\left(50\text{ N/m}\right)\left(0.02\text{ m}\right)}{10\text{ m/s}^2} = 0.1\text{ kg}$$

35. **C** The boy's push transfers energy to his sister, changing her speed according to the Work–Energy Theorem.

$$W = \Delta K \rightarrow W = \frac{1}{2}mv_1^2 - \frac{1}{2}mv_0^2 \rightarrow v_1 = \sqrt{\frac{2W}{m} + v_0^2} = \sqrt{\frac{2(50\text{ J})}{18\text{ kg}} + (3\text{ m/s})^2} = 3.82\text{ m/s}$$

The Conservation of Mechanical Energy determines how high she swings above her lowest point, where $h = 0$. At her highest point, her kinetic energy is zero as she slows to a stop before turning around.

$$K_i + U_i = K_f + U_f \rightarrow \frac{1}{2}mv_1^2 = mgh_f \rightarrow h_f = \frac{v_1^2}{2g} = \frac{(3.82\text{ m/s})^2}{2(10\text{ m/s}^2)} = 0.73\text{ m}$$

36. **A** By Conservation of Mechanical Energy, $K + U_s$ is a constant for the motion of the block. At the endpoints of the oscillation region, the block's displacement, x, is equal to $\pm A$. Since $K = 0$ here, all the energy is in the form of potential energy of the spring, $\frac{1}{2}KA^2$. Because $\frac{1}{2}KA^2$ gives the total energy at these positions, it also gives the total energy at any other position.

Using the equation $U_s(x) = \frac{1}{2}kx^2$, find that at $x = \frac{1}{4}A$,

$$U_s(x) = \frac{1}{2}k\left(\frac{1}{4}A\right)^2 = \frac{1}{32}kA^2$$

The potential energy at this point is 1/16 of the total energy.

37. **A** To stay on the track, the centripetal force at the top of the track must be greater than the force of gravity, $\frac{mv^2}{R} > mg$. Relate the speed of the rollercoaster to H and h using Conservation of Energy. $K_i + U_i = K_f + U_f \rightarrow 0 + mgH = \frac{1}{2}mv^2 + mgh \rightarrow mv^2 = 2mg(H - h)$. Substitute into the centripetal force relation, and solve for H:

$$\frac{2mg(H-h)}{R} > mg \rightarrow 2H - 2h > R \rightarrow H > h + \frac{R}{2}$$

38. **B** The rotational inertia of the system increases when the clay is dropped on the wheel. No net external torque acts on the system, so apply the Conservation of Angular Momentum:

$$L_i = L_f \rightarrow I_i\omega_i = I_f\omega_f \rightarrow I_{wheel}\omega_i = \left(I_{wheel} + I_{clay}\right)\omega_f \rightarrow \omega_f = \frac{I_{wheel}\omega_i}{I_{wheel} + mr^2}$$

$$\omega_f = \frac{\left(7 \text{ kg}\cdot\text{m}^2\right)\left(30 \text{ rad/s}\right)}{7 \text{ kg}\cdot\text{m}^2 + \left(2.5 \text{ kg}\right)\left(0.3 \text{ m}\right)^2} = 29.1 \text{ rad/s}$$

39. **A** Apply Newton's 2nd Law the system to solve for acceleration,

$$F_{net} = ma \rightarrow F_{eng} = \left(m_A + m_B\right)a \rightarrow a = \frac{F_{eng}}{m_A + m_B}$$

The net force on car B is the tension in the connector, so Newton's 2nd Law for car B yields,

$$T = m_B a = m_B\left(\frac{F_{eng}}{m_A + m_B}\right) = \left(0.3 \text{ kg}\right)\left(\frac{0.6 \text{ N}}{0.5 \text{ kg} + 0.3 \text{ kg}}\right) = 0.23 \text{ N}$$

40. **A** The frame is stationary, so $F_{net} = 0$. In the y-direction, $F_{net,y} = F_{app,y} + F_{f,s} - F_g = 0$. Therefore,

$$F_{f,s} = F_g - F_{app,y} = mg - F_{app}\sin\theta = \left(1.3 \text{ kg}\right)\left(9.8 \text{ m/s}^2\right) - \left(15 \text{ N}\right)\sin 25° = 6.4 \text{ N}$$

Note that the force of friction can take on any value up to its maximum value of $F_{f,s,max} = \mu_s F_N$.

41. **D** The friction from the rubber exerts torque on the spinning basket, which causes an angular acceleration on it. Note that the lever arm causing the torque equals the radius of the basket since the friction is applied to the outer edge.

$$\alpha = \frac{\tau}{I} = \frac{r_{\perp}F_f}{I} = \frac{r_{\perp}\mu_k F_n}{I} = \frac{\left(0.15 \text{ m}\right)\left(0.35\right)\left(5 \text{ N}\right)}{0.1 \text{ kg}\cdot\text{m}^2} = 2.625 \text{ rad/s}^2$$

Apply Big 5 #2 for rotational motion to solve for time. Use a negative value for angular acceleration since it slows the spinner down:

$$\omega = \omega_0 + \alpha t \rightarrow t = \frac{\omega - \omega_0}{\alpha} = \frac{0 \text{ rad/s} - 26 \text{ rad/s}}{-2.625 \text{ rad/s}^2} = 9.9 \text{ s}$$

42. **A** Moving the arms farther from the body's center of mass increases its rotational inertia. As a result, any torque on the body will result in a lower angular acceleration than if the arms were at their sides. Raising one's arms does raise the location of the body's center of mass, but this would increase the lever arm for forces exerted at the feet, so eliminate (B). Although having one arm outstretched at either side of the body does balance the torque, this can happen even if the arms are not outstretched, so eliminate (C). Increasing the body's cross-sectional area would increase the impact of air currents, so eliminate (D).

43. **D** Apply Newton's 2nd Law $F_{net} = ma$ to each mass. Assume that the pulley rotates clockwise, and use this direction of motion to define the positive direction for each mass. So, up is positive for m_1, and $T = m_1g = m_1a$. Down is positive for m_2, and $m_2g - T = m_2a$. The magnitudes of tension and acceleration are the same for each mass, so solve each equation for a, and set them equal to each other to solve for T:

$$\frac{T}{m_1} - g = g - \frac{T}{m_2} \rightarrow \frac{T}{m_1} + \frac{T}{m_2} = 2g \rightarrow T\left(\frac{1}{m_1} + \frac{1}{m_2}\right) = 2g \rightarrow T = \frac{2m_1m_2}{m_1 + m_2}g$$

44. **B** The two runners both start and finish their runs at rest, so the total work done on each is zero. The work done by each runner and the work done by gravity contribute to the total work done on each runner. The work done by gravity only depends on the change in height. As the runners experience the same change in height, gravity does the same work on each. Therefore, each runner does the same amount of work on their runs.

45. **A** Apply the definition for work done by a force, $W = Fd\cos\theta$. The elevator is not accelerating, so the net force on the woman is zero. Therefore, $F_N = mg$. F_N points up, and the displacement is down, so $\theta = 180°$. Therefore,

$$W = mgd\cos 180° = -(60\text{ kg})(9.8\text{ m/s}^2)(8\text{ m}) = -4700\text{ J}$$

46. **B, C** A particle slows down when the direction of its acceleration is opposite to the direction of its velocity. As the sign of a vector indicates its direction, the particle slows down if its velocity and acceleration have opposite signs, making (B) and (C) correct.

47. **A, D** The work done by a force is $W = Fd\cos\theta$, where θ is the angle between the force and displacement vectors. Since the normal force is perpendicular to the surface of the ramp and the displacement is along the surface, $\theta = 90°$ for the normal force, so it does no work. Therefore, (A) is correct. The work done gravity is also equal to the negative change in potential energy of the box. As the box slides from a given height, the change in potential energy is independent of ramp angle, so (B) is incorrect. Gravity does positive work on the box, friction does negative work on the box, and the sum of the work done by gravity and friction is the total work done on the box. As the box gains kinetic energy, the total work done on the box must be positive. Therefore, the work done by gravity must be greater than the magnitude of the work done by friction, making (C) incorrect. Work done by gravity is independent of ramp angle, but the force of friction is proportional to the normal force, which does depend on ramp angle, making (D) correct.

48. **A, D** When springs are connected in parallel as in (A) and (B), the effective spring constant can be calculated with the equation:

$$k_{eff} = k_1 + k_2$$

Choice (A) is correct, as $k_{eff} = k_1 + k_2 = k + k = 2k$.

When springs are connected in series, as in (C) and (D), the effective spring constant can be calculated with the equation:

$$\frac{1}{k_{\text{eff}}} = \frac{1}{k_1} + \frac{1}{k_2}$$

Choice (D) is correct, as $\frac{1}{k_{\text{eff}}} = \frac{1}{k_1} + \frac{1}{k_2} \Rightarrow k_{\text{eff}} = \frac{k_1 k_2}{k_1 + k_2} = \frac{4k(4k)}{4k + 4k} = \frac{16k^2}{8k} = 2k$.

49. **B, C** An object is defined to be in translational equilibrium when the sum of the forces acting on it is zero. If the net force on the object is equal to 0, the object's acceleration is also 0. An object at translational equilibrium can have a nonzero kinetic energy and net torque.

50. **B, C** The forearm moves with a constant angular speed, so the net torque on the forearm is zero throughout the motion, making (C) correct. Furthermore, the torque from the bicep to bend the elbow is equal in magnitude to the sum of the torques from the weight of the forearm and the object to straighten the elbow. The weight of the forearm exerts a non-zero torque on the forearm, so (D) is incorrect. Torque is given by $\tau = rF\sin\theta$, so since the r for the biceps is much smaller than the object and since the biceps produce a greater magnitude of torque than the object, the force from the biceps is greater than the force from the object, making (B) correct. Although the forearm doesn't experience angular acceleration, it does experience a centripetal acceleration as it takes a circular path around the elbow, so (A) is incorrect.

Section II: Free-Response Questions

1.

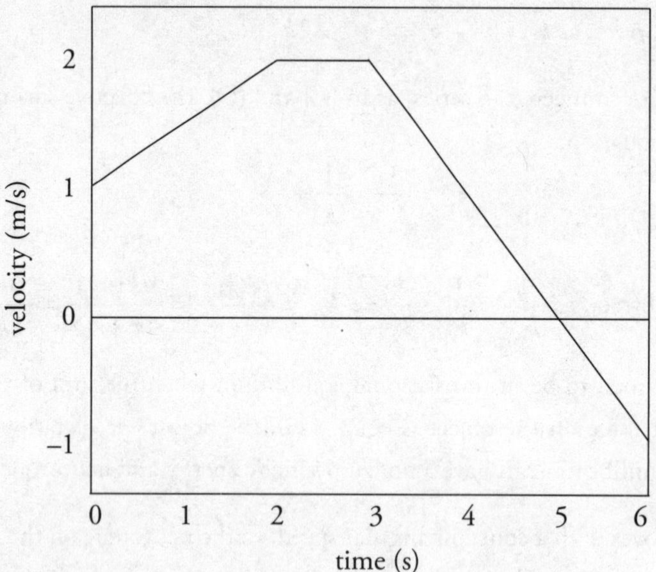

(A) On a velocity-versus-time graph, the displacement is the area between the graph and the x-axis. Area above the x-axis is positive displacement and area below the x-axis is negative displacement. The displacement from time $t = 0$ s to $t = 6$ s can be evaluated as

$t = 0$ s to $t = 2$ s: $(2 \text{ s})(1 \text{ m/s}) + \frac{1}{2}(2 \text{ s})(1 \text{ m/s}) = 3 \text{ m}$

$t = 2$ s to $t = 3$ s: $(1 \text{ s})(2 \text{ m/s}) = 2 \text{ m}$

$t = 3$ s to $t = 4$ s: $(1 \text{ s})(1 \text{ m/s}) + \frac{1}{2}(1 \text{ s})(1 \text{ m/s}) = 1.5 \text{ m}$

$t = 4$ s to $t = 5$ s: $\frac{1}{2}(1 \text{ s})(1 \text{ m/s}) = 0.5 \text{ m}$

$t = 5$ s to $t = 6$ s: $\frac{1}{2}(1 \text{ s})(-1 \text{ m/s}) = -0.5 \text{ m}$

The total displacement from time $t = 2$ s to $t = 6$ s is 3 m + 2 m + 1.5 m + 0.5 m – 0.5 m = 6.5 m.

(B) The speed of the object is increasing when the magnitude of the velocity is increasing. This occurs from time $t = 0$ s to $t = 2$ s and $t = 5$ s to $t = 6$ s.

(C)

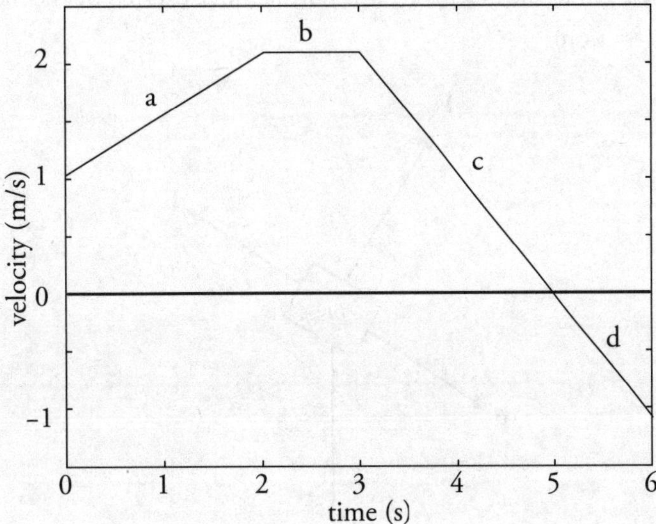

Section (A) tells you the object is speeding up in the positive direction (creating a parabola with an increasingly steeper slope). Section (B) tells you the object is traveling with constant speed in the positive direction (linear line). Section (C) tells you the object is slowing down but still moving in the positive direction (parabola with a decreasingly steeper slope). Section (D) tells you the object is moving in the negative direction and speeding up (parabola with an increasingly steeper slope).

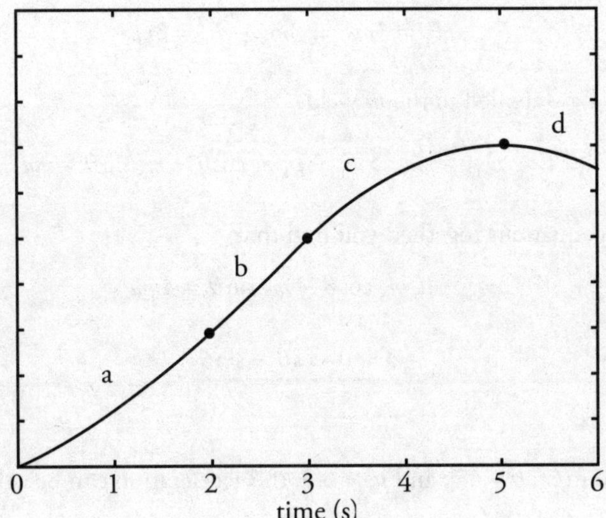

2. (A) The forces acting on the small m mass are $\mathbf{F}_T$ (the tension in the string connecting it to the $3m$ block), $\mathbf{F}_w$ (the weight of the block), $\mathbf{F}_N$ (the normal force exerted by the inclined plane), and $\mathbf{F}_f$ (the force of kinetic friction).

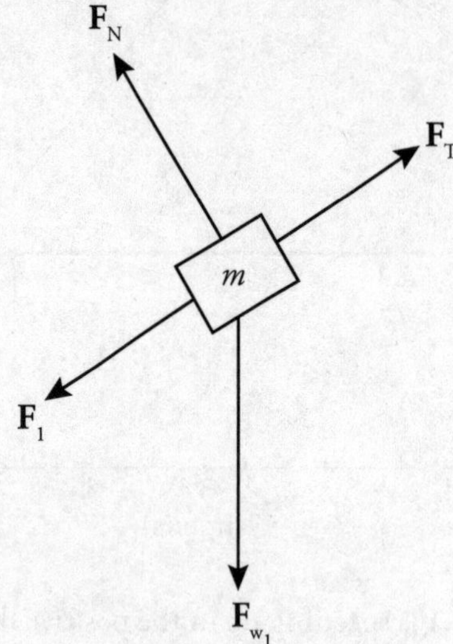

The forces acting on the larger $3m$ mass are $\mathbf{F}_T$ (the tension in the string connecting it to the m block) and $\mathbf{F}_w$ (the weight of the block).

(B) Newton's Second Law applied to the $3m$ mass yields

$$F_{w2} - F_T = 3ma \Rightarrow 3mg - F_T = 3ma$$

Newton's Second Law applied to the m yields

$$F_T - F_f - F_{w1,\parallel} = ma \Rightarrow F_T - \mu_k mg\cos\theta - mg\sin\theta = ma$$

Adding these two equations together, you find that

$$3mg - \mu_k mg\cos\theta - mg\sin\theta = 4ma$$

$$a = \frac{3 - \mu_k\cos\theta - \sin\theta}{4}g$$

(C) Plugging in $g = 10\text{ m/s}^2$, $\theta = 45°$, and $\mu_k = 0.3$, the acceleration can be calculated as follows:

$$a = \frac{3 - \mu_k\cos\theta - \sin\theta}{4}g$$

$$a = \frac{3 - (0.3)\cos 45° - \sin 45°}{4}\left(10\text{ m/s}^2\right)$$

$$a = 5.20\text{ m/s}^2$$

Apply Big Five #3:

$$d = v_0 t + \frac{1}{2}at^2 \Rightarrow d = 0 + \frac{1}{2}(5.20 \text{ m/s}^2)(3 \text{ s})^2 = 23.4 \text{ m}$$

3. (A) The energy lost is the difference in potential energy at the maximum heights before and after the bounce, $\Delta E = U_f - U_i = mg(h_f - h_i)$. The percentage of the initial energy lost in the bounce is therefore,

$$\frac{\Delta E}{U_i} = \frac{mg(h_f - h_i)}{mgh_i} = \frac{h_f - h_i}{h_i} = \frac{1.8 \text{ m} - 2 \text{ m}}{2 \text{ m}} = -0.10 = 10\% \text{ lost}$$

(B) Use impulse to determine the average force, $\vec{F}\Delta t = \Delta\vec{p} \rightarrow \vec{F} = \frac{\Delta\vec{p}}{\Delta t}$. Use Conservation of Energy to determine the speed before the bounce:

$$K_i + U_i = K_f + U_f \rightarrow 0 + mgh_i = \frac{1}{2}mv_f^2 + 0 \rightarrow v_f = \sqrt{2gh_i} = \sqrt{2(9.8 \text{ m/s}^2)(2 \text{ m})} = 6.26 \text{ m/s}$$

After the bounce, the kinetic energy is 90% the initial potential energy:

$$K' = 0.9U_i \rightarrow \frac{1}{2}mv'^2 = 0.9mgh_i \rightarrow v' = \sqrt{1.8gh_i} = \sqrt{1.8(9.8 \text{ m/s}^2)(2 \text{ m})} = 5.94 \text{ m/s}$$

Substitute into the impulse equation, remembering that the velocity before the bounce is down and after the bounce is up.

$$\vec{F} = \frac{m(v_f - v_i)}{\Delta t} = \frac{(0.25 \text{ kg})[5.94 \text{ m/s} - (-6.26 \text{ ms})]}{0.1 \text{ s}} = 15.3 \text{ N}$$

(C) Apply Conservation of Energy to the volleyball after it collides with the tennis ball to determine its speed after the collision,

$$K_1 + U_1 = K_2 + U_2 \rightarrow \frac{1}{2}mv'^2 + 0 = 0 + mgh_2 \rightarrow v' = \sqrt{2gh_2} = \sqrt{2(9.8 \text{ m/s}^2)(0.3 \text{ m})} = 2.42 \text{ m/s}$$

The velocity of the tennis ball before the collision is approximately the same as that of the volleyball before the volleyball's bounce off the ground as they've fallen approximately the same height.

Apply Conservation of Momentum to the collision between the volleyball and the tennis ball,

$$m_v v_{v,1} + m_t v_{t,1} = m_v v_{v,2} + m_t v_{t,2}$$

$$(0.25 \text{ kg})(5.94 \text{ m/s}) + (0.06 \text{ kg})(-6.26 \text{ m/s}) = (0.25 \text{ kg})(2.42 \text{ m/s}) + (0.06 \text{ kg})v_{t,2}$$

$$v_{t,2} = 8.41 \text{ m/s}$$

Apply Conservation of Energy to the tennis ball after it collides with the volleyball,

$$K_i + U_i = K_f + U_f \rightarrow \frac{1}{2}mv_{t,2}^2 + 0 = 0 + mgh_t \rightarrow h_t = \frac{v_{t,2}^2}{2g} = \frac{(8.41 \text{ m/s})^2}{2(9.8 \text{ m/s}^2)} = 3.61 \text{ m}$$

4. (A) The speed v_2 is related to the centripetal force experienced by the star with mass m_2, $F_{c,1} = m_2 \frac{v_2^2}{r_2}$.

The centripetal force is provided by the force of gravity between the two stars. Therefore,

$$m_2 \frac{v_2^2}{r_2} = G \frac{m_1 m_2}{(r_1 + r_2)^2} \rightarrow m_1 = \frac{v_2^2 (r_1 + r_2)^2}{G r_2}$$

(B) Derive an equation for m_2, with the same method as in part (A), $m_2 = \frac{v_1^2 (r_1 + r_2)^2}{G r_1}$. Divide the equation for m_1 by the equation for m_2.

$$\frac{m_1}{m_2} = \frac{v_2^2}{r_2} \frac{r_1}{v_1^2}$$

Eliminate the *r*'s using the relation $v = \frac{2\pi r}{T} \rightarrow r = \frac{vT}{2\pi}$. As the stars orbit each other, they have the same period *T*. Therefore,

$$\frac{m_1}{m_2} = \frac{v_2^2}{\frac{v_2 T}{2\pi}} \frac{\frac{v_1 T}{2\pi}}{v_1^2} = \frac{v_2}{v_1}$$

(C) Express m_2 as the total mass M minus m_1, and note that the ratio $\frac{v_2}{v_1} = 4$.

$$\frac{m_1}{M - m_1} = 4 \rightarrow m_1 = 4M - 4m_1 \rightarrow 5m_1 = 4M \rightarrow m_1 = \frac{4}{5} M = 8 \times 10^{30} \text{ kg}$$

$$m_2 = M - m_1 = 2 \times 10^{30} \text{ kg}$$

5. (A) In order for the objects to roll down the incline, there must be a force providing a torque on the two objects. A free-body diagram shows that there are three forces acting on the spheres: the weight of the sphere, the normal force, and the force of friction. Torque is given by the equation $\tau = rF\sin(\theta)$, where *F* is the magnitude of the force, *r* is the radius, and θ is the angle between *r* and *F*. As both the normal force and the weight of the sphere have a radius of 0, the torque must be provided by the force of friction.

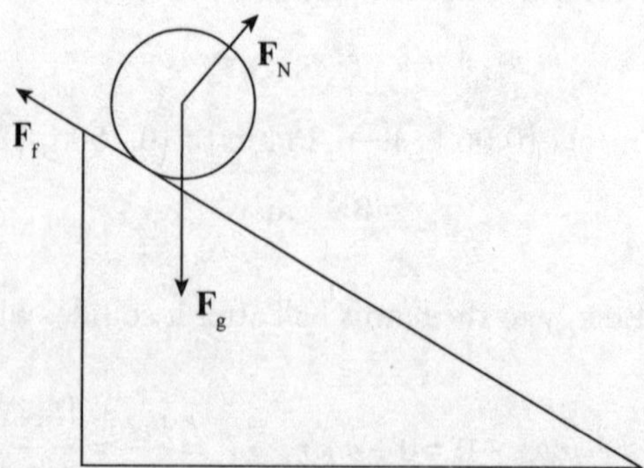

(B) As both objects have the same mass, both objects experience the same normal force and force of friction. This means both objects experience the same torque. Torque in terms of rotational inertia is given by the equation $\tau = I\alpha$, where I is the rotational inertia and α is the angular acceleration. Because the mass of the hollow sphere is distributed farther from the axis that it rotates around, the hollow sphere has a greater rotational inertia than the solid sphere. With a smaller rotational inertia, the solid sphere experiences a greater angular acceleration and reaches the bottom of the inclined plane first.

(C) As both objects have the same mass and start at the same height (on the top of the inclined plane), they both start with the same amount of potential energy. As both objects also start at rest, they both start with 0 kinetic energy. As the objects reach the bottom of the inclined plane, the potential energy of both objects is converted into kinetic energy. By Conservation of Energy, both objects have the same kinetic energy at the bottom of the inclined plane.

HOW TO SCORE PRACTICE TEST 3

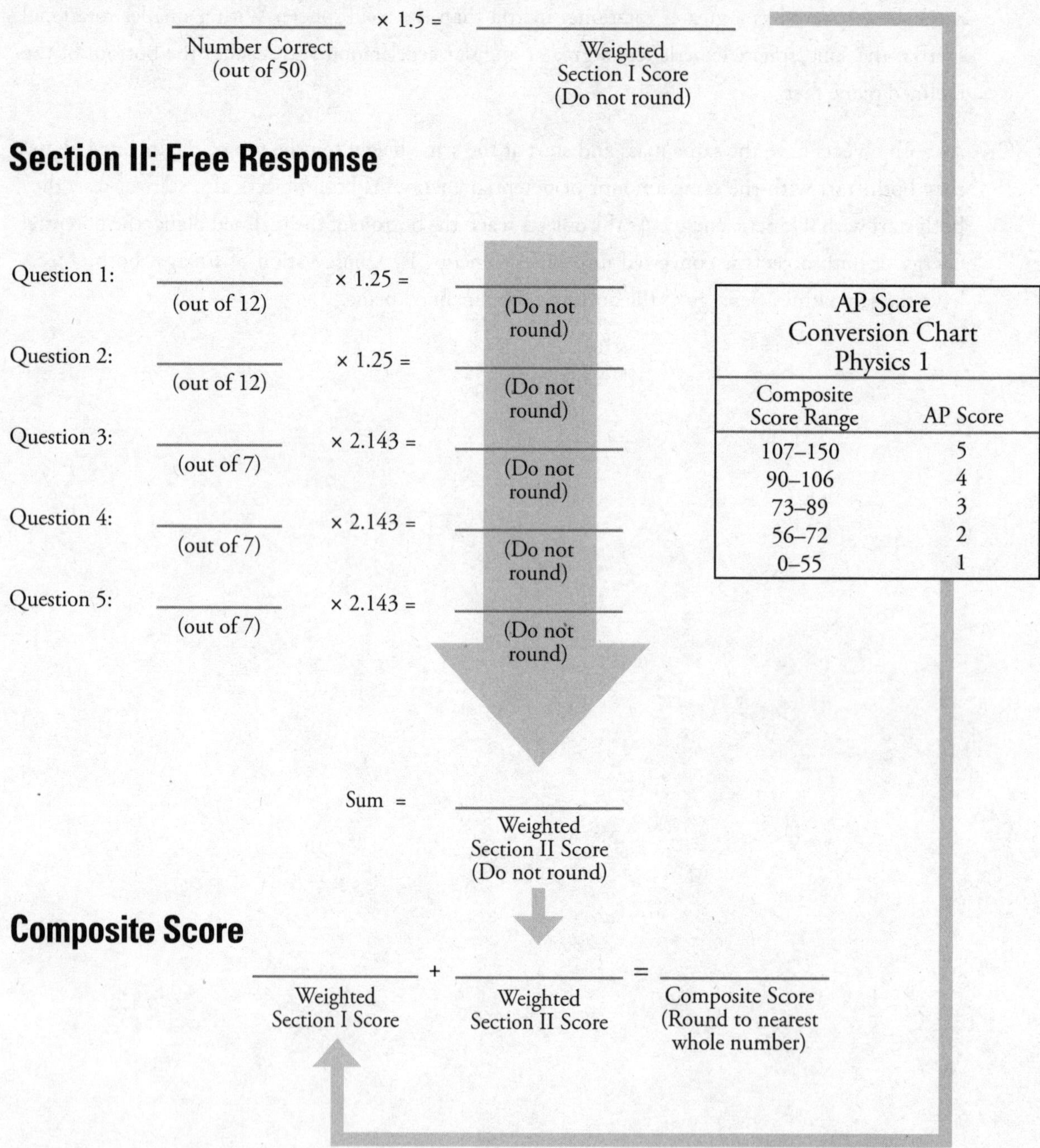

AP Score Conversion Chart
Physics 1

Composite Score Range	AP Score
107–150	5
90–106	4
73–89	3
56–72	2
0–55	1

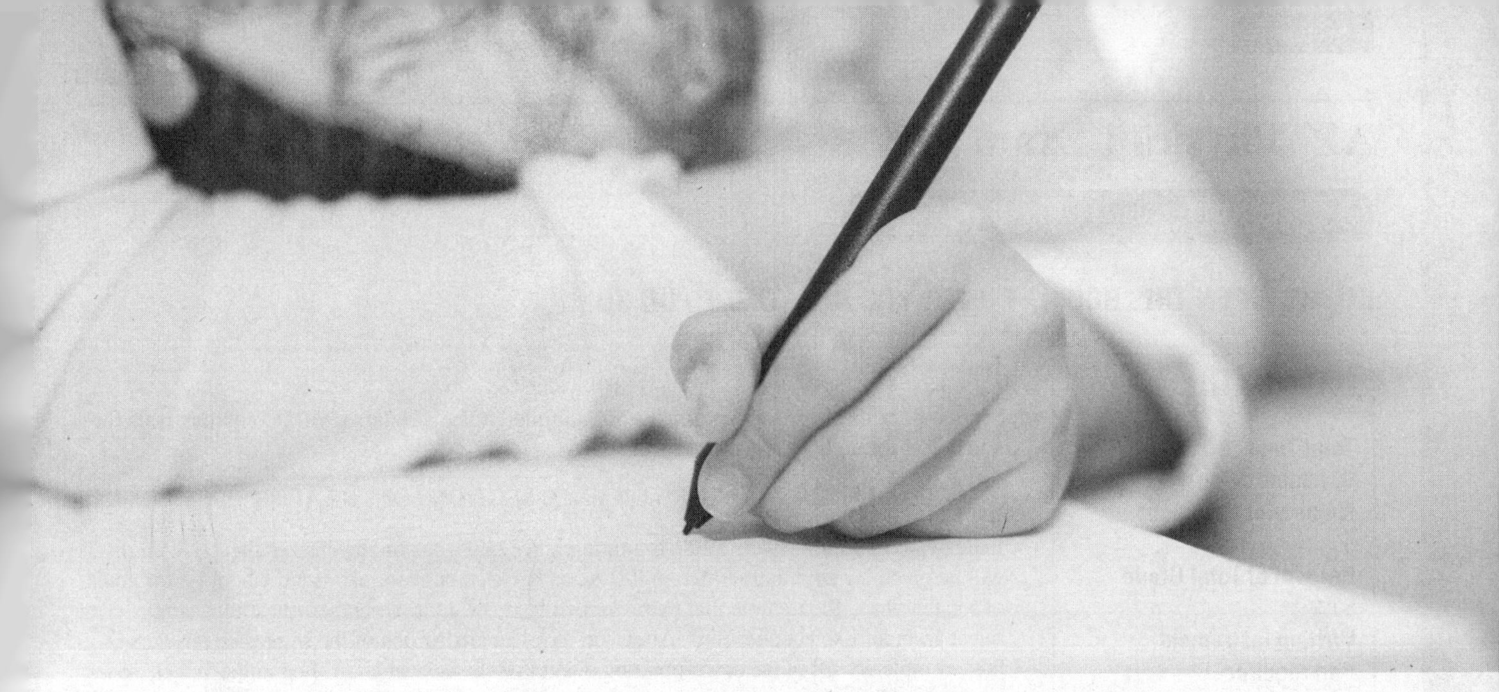

Practice Test 4

AP® Physics 1 Exam

SECTION I: Multiple-Choice Questions

DO NOT OPEN THIS BOOKLET UNTIL YOU ARE TOLD TO DO SO.

At a Glance

Total Time
90 minutes
Number of Questions
50
Percent of Total Grade
50%
Writing Instrument
Pen required

Instructions

Section I of this examination contains 50 multiple-choice questions. Fill in only the ovals for numbers 1 through 50 on your answer sheet.

CALCULATORS MAY BE USED ON BOTH SECTIONS OF THE AP PHYSICS 1 EXAM.

Indicate all of your answers to the multiple-choice questions on the answer sheet. No credit will be given for anything written in this exam booklet, but you may use the booklet for notes or scratch work. Please note that there are two types of multiple-choice questions: single-select and multi-select questions. After you have decided which of the suggested answers is best, completely fill in the corresponding oval(s) on the answer sheet. For single-select, you must give only one answer; for multi-select you must give BOTH answers in order to earn credit. If you change an answer, be sure that the previous mark is erased completely. Here is a sample question and answer.

Sample Question	Sample Answer
Chicago is a (A) state (B) city (C) country (D) continent	Ⓐ ● Ⓒ Ⓓ

Use your time effectively, working as quickly as you can without losing accuracy. Do not spend too much time on any one question. Go on to other questions and come back to the ones you have not answered if you have time. It is not expected that everyone will know the answers to all the multiple-choice questions.

About Guessing

Many candidates wonder whether or not to guess the answers to questions about which they are not certain. Multiple-choice scores are based on the number of questions answered correctly. Points are not deducted for incorrect answers, and no points are awarded for unanswered questions. Because points are not deducted for incorrect answers, you are encouraged to answer all multiple-choice questions. On any questions you do not know the answer to, you should eliminate as many choices as you can, and then select the best answer among the remaining choices.

GO ON TO THE NEXT PAGE.

ADVANCED PLACEMENT PHYSICS 1 TABLE OF INFORMATION

CONSTANTS AND CONVERSION FACTORS		
Proton mass, $m_p = 1.67 \times 10^{-27}$ kg	Electron charge magnitude,	$e = 1.60 \times 10^{-19}$ C
Neutron mass, $m_n = 1.67 \times 10^{-27}$ kg	Coulomb's law constant,	$k = 1/4\pi\varepsilon_0 = 9.0 \times 10^9\ \text{N}\cdot\text{m}^2/\text{C}^2$
Electron mass, $m_e = 9.11 \times 10^{-31}$ kg	Universal gravitational constant,	$G = 6.67 \times 10^{-11}\ \text{m}^3/\text{kg}\cdot\text{s}^2$
Speed of light, $c = 3.00 \times 10^8$ m/s	Acceleration due to gravity at Earth's surface,	$g = 9.8\ \text{m/s}^2$

UNIT SYMBOLS								
	meter,	m	kelvin,	K	watt,	W	degree Celsius,	°C
	kilogram,	kg	hertz,	Hz	coulomb,	C		
	second,	s	newton,	N	volt,	V		
	ampere,	A	joule,	J	ohm,	Ω		

PREFIXES		
Factor	Prefix	Symbol
10^{12}	tera	T
10^9	giga	G
10^6	mega	M
10^3	kilo	k
10^{-2}	centi	c
10^{-3}	milli	m
10^{-6}	micro	μ
10^{-9}	nano	n
10^{-12}	pico	p

VALUES OF TRIGONOMETRIC FUNCTIONS FOR COMMON ANGLES

θ	0°	30°	37°	45°	53°	60°	90°
$\sin\theta$	0	1/2	3/5	$\sqrt{2}/2$	4/5	$\sqrt{3}/2$	1
$\cos\theta$	1	$\sqrt{3}/2$	4/5	$\sqrt{2}/2$	3/5	1/2	0
$\tan\theta$	0	$\sqrt{3}/3$	3/4	1	4/3	$\sqrt{3}$	∞

The following conventions are used in this exam.

I. The frame of reference of any problem is assumed to be inertial unless otherwise stated.
II. Assume air resistance is negligible unless otherwise stated.
III. In all situations, positive work is defined as work done <u>on</u> a system.
IV. The direction of current is conventional current: the direction in which positive charge would drift.
V. Assume all batteries and meters are ideal unless otherwise stated.

GO ON TO THE NEXT PAGE.

ADVANCED PLACEMENT PHYSICS 1 EQUATIONS

MECHANICS

$v_x = v_{x0} + a_x t$

$x = x_0 + v_{x0}t + \frac{1}{2}a_x t^2$

$v_x^2 = v_{x0}^2 + 2a_x(x - x_0)$

$\vec{a} = \frac{\sum \vec{F}}{m} = \frac{\vec{F}_{net}}{m}$

$|\vec{F}_f| \le \mu |\vec{F}_n|$

$a_c = \frac{v^2}{r}$

$\vec{p} = m\vec{v}$

$\Delta\vec{p} = \vec{F}\,\Delta t$

$K = \frac{1}{2}mv^2$

$\Delta E = W = F_{\parallel}d = Fd\cos\theta$

$P = \frac{\Delta E}{\Delta t}$

$\theta = \theta_0 + \omega_0 t + \frac{1}{2}\alpha t^2$

$\omega = \omega_0 + \alpha t$

$x = A\cos(2\pi ft)$

$\vec{\alpha} = \frac{\sum \vec{\tau}}{I} = \frac{\vec{\tau}_{net}}{I}$

$\tau = r_{\perp}F = rF\sin\theta$

$L = I\omega$

$\Delta L = \tau\,\Delta t$

$K = \frac{1}{2}I\omega^2$

$|\vec{F}_s| = k|\vec{x}|$

$U_s = \frac{1}{2}kx^2$

$\rho = \frac{m}{V}$

$\Delta U_g = mg\,\Delta y$

$T = \frac{2\pi}{\omega} = \frac{1}{f}$

$T_s = 2\pi\sqrt{\frac{m}{k}}$

$T_p = 2\pi\sqrt{\frac{\ell}{g}}$

$|\vec{F}_g| = G\frac{m_1 m_2}{r^2}$

$\vec{g} = \frac{\vec{F}_g}{m}$

$U_G = -\frac{Gm_1 m_2}{r}$

a = acceleration
A = amplitude
d = distance
E = energy
f = frequency
F = force
I = rotational inertia
K = kinetic energy
k = spring constant
L = angular momentum
ℓ = length
m = mass
P = power
p = momentum
r = radius or separation
T = period
t = time
U = potential energy
V = volume
v = speed
W = work done on a system
x = position
y = height
α = angular acceleration
μ = coefficient of friction
θ = angle
ρ = density
τ = torque
ω = angular speed

ELECTRICITY

$|\vec{F}_E| = k\left|\frac{q_1 q_2}{r^2}\right|$

$I = \frac{\Delta q}{\Delta t}$

$R = \frac{\rho\ell}{A}$

$I = \frac{\Delta V}{R}$

$P = I\,\Delta V$

$R_s = \sum_i R_i$

$\frac{1}{R_p} = \sum_i \frac{1}{R_i}$

A = area
F = force
I = current
ℓ = length
P = power
q = charge
R = resistance
r = separation
t = time
V = electric potential
ρ = resistivity

WAVES

$\lambda = \frac{v}{f}$

f = frequency
v = speed
λ = wavelength

GEOMETRY AND TRIGONOMETRY

Rectangle
$A = bh$

Triangle
$A = \frac{1}{2}bh$

Circle
$A = \pi r^2$
$C = 2\pi r$

Rectangular solid
$V = \ell wh$

Cylinder
$V = \pi r^2\ell$
$S = 2\pi r\ell + 2\pi r^2$

Sphere
$V = \frac{4}{3}\pi r^3$
$S = 4\pi r^2$

A = area
C = circumference
V = volume
S = surface area
b = base
h = height
ℓ = length
w = width
r = radius

Right triangle
$c^2 = a^2 + b^2$
$\sin\theta = \frac{a}{c}$
$\cos\theta = \frac{b}{c}$
$\tan\theta = \frac{a}{b}$

c
a
θ
90°
b

GO ON TO THE NEXT PAGE.

THIS PAGE IS LEFT INTENTIONALLY BLANK.

GO ON TO THE NEXT PAGE.

AP PHYSICS 1

SECTION I

Note: To simplify calculations, you may use $g = 10\ m/s^2$ in all problems.

Directions: Each of the questions or incomplete statements is followed by four suggested answers or completions. Select the one that is best in each case and then fill in the corresponding circle on the answer sheet.

Questions 1–3 refer to the following scenario:

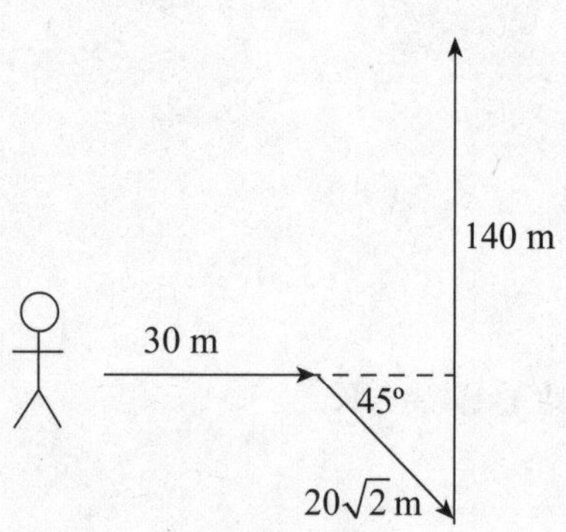

An explorer travels 30 m east, then $20\sqrt{2}$ m in a direction 45° south of east, and then 140 m north.

1. What is the distance traveled by the explorer?
   (A) 167.2 m
   (B) 169 m
   (C) 170 m
   (D) 198.2 m

2. What is the displacement of the explorer?
   (A) 130 m
   (B) 169 m
   (C) 170 m
   (D) 215 m

3. The explorer took 60 s, 130 s, and 70 s to travel the 30 m, $20\sqrt{2}$ m, and 140 m north distances, respectively. What is the average velocity of the explorer over the total distance traveled?
   (A) 0.50 m/s
   (B) 33.3 m/min
   (C) 0.76 m/s
   (D) 100 m/min

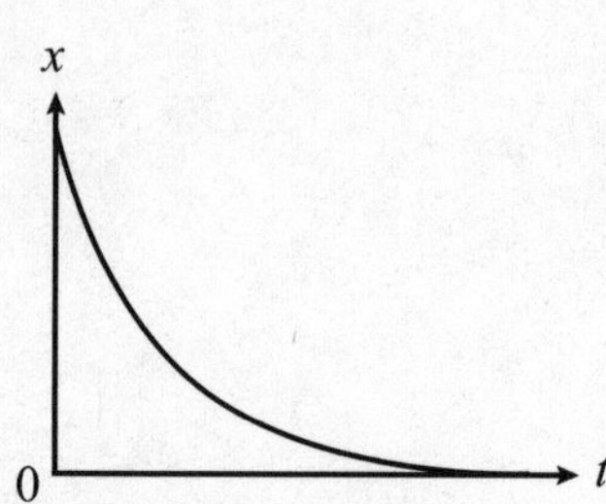

4. The graph above is the position-versus-time graph of an object. Which of the following is the velocity-versus-time graph of the object?

(A)
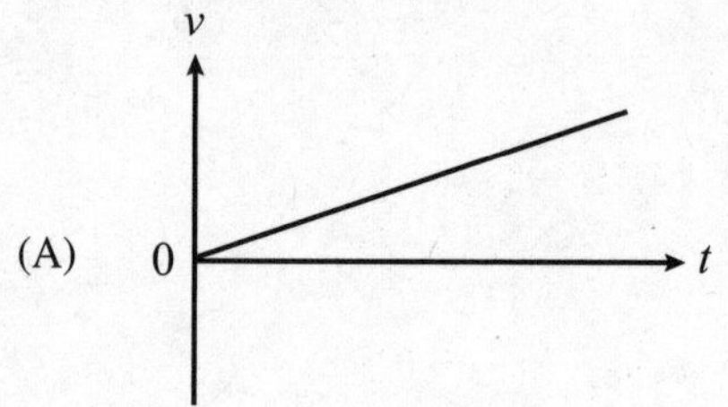

(B)
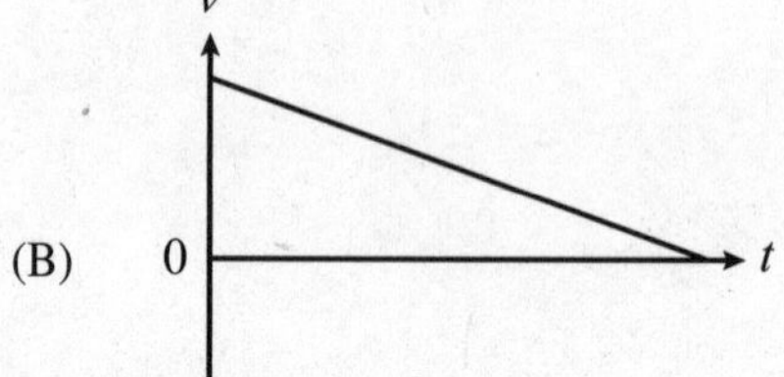

(C)
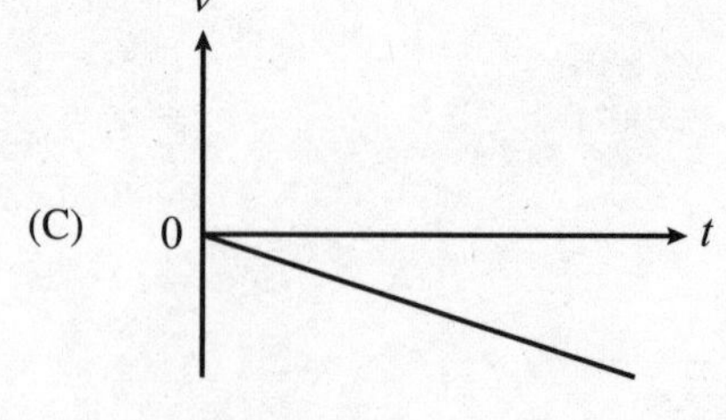

(D)
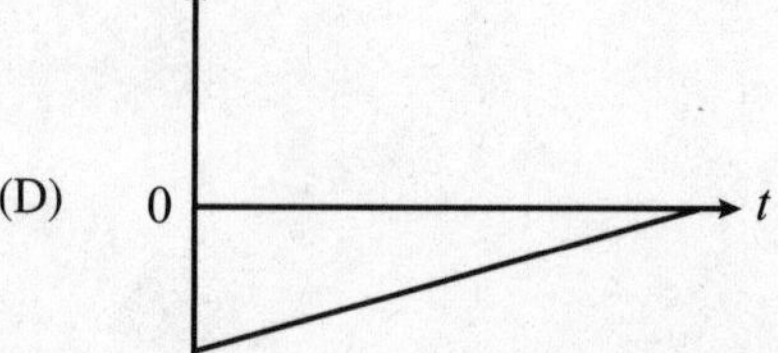

GO ON TO THE NEXT PAGE.

5. In which section of the following velocity-versus-time graph is the object slowing down and moving in the negative direction?

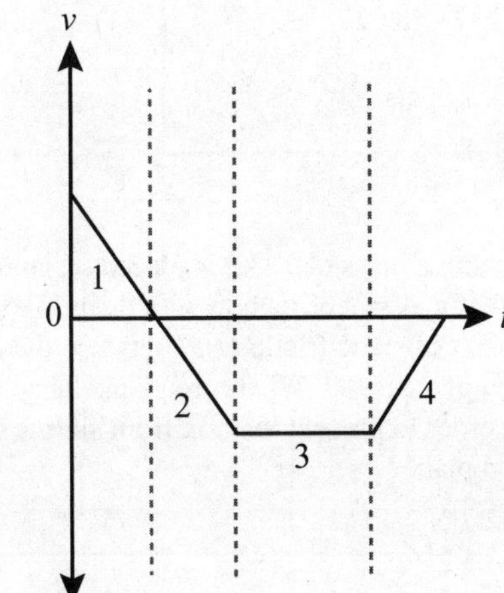

(A) Section 1
(B) Section 2
(C) Section 3
(D) Section 4

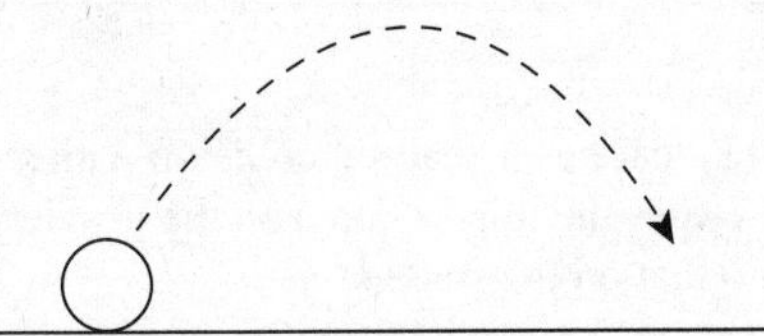

6. Which of the following is a true statement regarding the motion of projectiles?

(A) The velocity of projectiles is smallest at the apex of the trajectory.
(B) All projectiles have zero velocity at the apex of the trajectory.
(C) The acceleration of projectiles is greatest at the apex of the trajectory.
(D) Projectiles have maximum kinetic energy at the apex of the trajectory.

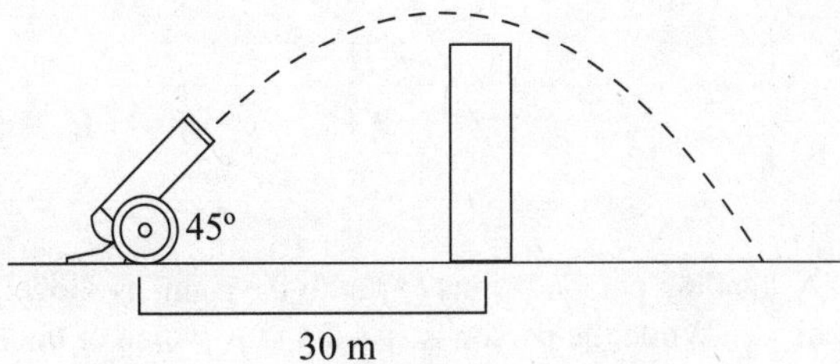

7. A cannonball is fired with an initial velocity of 20 m/s and a launch angle of 45° at a wall 30 m away. If the cannonball just barely clears the wall, what is the maximum height of the wall?

(A) 5.92 m
(B) 6.34 m
(C) 7.51 m
(D) 8.32 m

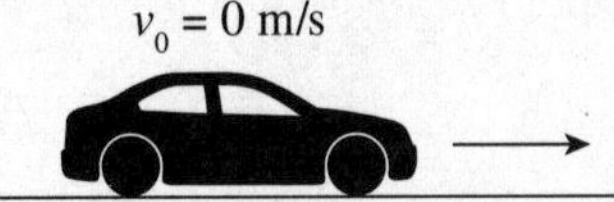

8. A car initially at rest accelerates linearly at a constant rate for 5 s. If the final speed of the car is 15 m/s, what was the displacement of the car?

(A) 20 m
(B) 37.5 m
(C) 75 m
(D) 150 m

GO ON TO THE NEXT PAGE.

9. A standing person begins to run to the right, as shown above. While the person is speeding up, which of the following could be the direction of the (total) contact force that the ground exerts on his/her foot?

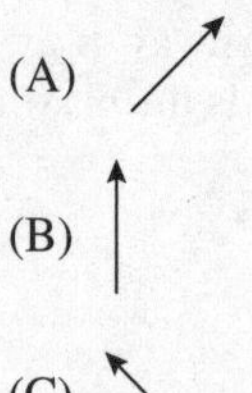

(D) None of these are possible.

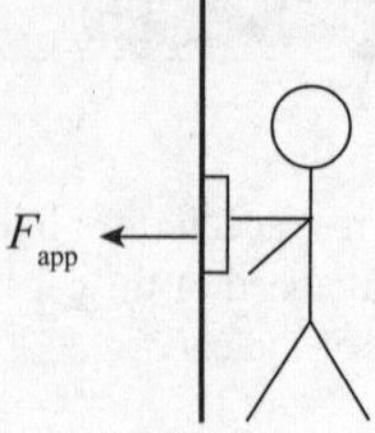

10. A student presses a 0.5 kg book against the wall. If the μ_s between the book and the wall is 0.2, what force must the student apply to hold the book in place?

(A) 0 N
(B) 15 N
(C) 25 N
(D) 35 N

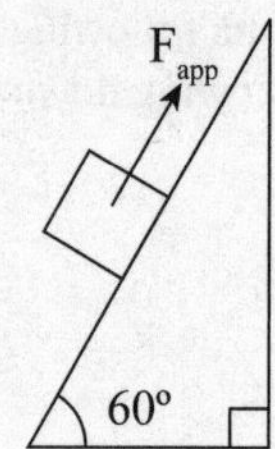

11. A box with a mass of 10 kg is placed on an inclined plane that makes a 60° angle with the horizontal. The coefficient of static friction (μ_s) between the box and the inclined plane is 0.2. What force must be applied on the box in order to prevent the box from sliding down the inclined plane?

(A) 32.7 N
(B) 48.3 N
(C) 56.2 N
(D) 76.6 N

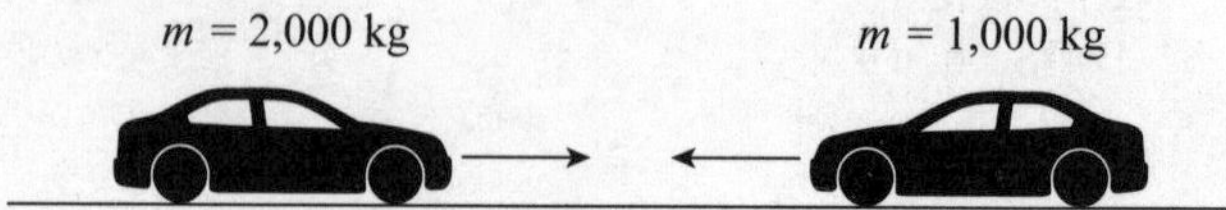

12. A 2,000 kg car has a head-on collision with a 1,000 kg car. How does the impact force on the heavier car compare with that on the smaller car?

(A) The heavier car experiences a greater impact force.
(B) The smaller car experiences a greater impact force.
(C) Both vehicles experience an impact force with the same magnitude.
(D) Cannot be determined

GO ON TO THE NEXT PAGE.

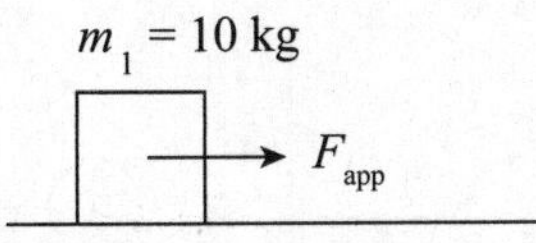

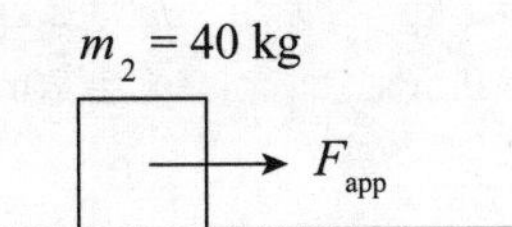

13. Two students push two different boxes across a frictionless floor. Both students exert a horizontal force of 50 N. If the first box has mass $m_1 = 10$ kg and the second box has mass $m_2 = 40$ kg, what is the value of a_1/a_2?

    (A) 1/4
    (B) 1/2
    (C) 2
    (D) 4

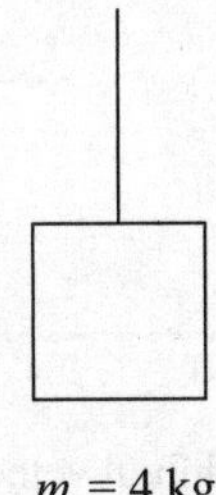

14. A 4 kg box is accelerated upwards by a string with a breaking strength of 80 N. What is the maximum upward acceleration that can be applied on the box without breaking the string?

    (A) 2.5 m/s^2
    (B) 5.0 m/s^2
    (C) 7.5 m/s^2
    (D) 10 m/s^2

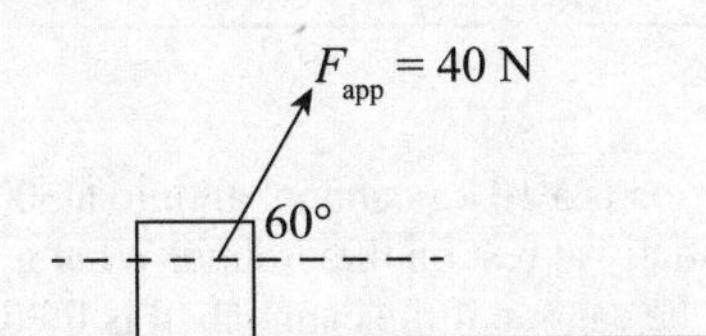

15. A worker moves a 30 kg box by pulling on it with a rope that makes a 60° angle with the horizontal. If the worker applies a force of 40 N and pulls the box over a distance of 20 m, how much work did the worker do?

    (A) 100 J
    (B) 200 J
    (C) 400 J
    (D) 800 J

16. A 2 kg rock is dropped off a cliff with a height of 20 m. What is the speed of the rock at the bottom of the hill?

    (A) 10 m/s
    (B) 14 m/s
    (C) 20 m/s
    (D) 40 m/s

17. How much work is done by a pitcher to throw a 0.2 kg ball at a speed of 30 m/s?

    (A) 3 J
    (B) 9 J
    (C) 30 J
    (D) 90 J

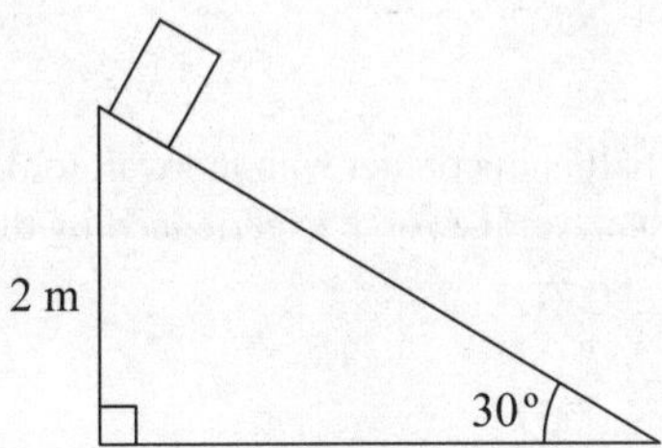

18. A 10 kg block is placed at the top of an inclined plane with an angle of incline of 30° and $\mu_k = 0.1$. If the height of the inclined plane is 2 m, what is the kinetic energy of the block when it is halfway down the incline?

    (A) 64 J
    (B) 72 J
    (C) 83 J
    (D) 100 J

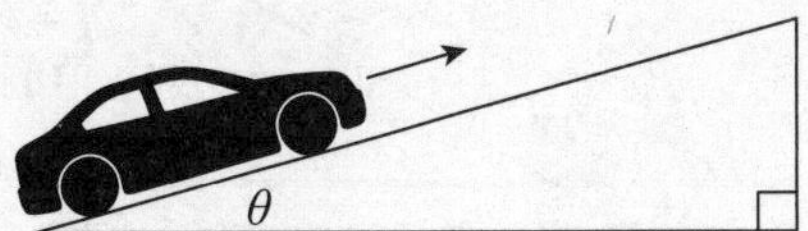

19. A car's engine must exert a force of 2,000 N to maintain a speed of 30 m/s up an incline. What is the power provided by the engine during this motion?

    (A) 67 W
    (B) 9,000 W
    (C) 60,000 W
    (D) 90,000 W

GO ON TO THE NEXT PAGE.

Questions 20–22 refer to the following scenario:

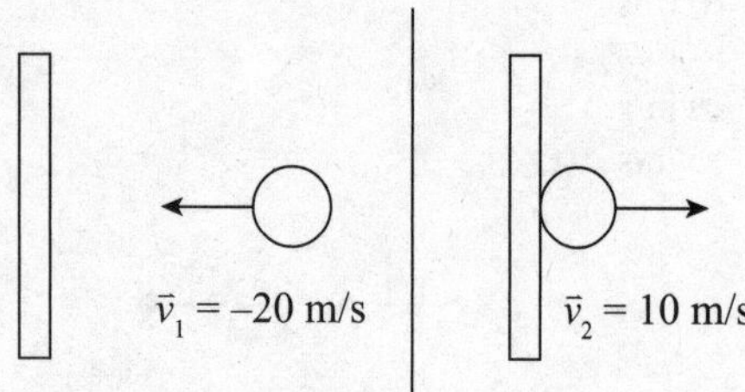

A 2 kg ball with a velocity of –20 m/s collides with the wall and bounces back with a velocity of 10 m/s.

20. What is the impulse during the collision?

(A) 10 N·s
(B) 20 N·s
(C) 30 N·s
(D) 60 N·s

21. If the ball is in contact with the wall for 0.002 s, determine the average force experienced by the ball.

(A) 3,000 N
(B) 5,000 N
(C) 30,000 N
(D) 50,000 N

22. How much work did the wall do on the ball?

(A) –300 J
(B) –400 J
(C) –750 J
(D) –1,000 J

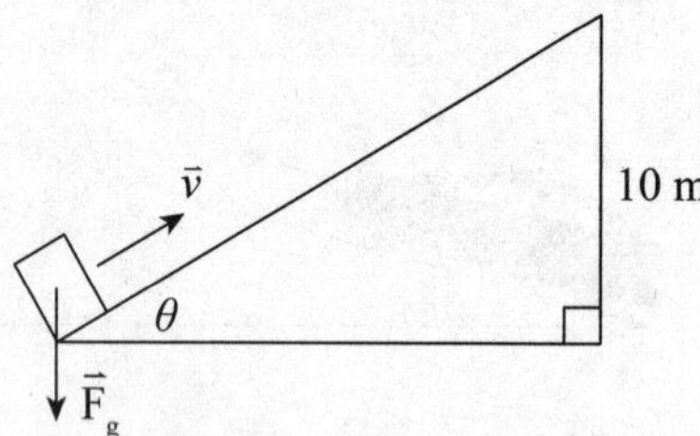

23. A student pushes a 6 kg box up an inclined plane with a height of 10 m. How much work does gravity do on the box during this process?

(A) –1,200 J
(B) –600 J
(C) 600 J
(D) 1,200 J

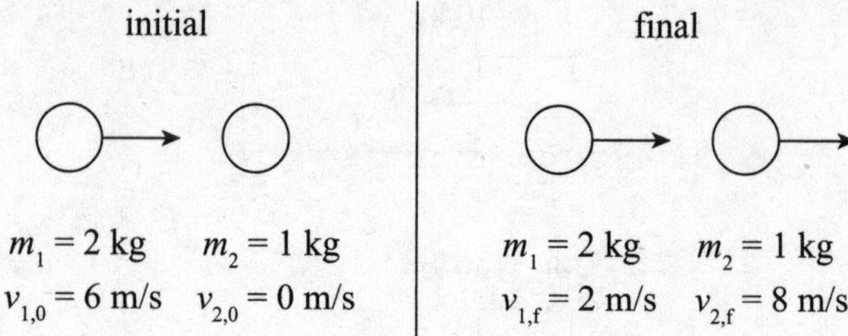

24. A 2 kg ball traveling to the right at 6 m/s collides head on with a 1 kg ball at rest. After impact, the 2 kg ball is traveling to the right at 2 m/s and the 1 kg ball is traveling to the right at 8 m/s. What type of collision occurred?

(A) Inelastic
(B) Perfectly inelastic
(C) Elastic
(D) Cannot be determined

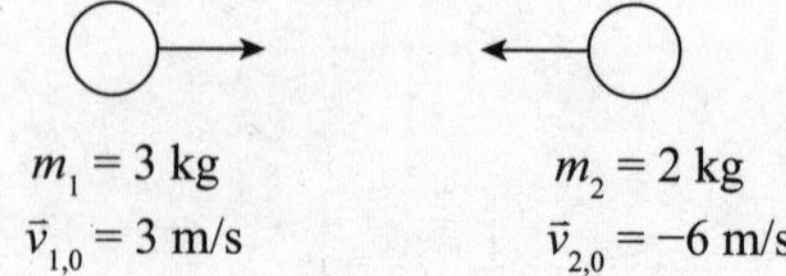

25. A 3 kg mass with an initial velocity of +3 m/s has a perfectly inelastic collision with a 2 kg mass with an initial velocity of –6 m/s. What is the final velocity after impact?

(A) –1.0 m/s
(B) –0.6 m/s
(C) 0.6 m/s
(D) 1.0 m/s

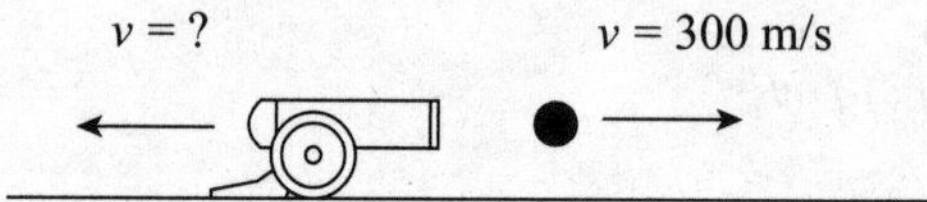

26. A soldier loads a 10 kg cannonball into a 300 kg cannon that is initially at rest on the ground. What is the recoil speed of the cannon if the cannonball is fired with a horizontal velocity of 300 m/s ?

(A) 5 m/s
(B) 7.5 m/s
(C) 10 m/s
(D) 15 m/s

GO ON TO THE NEXT PAGE.

27. A student attaches a 1.5 kg mass to a 40 cm long string. The student then spins the mass in a horizontal circle with increasing speed. If the string snaps when the mass has a speed of 6 m/s, what is the breaking strength of the string?

(A) 22.5 N
(B) 135 N
(C) 240 N
(D) 360 N

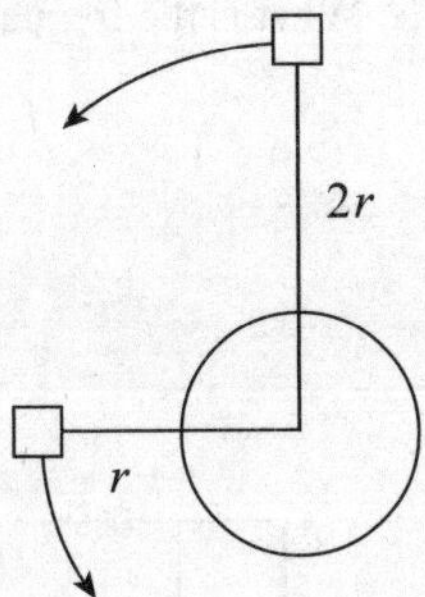

28. If the speed of a satellite orbiting the Earth at a distance r from the center of the Earth is v, what is the speed of a second satellite orbiting the Earth at a distance $2r$ from the center of the Earth?

(A) $2v$

(B) $\sqrt{2}v$

(C) $\frac{1}{\sqrt{2}}v$

(D) $\frac{1}{2}v$

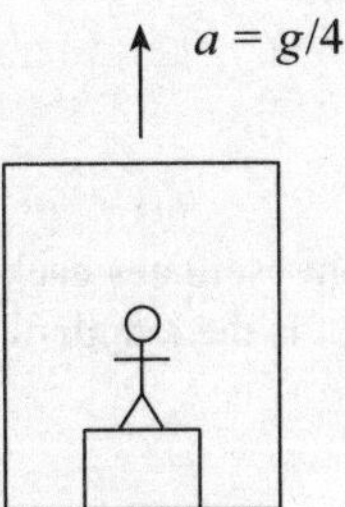

29. A 50 kg man stands on a scale that measures force in an elevator traveling up with an acceleration of $g/4$. What will the scale read?

(A) 375 N
(B) 500 N
(C) 575 N
(D) 625 N

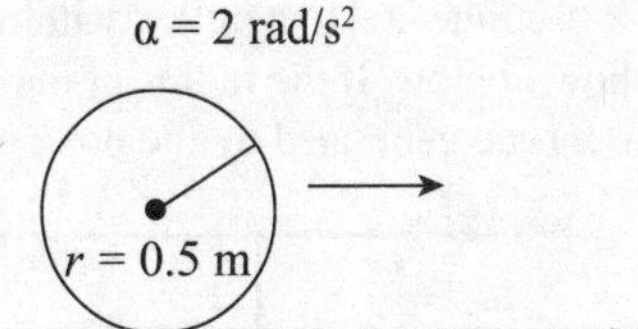

30. A mechanical wheel initially at rest on the floor begins rolling forward with an angular acceleration of 2 rad/s². If the radius of the wheel is 0.5 m, what is the linear velocity of the wheel after 5 s?

(A) 0.5 m/s
(B) 1 m/s
(C) 5 m/s
(D) 10 m/s

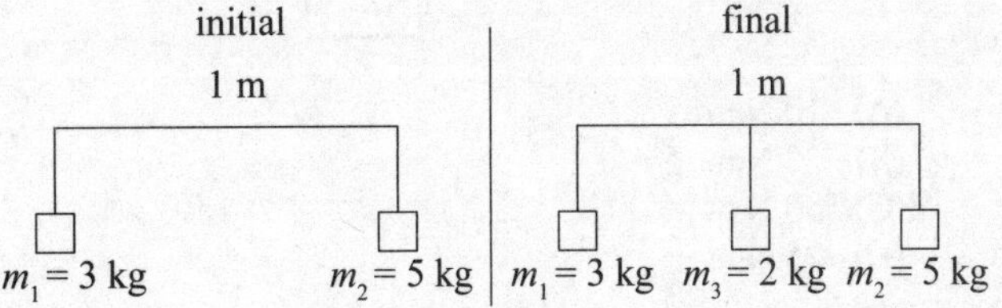

31. Two masses are attached to a 1 m long massless bar. Mass 1 is 3 kg and is attached to the far left side of the bar. Mass 2 is 5 kg and is attached to the far right side of the bar. If a third mass that is 2 kg is added to the middle of the bar, how does the center of mass of the system change?

(A) The center of mass shifts to the left by 0.025 m.
(B) The center of mass shifts to the right by 0.025 m.
(C) The center of mass shifts to the left by 0.075 m.
(D) The center of mass shifts to the right by 0.075 m.

GO ON TO THE NEXT PAGE.

32. A 5 kg box is connected to a pulley with rope in the diagram shown below. If the radius of the pulley is 0.5 m, what is the torque generated by the box on the pulley?

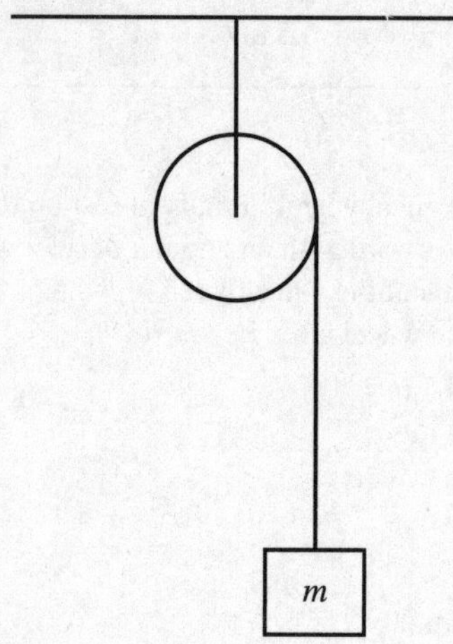

(A) 10 N·m
(B) 25 N·m
(C) 50 N·m
(D) 75 N·m

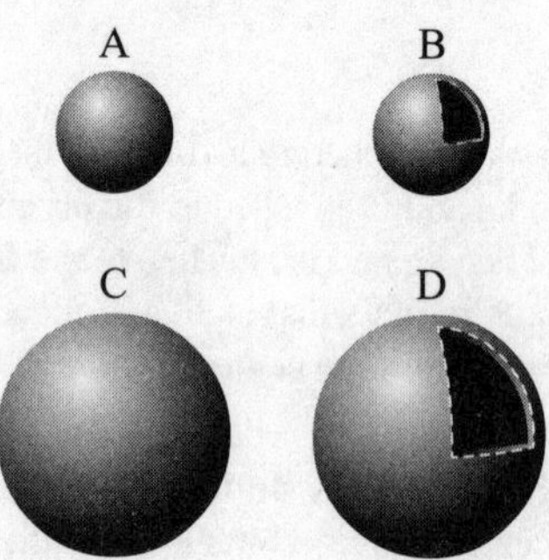

33. Which of the following objects has the greatest rotational inertia?

(A) A 1 kg solid ball with radius of 5 cm
(B) A 1 kg hollow ball with radius of 5 cm
(C) A 5 kg solid ball with radius of 5 cm
(D) A 5 kg hollow ball with radius 5 cm

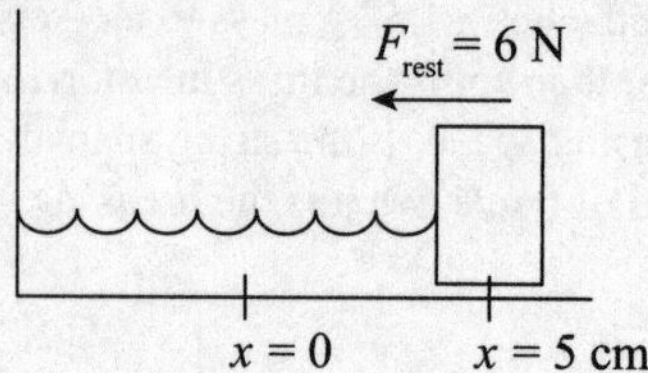

34. A horizontal spring is attached to a 5 kg block. When the block is pulled 5 cm to the right, the restoring force has a magnitude of 6 N. What is the frequency of the spring?

(A) 0.32 Hz
(B) 0.56 Hz
(C) 0.78 Hz
(D) 0.98 Hz

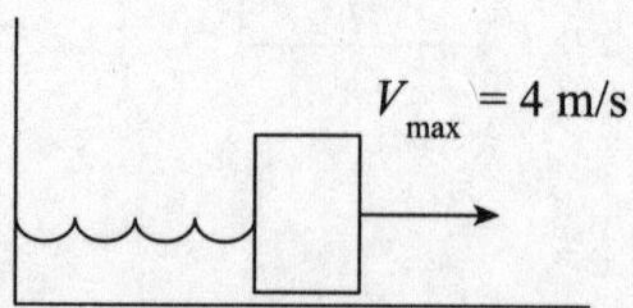

35. A 3 kg block is attached to a horizontal spring with a force constant of 10 N/m. If the maximum speed of the block is 4 m/s, what is the amplitude of the block?

(A) 0.55 m
(B) 1.1 m
(C) 2.2 m
(D) 4.4 m

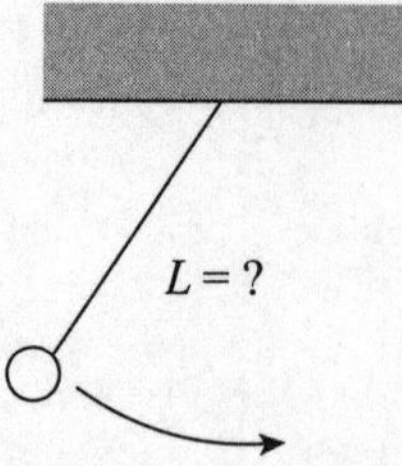

36. A simple pendulum oscillates back and forth with a period of 2 s. What is the length of the string of the pendulum?

(A) 0.25 m
(B) 0.5 m
(C) 1 m
(D) 2 m

GO ON TO THE NEXT PAGE.

37. An object of mass m is attached to a horizontal spring with spring constant k, oscillating on a frictionless horizontal surface with an amplitude A. What is the speed of the object when its displacement from its equilibrium position is $\frac{A}{2}$?

(A) $\frac{A}{2}\sqrt{\frac{k}{m}}$

(B) $\frac{A}{2}\sqrt{\frac{2k}{m}}$

(C) $\frac{A}{2}\sqrt{\frac{3k}{m}}$

(D) $A\sqrt{\frac{k}{m}}$

38. A planet forms from a cloud of gas and dust that collapses under gravity. Why does the planet rotate faster than the cloud of gas and dust it was formed from?

(A) As the cloud collapses, the gravitational forces between particles increase, causing more torque on the cloud.
(B) As the cloud collapses, the gravitational forces between particles increase, increasing the angular momentum of the system.
(C) As the cloud collapses, its rotational inertia decreases, so its angular velocity increases to conserve angular momentum.
(D) As the cloud collapses, its rotational inertia decreases, so torques acting on it result in higher angular acceleration.

39. An object (mass = m) above the surface of the Moon (mass = M) is dropped from an altitude h equal to the Moon's radius (R). What is the object's impact speed?

(A) $\sqrt{GM/R}$
(B) $\sqrt{GM/(2R)}$
(C) $\sqrt{2GM/R}$
(D) $\sqrt{2GMm/R}$

40. While loading identical boxes onto a truck, some boxes are lifted from ground level directly onto the truck, while others are pushed up a ramp with a somewhat rough surface. Which of the following MUST be true?

(A) The work required to push a box up the ramp is greater than the work required to lift a box onto the truck.
(B) The work required to push a box up the ramp less than the work required to lift a box onto the truck.
(C) The force required to push a box up the ramp is greater than the force required to lift a box onto the truck.
(D) The force required to push a box up the ramp is less than the force required to lift a box onto the truck.

41. A ball of mass m hangs on a string of length L. A child pushes the ball so that it starts to swing back-and-forth. When the string is at an angle θ from the vertical, what torque does gravity apply to this system?

(A) mgL
(B) $mgL \sin \theta$
(C) $mgL \cos \theta$
(D) $mgL(1 - \cos \theta)$

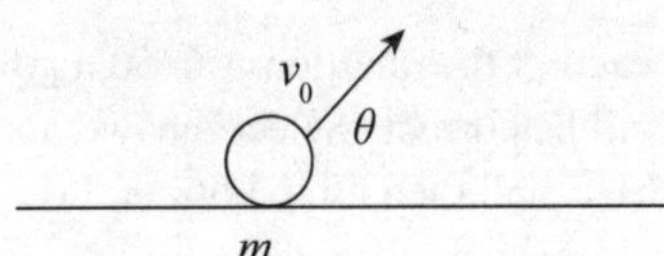

42. A projectile of mass m is fired into the air at some angle θ. If its launch velocity is v_0, which of the following expressions represents the object's kinetic energy at the peak of its motion?

(A) 0
(B) $(1/2)m(v_0 \sin \theta)^2$
(C) $(1/2)m(v_0 \cos \theta)^2$
(D) $(1/2)m[(v_0 \sin \theta)^2 + (v_0 \cos \theta)^2]$

43. A spider of mass m accelerates at a rate of a down a vertical strand of its web. What is the magnitude of tension in the strand?

(A) mg
(B) $m(g - a)$
(C) $m(a - g)$
(D) $m(g + a)$

GO ON TO THE NEXT PAGE.

44. An object slides from rest down a 10.0-m long, frictionless incline that is at 30° to the horizontal, then it skids to a stop on a rough horizontal surface over a distance of 12.5 m. What is the coefficient of kinetic friction between the object and the rough horizontal surface?

(A) 0.400
(B) 0.462
(C) 0.693
(D) 0.800

45. What is the power delivered by gravity as it pulls an object of mass *m* that is dropped from a height *h* down to the ground?

(A) mgh
(B) $\frac{mg^2}{2}$
(C) $\frac{m\sqrt{2g^3h}}{2}$
(D) $m\sqrt{2g^3h}$

Directions: For each of the questions 46–50, two of the suggested answers will be correct. Select the two answers that are best in each case, and then fill in both of the corresponding circles on the answer sheet.

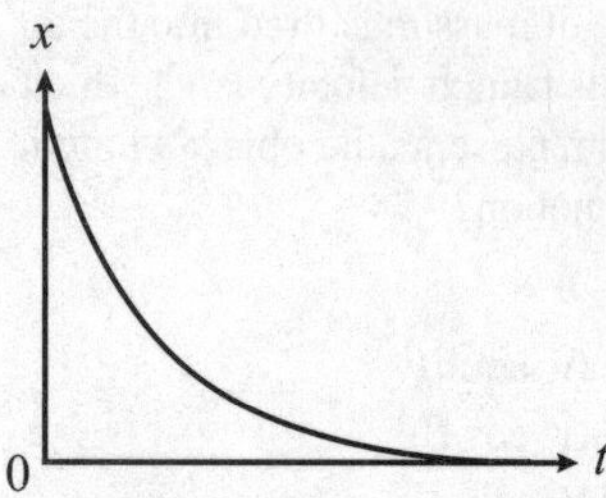

46. The graph above is the position-versus-time graph of an object. Which of the following is true regarding the motion of the object? Select two answers.

(A) The object is moving in the positive direction.
(B) The object is moving in the negative direction.
(C) The acceleration of the object is decreasing.
(D) The speed of the object is decreasing.

47. In which of the following scenarios is the total work done on the box equal to zero? Select two answers.

(A) A box is lifted from the floor and placed on a shelf.
(B) A box falls some distance to the ground.
(C) A box is pushed along the ground at a constant speed.
(D) A box slides along a horizontal surface, slowing from an initial speed to a stop.

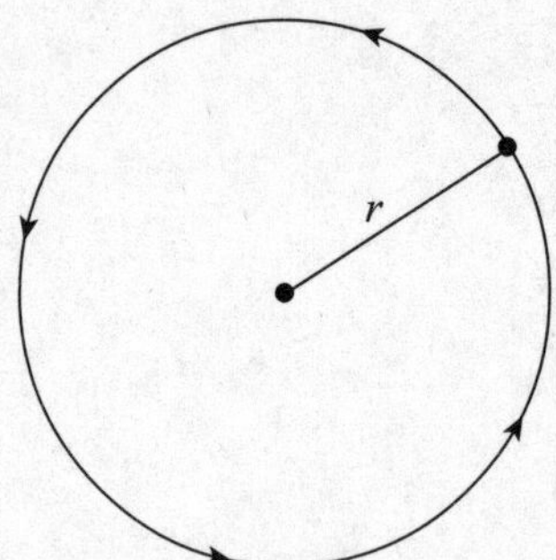

48. Which of the following statements is true regarding an object undergoing uniform circular motion? Select two answers.

(A) The velocity of the object is constant.
(B) The centripetal force is always the net force acting in the plane of the circular path.
(C) The object must feel a force directed radially outward.
(D) The acceleration of the object must point toward center of the circle.

49. Which of the following is true regarding conservative forces? Select two answers.

(A) The work done by conservative forces is path dependent.
(B) The work done by conservative forces is path independent.
(C) Conservative forces depend only on the position of an object.
(D) When a conservative force does positive work on an object, its potential energy increases.

GO ON TO THE NEXT PAGE.

50. In which of the following scenarios is the normal force acting on the object equal to mg? Select two answers.

(A) 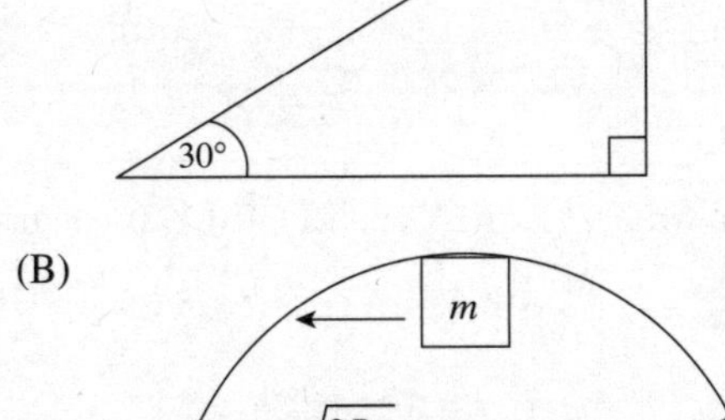

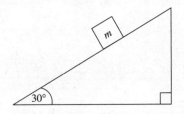

An object of mass m rests on an incline plane at an angle of 30°.

(B) 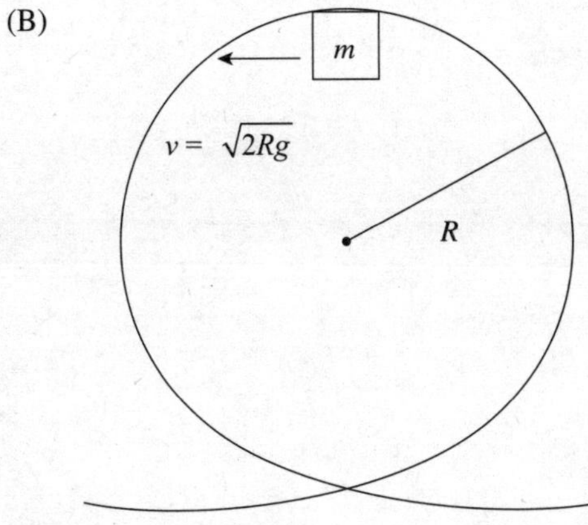

An object of mass m slides upside down on a curved track of radius R with a speed $v = \sqrt{2Rg}$.

(C) 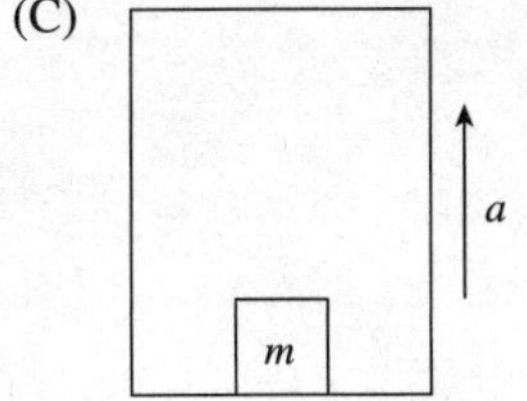

An object of mass m rests on an elevator that accelerates upward.

(D) 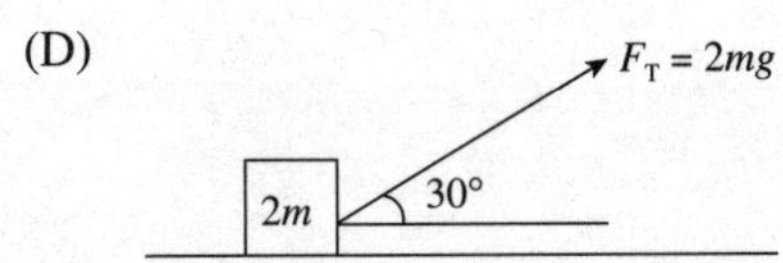

An object of mass $2m$ is pulled along a horizontal surface by a tension force $F_T = 2mg$ at an angle of 30° above the horizontal.

END OF SECTION I

DO NOT CONTINUE UNTIL INSTRUCTED TO DO SO.

AP PHYSICS 1

SECTION II

Free-Response Questions

Time—90 minutes

Percent of total grade—50

General Instructions

Use a separate piece of paper to answer these questions. Show your work. Be sure to write CLEARLY and LEGIBLY. If you make an error, you may save time by crossing it out rather than trying to erase it.

GO ON TO THE NEXT PAGE.

AP PHYSICS 1

SECTION II

Directions: Questions 1–5 here are as follows: one experimental design question (worth 12 points), one quantitative/qualitative translation question (worth 12 points), one paragraph argument short answer question (worth 7 points), and two additional short answer questions (worth 7 points each). You have a total of 90 minutes to complete this section. Show your work for each part in the space provided after that part.

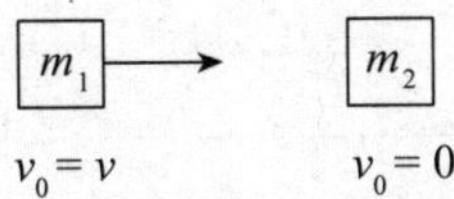

1. A mass m_1 traveling with an initial velocity of v has an elastic collision with a mass m_2 initially at rest.

(A) Determine the final velocity v_1 of m_1 in terms of m_1, m_2, and v.

(B) Determine the final velocity v_2 of m_2 in terms of m_1, m_2, and v.

(C) For what values of m_1 and m_2 would the final velocities of the two masses be in the same direction? The opposite direction?

GO ON TO THE NEXT PAGE.

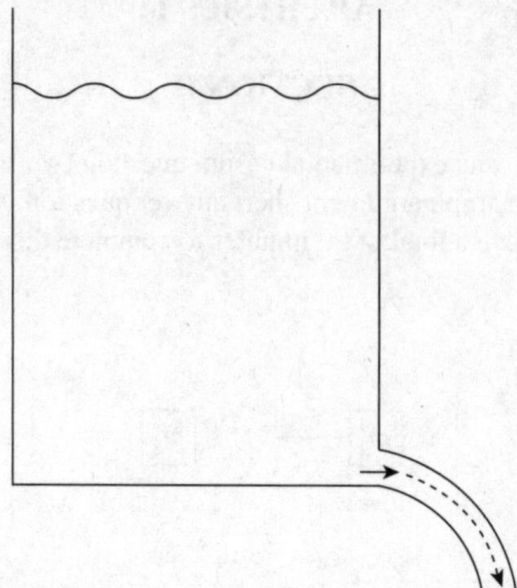

2. A student observes water flowing out of a small hole at the bottom of a large tank (onto a floor with a drain). The water flows out of the hole horizontally.

The student hypothesizes that the water flowing out of the hole undergoes projectile motion, so that its range can be determined by the equations of projectile motion.

(A) The student plans to test the hypothesis by observing the water flowing out of the tank. They are able to open the top of the tank and refill it as needed.

i. State the basic physics principles or laws the student could use in designing an experiment to test the hypothesis.

ii. Design an experimental procedure to test whether or not the water undergoes projectile motion. Assume equipment usually found in a school physics laboratory is available. In the table below, list the quantities and associated symbols that would be measured in your experiment. Also list the equipment that would be used to measure each quantity. You do not need to fill in every row. If you need additional rows, you may add them to the space just below the table.

Quantity to be Measured	Symbol for Quantity	Equipment for Measurement

(B) Describe the overall procedure to be used to determine whether the range of the water is described by projectile motion, referring to the table. Provide enough detail so that another student could replicate the experiment, including any steps necessary to reduce experimental uncertainty. As needed, use the symbols defined in the table and/or include a simple diagram of the setup.

(C) Describe how the experimental data could be analyzed to confirm or disconfirm the hypothesis that the water's range can be calculated with the equations of projectile motion.

GO ON TO THE NEXT PAGE.

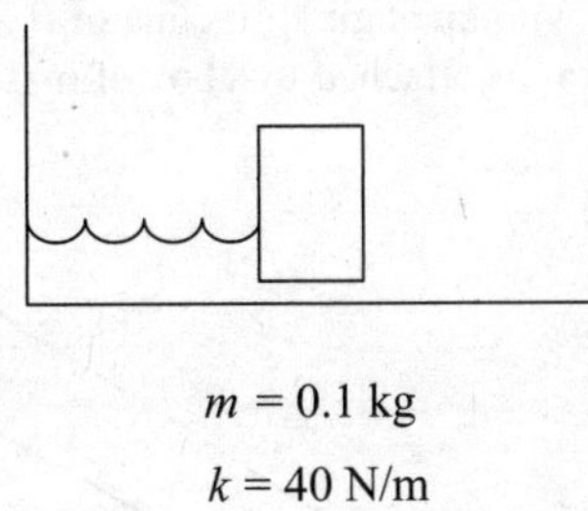

$m = 0.1$ kg

$k = 40$ N/m

3. A horizontal spring with a spring constant of 40 N/m is attached a 0.1 kg block.

(A) If the block is pulled to a distance of 0.5 m from equilibrium and released, what is the maximum speed of the block?

(B) What is the frequency of the oscillations?

(C) If the spring were flipped vertically and attached to the ground with the block placed on top, how would the natural length of the spring change?

(D) How does the frequency of the oscillations of the vertical spring-block oscillator compare with the frequency when it was placed horizontally?

GO ON TO THE NEXT PAGE.

4. A massless tray is placed on an inclined plane with an angle of incline of θ. There is a coefficient of static friction μ_s between the inclined plane and the massless tray. The tray is attached to a box of mass M by the pulley system shown below.

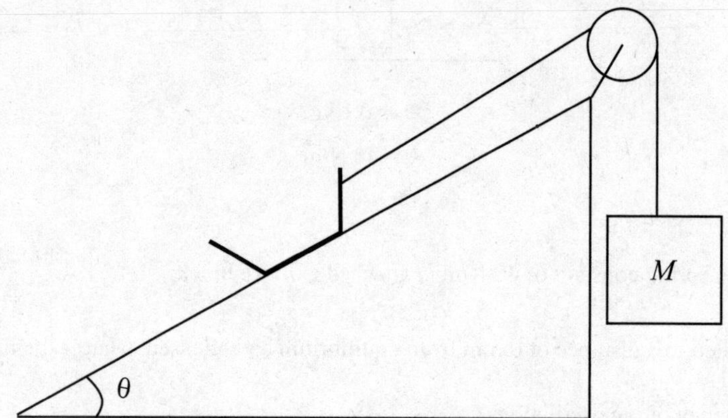

(A) If mass can be loaded onto the massless tray, how much mass m has to be loaded to stop the tray from being pulled up the inclined plane by M?

(B) How much mass m has to be loaded until the tray starts sliding down the inclined plane?

(C) If $\mu_k = 0.3$ and $\theta = 45°$, what is the acceleration of the massless tray if $m = 4M$?

GO ON TO THE NEXT PAGE.

5. A machine launches a 2 kg ball to the right with an initial velocity 16 m/s at an angle of 30° to a student standing 20 m away with a baseball bat.

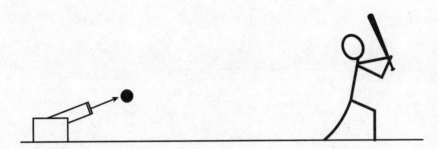

(A) What height must the student swing the bat to hit the ball?

(B) What is the magnitude of the velocity of the ball just before impact?

(C) If the student hits the ball with an upward vertical velocity of 5 m/s and horizontal velocity of 12 m/s to the left, what are the horizontal and vertical components of the impulse of the ball from the collision?

(D) If the impact time with the bat was 0.05 s, what is the average force experienced by the ball during impact?

STOP

END OF EXAM

Practice Test 4: Answers and Explanations

PRACTICE TEST 4: ANSWER KEY

1.	D	26.	C
2.	A	27.	B
3.	A	28.	C
4.	D	29.	D
5.	D	30.	C
6.	A	31.	A
7.	C	32.	B
8.	B	33.	D
9.	A	34.	C
10.	C	35.	C
11.	D	36.	C
12.	C	37.	C
13.	D	38.	C
14.	D	39.	A
15.	C	40.	A
16.	C	41.	B
17.	D	42.	C
18.	C	43.	B
19.	C	44.	A
20.	D	45.	C
21.	C	46.	B, D
22.	A	47.	A, C
23.	B	48.	B, D
24.	C	49.	B, C
25.	B	50.	B, D

PRACTICE TEST 4: ANSWERS AND EXPLANATIONS

Section I: Multiple-Choice Questions

1. **D** The total distance traveled is equal to the sum of the individual distances traveled by the explorer: $d = 30\text{ m} + 20\sqrt{2}\text{ m} + 140\text{ m} = 198.2\text{ m}$.

2. **A** The displacement is equal to the change in position of the explorer. The horizontal and vertical components of the explorer's displacement can be calculated as follows:

$$\Delta x = 30\text{ m} + 20\sqrt{2}\cos 45° = 50\text{ m}$$

$$\Delta y = -20\sqrt{2}\sin 45° + 140\text{ m} = 120\text{ m}$$

The displacement of the explorer then is the magnitude of the vector $(50\mathbf{i} + 120\mathbf{j})\text{m}$:

$$\Delta s = \sqrt{(\Delta x)^2 + (\Delta y)^2} = \sqrt{(50\text{ m})^2 + (120\text{ m})^2} = 130\text{ m}$$

3. **A** As the displacement of the explorer was 130 m, the average velocity of the explorer is equal to

$$\text{average velocity} = \frac{\text{displacement}}{\text{time}} = \frac{130\text{ m}}{260\text{ s}} = 0.50\text{ m/s}$$

4. **D** The position of the object is decreasing, so the object must be moving in the negative direction with a negative velocity. Eliminate (A) and (B). The position of the object is decreasing at a slower rate over time, so the magnitude of the velocity must be decreasing. Eliminate (C). Choice (D) is correct.

5. **D** In section 1, the object is slowing down and moving in the positive direction. Eliminate (A). In section 2, the object is speeding up and moving in the negative direction. Eliminate (B). In section 3, the object is moving with constant speed in the negative direction. Eliminate (C). In section 4, the object is slowing down and moving in the negative direction. Choice (D) is correct.

6. **A** When a projectile reaches the top of its trajectory, the vertical component of its velocity is momentarily zero. As the horizontal velocity in standard parabolic motion is always constant, this means that the velocity of projectiles is smallest at the apex of the trajectory. Choice (A) is correct. Choice (B) is incorrect as the projectile can still have a horizontal velocity at the apex. The acceleration experienced by projectiles in flight is gravity, which is constant at all points of the trajectory. Eliminate (C). Projectiles have the smallest velocity at the apex of the trajectory, so they cannot have the maximum kinetic energy at this same point. Eliminate (D).

7. **C** First, calculate how long it takes for the cannonball to reach the wall. This is a horizontal question. As the horizontal velocity is constant, the time it takes for the cannonball to reach the wall can be calculated as follows:

$$\Delta x = v_{0x}t \Rightarrow t = \frac{\Delta x}{v_{0x}} = \frac{30\text{ m}}{20\cos 45°} = 2.12\text{ s}$$

Next, calculate the height of the projectile at this time by applying Big Five #3:

$$\Delta y = v_{0y}t - \frac{1}{2}gt^2 = 20\sin 45°(2.12\text{ s}) - \frac{1}{2}(10\text{ m/s}^2)(2.12\text{ s})^2 = 7.51\text{ m}$$

This is the maximum height of a wall that the cannonball can clear.

8. **B** Use Big Five #1:

$$\Delta x = \frac{1}{2}(v_0 + v)t = \frac{1}{2}(0\text{ m/s} + 15\text{ m/s})(5\text{ s}) = 37.5\text{ m}$$

9. **A** The forces on the person's foot are shown in the diagram below:

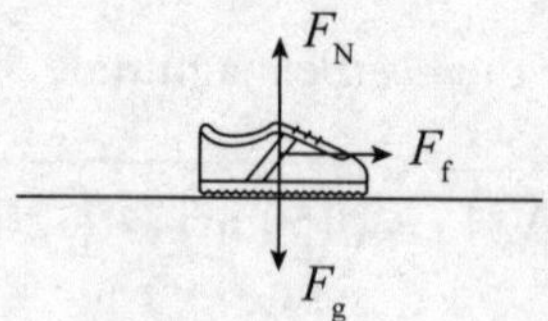

Since the person is accelerating to the right, the net force must be to the right. The interaction causing this force is friction between the ground and the person's shoe. But the upward normal force also comes from the ground. Thus, the total contact force from the ground points up and to the right.

Of these forces, both the normal force and the friction force are coming from the ground, so the sum of those vectors is the correct answer.

10. **C** In the horizontal direction, there are two forces acting on the book: the force applied by the student and the normal force from the wall. In the vertical direction, there are also two forces acting on the book: the force of friction and the weight of the book. In order for the student to hold the book in place, the net force on the book must equal zero in both the horizontal and vertical directions.

$$\Sigma F_x = F_N - F_{app} = 0 \Rightarrow F_N = F_{app}$$

$$\Sigma F_y = F_f - F_g = 0 \Rightarrow F_f = F_g \Rightarrow \mu_s F_N = mg \Rightarrow F_N = \frac{mg}{\mu_s} = \frac{0.5\text{ kg}(10\text{ m/s}^2)}{0.2} = 25\text{ N}$$

As the normal force is equal to the force applied, the student must apply a force of 25 N to hold the book in place.

11. **D** There are four forces acting on the box: the weight of the box, the normal force of the inclined plane on the box, the force applied, and the force of friction. The weight of the box can be separated into its parallel and perpendicular components.

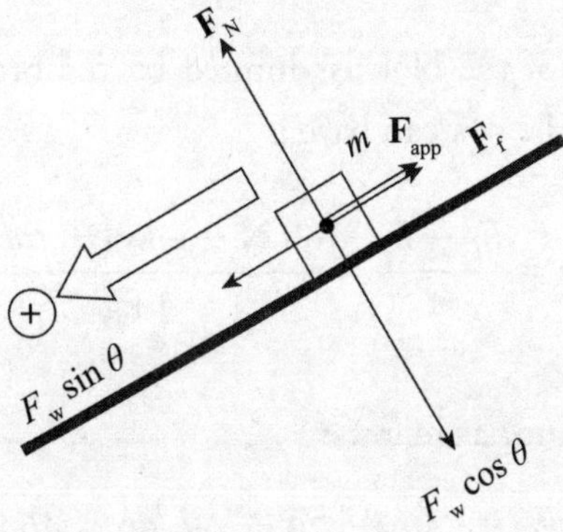

In order for the box to be at rest, the net force in both the parallel and perpendicular directions must be zero:

$$F_N - F_w \cos(\theta) = 0 \Rightarrow F_N = F_w \cos(\theta) \Rightarrow F_N = mg\cos(\theta)$$

$$F_w \sin(\theta) - F_f - F_{app} = 0 \Rightarrow F_{app} = F_w \sin(\theta) - F_f = mg\sin(\theta) - \mu_s F_N$$

$$= mg\sin(\theta) - \mu_s mg\cos(\theta)$$

$$= (10 \text{ kg})(10 \text{ m/s}^2)(\sin 60°) - 0.2(10 \text{ kg})(10 \text{ m/s}^2)(\cos 60°) = 76.6 \text{ N}$$

A force of 76.6 N must be applied to prevent the box from sliding down the inclined plane.

12. **C** By Newton's Third Law, when the heavier car exerts a force on the smaller car, that small car exerts an equal but opposite force back onto the heavy car. While the forces are in opposite direction, both forces have the same magnitude.

13. **D** By Newton's Second Law, the acceleration of each mass can be evaluated using $F = ma$. As both objects experience the same magnitude of force, the a_1/a_2 can be evaluated as follows:

$$F_1 = F_2$$

$$m_1 a_1 = m_2 a_2$$

$$\frac{a_1}{a_2} = \frac{m_2}{m_1} = \frac{40 \text{ kg}}{10 \text{ kg}} = \frac{4}{1}$$

14. **D** There are two forces acting on the box: the weight of the box and the tension force from the string. Newton's Second Law gives the acceleration of the box:

$$F_{net} = ma = F_T - F_g$$

The maximum acceleration of the block is limited by the breaking strength of the string. The maximum acceleration that the box can have is

$$ma = F_T - F_g \Rightarrow a = \frac{F_T - F_g}{m} = \frac{80\text{ N} - 4\text{ kg}\left(10\text{ m/s}^2\right)}{4\text{ kg}} = 10\text{ m/s}^2$$

15. **C** The work done can be calculated as follows:

$$W = Fd\cos\theta = 40\text{ N} \cdot 20\text{ m} \cdot \cos 60° = 400\text{ J}$$

16. **C** Apply Conservation of Mechanical Energy:

$$K_i + U_i = K_f + U_f$$

$$0 + mgh = \frac{1}{2}mv_f^2 + 0$$

$$v_f^2 = 2gh$$

$$v_f = \sqrt{2gh} = \sqrt{2\left(10\text{ m/s}^2\right)\left(20\text{ m}\right)} = 20\text{ m/s}$$

17. **D** The work done can be calculated using the Work–Energy Theorem:

$$W = \Delta K = K_{\text{final}} - K_{\text{initial}} = \frac{1}{2}mv^2 - \frac{1}{2}mv_0^2 = \frac{1}{2}m\left(v^2 - v_0^2\right)$$

$$= \frac{1}{2}\left(0.2\text{ kg}\right)\left(\left(30\text{ m/s}\right)^2 - 0\right) = 90\text{ J}$$

18. **C** Initially, all of the energy of the block is potential energy. At the halfway point of the block, half of the potential energy of the block has been converted to kinetic energy. However, the force of friction does negative work on the block, which reduces the kinetic energy of the block. As a result, the kinetic energy of the block at the halfway point is equal to

$$K = \frac{mgh}{2} - W_f$$

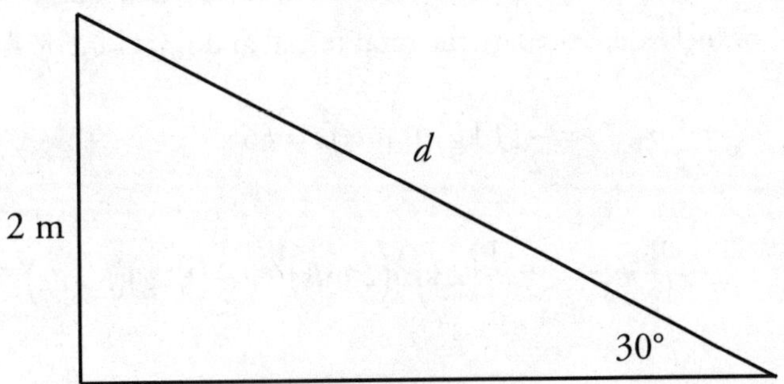

The length of the inclined plane (d) can be calculated as follows: $\sin\theta = \frac{2\text{ m}}{\text{d}} \Rightarrow d = \frac{2\text{ m}}{\sin 30°} = 4\text{ m}.$

As the block is halfway down, the block has traveled a distance of 2 m.

$$K = \frac{mgh}{2} - F_f d = \frac{mgh}{2} - \mu_k mg\cos\theta \cdot d$$

$$= \frac{(10\text{ kg})(10\text{ m/s}^2)(2\text{ m})}{2} - (0.1)(10\text{ kg})(10\text{ m/s}^2)(\cos 30°)(2\text{ m}) = 83\text{ J}$$

19. **C** The power can be calculated as follows:

$$P = Fv = 2{,}000\text{ N}(30\text{ m/s}) = 60{,}000\text{ W}$$

20. **D** The impulse is equal to the change in linear momentum, which can be calculated as follows:

$$J = \Delta\mathbf{p} = m\Delta\mathbf{v} = m(\mathbf{v}_f - \mathbf{v}_i) = 2\text{ kg}(10\text{ m/s} - (-20\text{ m/s})) = 60\text{ N·s}$$

21. **C** The average force experienced by the ball can be calculated using

$$\boldsymbol{F} = \frac{\Delta\mathbf{p}}{\Delta t} = \frac{60\text{ N}\cdot\text{s}}{0.002\text{ s}} = 30{,}000\text{ N}$$

22. **A** The work done can be calculated using the Work–Energy Theorem:

$$W = \Delta K = K_{\text{final}} - K_{\text{initial}} = \frac{1}{2}mv^2 - \frac{1}{2}mv_0^2 = \frac{1}{2}m\left(v^2 - v_0^2\right)$$
$$= \frac{1}{2}(2\text{ kg})\left((10\text{ m/s})^2 - (20\text{ m/s})^2\right) = -300\text{ J}$$

23. **B** As gravity is a conservative force, the work done by gravity is path independent and can be calculated as

$$\Delta U_g = -W_g \Rightarrow W_g = -\Delta U_g = -mg(h_f - h_i) = -6\text{ kg}(10\text{ m/s}^2)(10\text{ m} - 0\text{ m}) = -600\text{ J}$$

24. **C** The objects did not stick together after the collision, so (B) is wrong. To determine whether the collision is elastic or inelastic, calculate the total initial and final kinetic energies:

$$K_i = \frac{1}{2}m_1v_{1,i^2} = \frac{1}{2}(2\text{ kg})(6\text{ m/s})^2 = 36\text{ J}$$

$$K_f = \frac{1}{2}m_{1,f}v_{1,f^2} + \frac{1}{2}m_2v_{2,f^2} = \frac{1}{2}(2\text{ kg})(2\text{ m/s})^2 + \frac{1}{2}(1\text{ kg})(8\text{ m/s})^2 = 36\text{ J}$$

As kinetic energy was conserved, the collision was elastic.

25. **B** The final velocity of a perfectly inelastic collision can be calculated using the equation:

$$m_1\mathbf{v}_{1,i} + m_2\mathbf{v}_{2,i} = (m_1 + m_2)\mathbf{v}_f$$

$$\mathbf{v}_f = \frac{m_1\mathbf{v}_{1,i} + m_2\mathbf{v}_{2,i}}{m_1 + m_2} = \frac{3\text{ kg}(3\text{ m/s}) + 2\text{ kg}(-6\text{ m/s})}{3\text{ kg} + 2\text{ kg}} = -0.6\text{ m/s}$$

26. **C** As the cannon and cannonball are both initially at rest, the total initial momentum is 0. By Conservation of Linear Momentum, the total final momentum must also be equal to 0.

$$m_{\text{cannon}}\mathbf{v}_{\text{cannon,i}} + m_{\text{cannonball}}\mathbf{v}_{\text{cannonball,i}} = m_{\text{cannon}}\mathbf{v}_{\text{cannon,f}} + m_{\text{cannonball}}\mathbf{v}_{\text{cannonball,f}}$$

$$0 = m_{\text{cannon}}\mathbf{v}_{\text{cannon,f}} + m_{\text{cannonball}}\mathbf{v}_{\text{cannonball,f}}$$

$$\mathbf{v}_{\text{cannon,f}} = \frac{-m_{\text{cannonball}}\mathbf{v}_{\text{cannonball,f}}}{m_{\text{cannon}}} = \frac{-10\text{ kg}(300\text{ m/s})}{300\text{ kg}} = -10\text{ m/s}$$

The final velocity of the cannon is –10 m/s. The final speed is the magnitude of velocity, which is 10 m/s.

27. **B** As the mass spins in a horizontal circle, it is the tension force that snapped the string, which is equal to the breaking strength when the mass is traveling at 6 m/s. As the tension force is the only force acting in the horizontal direction, it is also producing centripetal force and can be calculated as follows:

$$\mathbf{F}_T = \mathbf{F}_c = \frac{mv^2}{r} = \frac{(1.5\text{ kg})(6\text{ m/s})^2}{0.4\text{ m}} = 135\text{ N}$$

28. **C** The gravitational pull of the planet provides the centripetal force on the satellite:

$$\mathbf{F}_g = \mathbf{F}_c$$

$$G\frac{Mm}{r^2} = \frac{mv^2}{r} \Rightarrow G\frac{M}{r} = v^2 \Rightarrow v = \sqrt{\frac{GM}{r}}$$

The speed of the second satellite is

$$v_2 = \sqrt{\frac{GM}{2r}} = \frac{1}{\sqrt{2}}\sqrt{\frac{GM}{r}} = \frac{1}{\sqrt{2}}v$$

29. **D** The reading on the scale is equal to the magnitude of the downward force exerted by the man on the scale. By Newton's Third Law, this is equivalent to the magnitude of the normal force exerted by the scale on the man. There are two forces acting on the man: the weight of the man and the normal force of the scale. Apply Newton's Second Law:

$$F_{net} = F_N - F_g = ma$$

$$F_N = ma + F_g = \frac{mg}{4} + mg = \frac{5mg}{4} = \frac{5(50\text{ kg})(10\text{ m/s}^2)}{4} = 625\text{ N}$$

30. **C** Apply Big Five #2 for rotational motion:

$$\omega = \omega_0 + \alpha t = 0 + 2\text{ rad/s}^2(5\text{ s}) = 10\text{ rad/s}$$

The angular velocity can be related to linear velocity with the equation:

$$v = r\omega = (0.5\text{ m})(10\text{ rad/s}) = 5\text{ m/s}$$

31. **A** Start by calculating the center of mass for the system with two masses:

$$x(c.m.) = \frac{x_1m_1 + x_2m_2}{m_1 + m_2} = \frac{0\text{ m}(3\text{ kg}) + 1\text{ m}(5\text{ kg})}{3\text{ kg} + 5\text{ kg}} = 0.625\text{ m}$$

Next, calculate the new center of the mass of the system with the addition of the third mass:

$$x(c.m.) = \frac{x_1m_1 + x_2m_2 + x_3m_3}{m_1 + m_2 + m_3} = \frac{0\text{ m}(3\text{ kg}) + 1\text{ m}(5\text{ kg}) + 0.5\text{ m}(2\text{ kg})}{3\text{ kg} + 5\text{ kg} + 2\text{ kg}} = 0.6\text{ m}$$

The center of mass of the system has shifted to the left by 0.025 m.

32. **B** The weight of the box pulls down on the rope, producing tension that creates a torque on the pulley. This torque is equal to

$$\tau = rF = rF_g = rmg = 0.5\text{ m}(5\text{ kg})(10\text{ m/s}^2) = 25\text{ N·m}$$

33. **D** Rotational inertia is directly proportional to the mass of the object. Eliminate (A) and (B) as they have smaller masses. The farther away the mass is from the axis of rotation, the greater the rotational inertia. The mass of a hollow ball is farther away from the axis of rotation than that of a solid ball. Eliminate (C). Choice (D) is correct.

34. **C** Apply Hooke's Law to calculate the force constant of the spring:

$$F = kx \Rightarrow k = \frac{F}{x} = \frac{6\text{ N}}{0.05\text{ m}} = 120\text{ N/m}$$

The frequency of a spring-block oscillator is given by

$$f = \frac{1}{2\pi}\sqrt{\frac{k}{m}} = \frac{1}{2\pi}\sqrt{\frac{120\text{ N/m}}{5\text{ kg}}} = 0.78\text{ Hz}$$

35. **C** When the block is at its amplitude, all of the kinetic energy has been converted to potential energy. The maximum potential energy is given by $U = \frac{1}{2}kA^2$. The amplitude can then be calculated by applying Conservation of Energy:

$$K_{max} = U_{max}$$

$$\frac{1}{2}mv_{max}^2 = \frac{1}{2}kA^2$$

$$mv_{max}^2 = kA^2$$

$$A^2 = \frac{mv_{max}^2}{k}$$

$$A = \sqrt{\frac{mv_{max}^2}{k}} = \sqrt{\frac{3\text{ kg}(4\text{ m/s})^2}{10\text{ N/m}}} = 2.2\text{ m}$$

36. **C** The period of a string is given by

$$T = 2\pi\sqrt{\frac{L}{g}} \Rightarrow \sqrt{\frac{L}{g}} = \frac{T}{2\pi} \Rightarrow \frac{L}{g} = \left(\frac{T}{2\pi}\right)^2 \Rightarrow L = g\left(\frac{T}{2\pi}\right)^2 = 10\frac{\text{m}}{\text{s}^2}\left(\frac{2\text{ s}}{2\pi}\right)^2 = 1\text{ m}$$

37. **C** Apply the Conservation of Mechanical Energy to the object since the system is frictionless, $K_i + U_i = K_f + U_f$. Let the object at equilibrium be the initial point, where the kinetic energy is zero, and let the final point be the object at $\frac{A}{2}$. Using these points, substitute the expressions for the kinetic and potential energies, recalling that for a spring, $U = \frac{1}{2}kx^2$.

$$0 + \frac{1}{2}kA^2 = \frac{1}{2}mv^2 + \frac{1}{2}k\left(\frac{A}{2}\right)^2 \rightarrow kA^2 - \frac{1}{4}kA^2 = mv^2 \rightarrow v^2 = \frac{3kA^2}{4m} \rightarrow v = \frac{A}{2}\sqrt{\frac{3k}{m}}$$

38. **C** As the mass within the cloud moves closer to its axis of rotation, the rotational inertia decreases. In space, the external forces acting on the cloud are negligible, so the external torque on the cloud is negligible, so eliminate (A) and (D). Note that gravity between particles in the cloud is an internal force, not an external force. In the absence of external torque, angular momentum is conserved, so eliminate (B).

39. **A** Because the initial height of the object is comparable to the radius of the Moon, you cannot simply use mgh for its initial potential energy. Instead, use the more general expression $U = -GMm/r$, where r is the distance from the center of the Moon. Note that since $h = R$, the object's initial distance from the Moon's center is $h + R = R + R = 2R$. Conservation of Energy then gives

$$K_i + U_i = K_f + U$$
$$0 - \frac{GMm}{2R} = \frac{1}{2}mv^2 - \frac{GMm}{R}$$
$$\frac{GMm}{2R} = \frac{1}{2}mv^2$$
$$v = \sqrt{\frac{GM}{R}}$$

40. **A** Simple machines, like the inclined plane of the ramp, do not reduce the amount of work required to lift an object, eliminating (B). In the ideal friction-free case, the work is equal with and without friction. With friction, the applied force on the box sliding up the ramp must be greater than it would be in the ideal case, so more work is required to slide the box up the ramp than lifting it, thus (A) is correct. Inclined planes are able to reduce the force at the expense of increasing the distance over which the force is applied. To lift a box of mass m straight up, a force of mg is required. To push such a box up a ramp, the applied force must oppose the component of gravity that points down the ramp and the force of friction, so a force of $mg \sin\theta + \mu_k mg\cos\theta = mg(\sin\theta + \mu_k \cos\theta)$. For many typical values of θ and k in such a situation, $\sin\theta + \mu_k \cos\theta < 1$. However, there are some values of θ and k for which $\sin\theta + \mu_k \cos\theta > 1$. The problem does not specify the values of θ and k, so neither (C) nor (D) must be true, eliminating (C) and (D).

41. **B** Torque is given by $\tau = rF\sin\theta$. The force is gravity, which acts on the ball. The ball is a distance L from the pivot point (where the string attaches to some support), so $r = L$. Gravity acts in the vertical direction, so the angle between $\vec{r}$ and $\vec{F}$ is the same as the angle between the string and the vertical. Therefore, $\tau = L(mg)\sin\theta = mgL\sin\theta$.

42. **C** At the peak of its motion, the object is only moving horizontally. Furthermore, because there is no horizontal acceleration, its horizontal velocity will be the same at the peak as it is at the launch. Finally, remember that $\cos\theta$ (not $\sin\theta$) is used for taking the horizontal component.

43. **B** Define the coordinate system to apply Newton's Second Law. Let up be the positive direction so that the tension that points up is positive, and the force of gravity and acceleration that point down are negative. $\vec{F}_{net} = m\vec{a} \rightarrow T - mg = m(-a) \rightarrow T = mg - ma = m(g - a)$.

44. **A** Friction acts during part of the object's motion, so apply the Conservation of Energy with Nonconservative Forces. $K_i + U_i + W_{friction} = K_f + U_f$. Define the bottom of the incline to be where the height is 0. Let the initial point be at the top of the incline of length L, which is at a height of $L \sin 30°$. The object slides from rest, so $K_i = 0$. Let the final point be on the horizontal surface where the object has stopped, so $U_f = 0$ and $K_f = 0$. Therefore, $0 + mgL\sin 30° + W_{friction} = 0 + 0 \rightarrow W_{friction} = -mgL\sin 30°$. The work done by friction is $W_{friction} = F_{friction} d \cos 180°$ The force of friction is the coefficient of kinetic friction times the normal force. Since the object experiences friction on a horizontal surface, $F_N = mg$. Therefore,

$$\mu_k F_N d \cos 180° = -mgL \sin 30° \rightarrow \mu_k mgd(-1) = -mgL\left(\frac{1}{2}\right) \rightarrow$$

$$\mu_k = \frac{L}{2d} = \frac{10.0 \text{ m}}{2(12.5 \text{ m})} = 0.400$$

45. **C** Power is work per unit time, $P = W/t$. The work done by gravity is the negative of the object's change in potential energy. $W = -\Delta U = -mg\Delta h = -mg(0 - h) = mgh$. Use one of the Big 5 to determine the time it takes for the object to reach the ground. The final velocity is unknown, so use $y = y_0 + v_0 t - \frac{1}{2}gt^2$. The object is dropped, so $v_0 = 0$. Therefore, $0 = h - \frac{1}{2}gt^2 \rightarrow t = \sqrt{\frac{2h}{g}}$. Substitute the expressions for W and t into the expression for power,

$$P = \frac{mgh}{\sqrt{\frac{2h}{g}}} = \frac{m\left(g\sqrt{g}\right)h}{\sqrt{2h}} = \frac{m\sqrt{2g^3h}}{2}$$

46. **B, D** Choices (B) and (D) are both true statements regarding objects undergoing uniform circular motion. As the objects are moving in a circle, the direction of their velocity is constantly changing, so eliminate (A). Although it is possible for the object to experience a force pointing radially outward, this is not a necessity for uniform circular motion, so eliminate (C).

47. **A,C** The total work done on an object is equal to the object's change in kinetic energy. In (A), the object starts and ends at rest, so $\Delta K = 0$, so (A) is correct. In (B), the box accelerates as it falls as gravity does positive work on the object, so (B) is incorrect. In (C), the box maintains a constant speed, so (C) is correct. In (D), the box decelerates, so the total work done on the box is negative, so (D) is incorrect.

48. **B, D** Choices (B) and (D) are both true statements regarding objects undergoing uniform circular motion. As the objects are moving in a circle, the direction of their velocity is constantly changing, so eliminate (A). Although it is possible for the object to feel a force pointing radially outward, this is not a necessity for uniform circular motion, so eliminate (C).

49. **B, C** The work done by conservative forces is path independent, so (B) is correct, while (A) is incorrect. Conservative forces, such as gravity and the spring force, only depend on the position of an object, so (C) is correct. When conservative forces do positive work, it decreases the potential energy of the system.

50. **B, D** For (A), the normal force is equal to the component of the force of gravity that is perpendicular to the surface, $mg\cos 30° = \frac{\sqrt{3}}{2}mg$, so (A) is incorrect. For (B), the centripetal force acting on the object is equal to the sum of the force of gravity and the normal force, $F_C = F_N + W$. Solving for normal force and substituting the equation for centripetal force yields $F_N = m\frac{v^2}{R} - mg = \frac{m}{R}\left(\sqrt{2Rg}\right)^2 - mg = mg$, so (B) is correct. For (C), the two forces acting on the object are the upward normal force and the downward force of gravity. In order for the object to accelerate upward, the net force must point upward, so $F_N < mg$. Therefore, (C) is incorrect. For (D), the net vertical force on the object is $F_{net,y} = F_N + F_{T,y} - W = 0$. Solving for the normal force and substituting the expressions for the y-component of tension and the weight of the object with mass $2m$ yields $F_N = W - F_{T,y} = (2m)g - (2mg)\sin 30° = 2mg - mg = mg$. Therefore, (D) is correct.

Section II: Free-Response Questions

1. (A) In a perfectly elastic collision, both momentum and kinetic energy are conserved:

Equation 1: $m_1v = m_1v_1 + m_2v_2$

Equation 2: $\frac{1}{2}m_1v^2 = \frac{1}{2}m_1v_1^2 + \frac{1}{2}m_2v_2^2 \Rightarrow m_1v^2 = m_1v_1^2 + m_2v_2^2$

Rearrange these equations into

$$m_1v = m_1v_1 + m_2v_2 \Rightarrow m_2v_2 = m_1(v - v_1)$$
$$m_2v_2^2 = m_1v^2 - m_1v_1^2 \Rightarrow m_2v_2^2 = m_1(v - v_1)(v + v_1)$$

Divide the second rearranged equation by the first rearranged equation to get

$$v_2 = v + v_1$$

Substitute this value back into Equation 1 to get

$$m_1 v = m_1 v_1 + m_2 v_2$$

$$m_1 v = m_1 v_1 + m_2 v + m_2 v_1$$

$$m_1 v_1 + m_2 v_1 = m_1 v - m_2 v$$

$$v_1 = \frac{m_1 v - m_2 v}{m_1 + m_2}$$

(B) Substitute the expression for v_1 from (a) back into $v_2 = v + v_1$:

$$v_2 = v + \frac{m_1 v - m_2 v}{m_1 + m_2} = \frac{m_1 v + m_2 v}{m_1 + m_2} + \frac{m_1 v - m_2 v}{m_1 + m_2} = \frac{2m_1 v}{m_1 + m_2}$$

(C) According to the expression for v_2 from (B), v_2 will always have a positive value. The two masses will be traveling in the same direction if v_1 is also positive. According to the expression for v_1 from (A), this is true if $m_1 > m_2$. If $m_1 < m_2$, the two masses will be traveling in the opposite direction. Note: if $m_1 = m_2$, the final velocity of m_1 would be zero, but m_2 would still be moving in the positive direction.

2. (A) (i) The range of a projectile is $R = v_{0x}t$, where time is determined by motion in the y-direction. The water flows out of the hole horizontally, so $v_{0y} = 0$, meaning that the height determines the flight time, $h = \frac{1}{2}gt^2 \rightarrow t = \sqrt{\frac{2h}{g}}$. Therefore, measuring the height and initial velocity would determine the expected range. The initial velocity of the water flowing out of the hole is determined by Bernoulli's equation. The surface of the water in the tank and the water coming out of the hole are both exposed to atmospheric pressure, so the speed of the water flowing out of the hole is determined by the height of the water in the tank.

(ii) Based on the principles involved, a meter stick will be needed to measure the depth of water in the tank, D, the height of the hole, h, and the range, R.

(B) First, use a meter stick to measure the height of the hole. Then, use a plumb line to align a meter stick to measure a range of zero right beneath the hole and to extend in the direction of water flow. Also place a meter stick vertically in the tank to measure the depth of the water. With the water flowing, record the depth of the water in the tank and then immediately record the range of the water flowing out. Make the same measurement for several depths to verify the relationship between these quantities in multiple situations.

(C) Plot the data on a graph of range vs. depth. According to Bernoulli's equation, with the area of the hole << area of the cross section of the tank, the speed of the water coming out of the hole is $\sqrt{2gD}$, so if the student's hypothesis is correct, $R = \left(\sqrt{2gD}\right)\sqrt{\frac{2h}{g}} = 2\sqrt{Dh}$. Add a plot of this expected relationship to compare with the measured data points.

3. (A) When pulled to a distance of 0.5 m, the block has maximum potential energy given by the equation $U = \frac{1}{2}kA^2$. All of this energy is converted to kinetic energy when the block reaches maximum speed:

$$K_{max} = U_{max}$$

$$\frac{1}{2}mv_{max}^2 = \frac{1}{2}kA^2$$

$$mv_{max}^2 = kA^2$$

$$v_{max}^2 = \frac{kA^2}{m}$$

$$v_{max} = \sqrt{\frac{kA^2}{m}} = A\sqrt{\frac{k}{m}} = 0.5\text{ m}\left(\sqrt{\frac{40\text{ N/m}}{0.1\text{ kg}}}\right) = 10\text{ m/s}$$

(B) The frequency of oscillations is given by the following equation:

$$f = \frac{1}{2\pi}\sqrt{\frac{k}{m}} = \frac{1}{2\pi}\sqrt{\frac{40\text{ N/m}}{0.1\text{ kg}}} = \frac{10}{\pi}\text{ Hz} = 3.2\text{ Hz}$$

(C) When the spring is placed vertically on the ground with the block on top, there are two forces acting on the block at the new natural length: the weight of the block and the upward restoring force. These two forces are equal to each other at the natural length:

$$F_{rest} = F_g$$

$$kx = mg$$

$$x = \frac{mg}{k} = \frac{(0.1\text{ kg})(10\text{ m/s}^2)}{40\text{ N/m}} = 0.025\text{ m}$$

The natural length of the spring would decrease by 0.025 m.

(D) The frequency of spring-block oscillator is given by the equation $f = \frac{1}{2\pi}\sqrt{\frac{k}{m}}$. Flipping the spring vertically does not affect k or m, so the frequency of oscillations will remain unchanged.

4. (A) The forces acting on the m mass are $\mathbf{F}_T$ (the tension in the string connecting it to M mass), $\mathbf{F}_w$ (the weight of the mass m), $\mathbf{F}_N$ (the normal force exerted by the inclined plane), and $\mathbf{F}_f$ (the force of static friction).

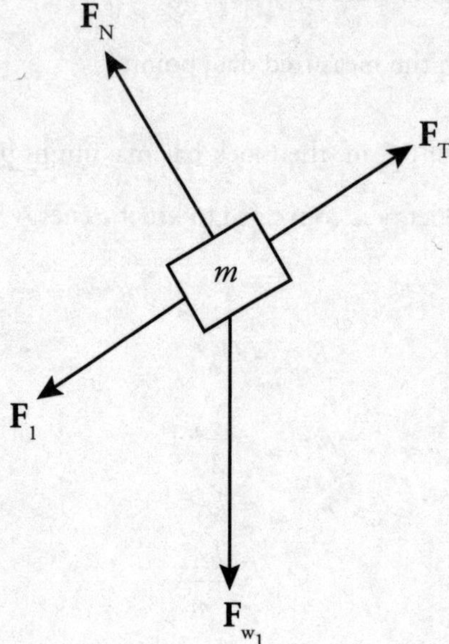

The forces acting on the M mass are $\mathbf{F}_T$ (the tension in the string connecting it to the m mass) and $\mathbf{F}_w$ (the weight of the block).

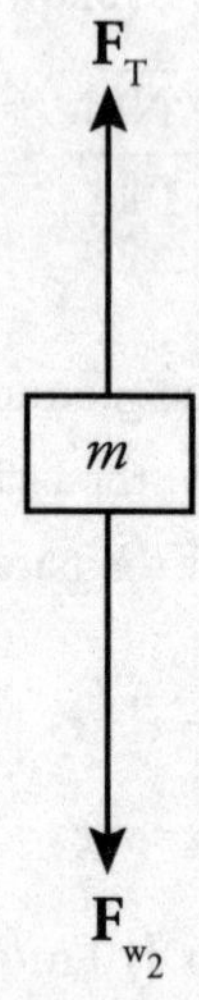

In order to stop the tray from moving, the net force on both the M mass and the mass m loaded on the tray must be zero. Newton's First Law applied to the M mass yields

$$F_{w_2} - F_T = 0 \Rightarrow F_T = F_{w_2} = Mg$$

Newton's First Law applied to the *m* yields

$$F_T - F_g - F_f = 0 \Rightarrow F_T - mg\sin(\theta) - \mu_s mg\cos(\theta) = 0$$

Substituting $F_T = Mg$, you get

$$Mg - mg\sin\theta - \mu_s mg\cos\theta = 0$$

$$mg\sin\theta + \mu_s mg\cos\theta = Mg$$

$$m = \frac{Mg}{g\sin\theta + \mu_s g\cos\theta}$$

(B) As the mass is trying to slide down the incline, the force of friction is acting in the opposite direction as in (A). There is no change to the *M* mass, so Newton's First Law applied to the *M* mass still yields

$$F_{w_2} - F_T = 0 \Rightarrow F_T = F_{w_2} = Mg$$

Newton's First Law applied to the *m* yields

$$F_T + F_g - F_f = 0 \Rightarrow F_T + mg\sin\theta - \mu_s mg\cos\theta = 0$$

Substituting $F_T = Mg$, you get

$$Mg + mg\sin\theta - \mu_s mg\cos\theta = 0$$

$$mg\sin\theta - \mu_s mg\cos\theta = Mg$$

$$m = \frac{Mg}{g\sin\theta + \mu_s g\cos\theta}$$

In order for the massless tray to start sliding down the inclined plane, *m* has to be greater than $\frac{Mg}{g\sin\theta - \mu_s g\cos\theta}$.

(C) Newton's First Law applied to the *M* mass yields

$$F_T - F_{w_2} = Ma \Rightarrow F_T = F_{w_2} + Ma \Rightarrow F_T = Mg + Ma$$

Newton's First Law applied to the *m* yields

$$F_g - F_f - F_T = ma \Rightarrow mg\sin\theta - \mu_k mg\cos\theta - F_T = ma$$

Substituting $F_T = Mg + Ma$, you get

$$mg\sin\theta - \mu_k mg\cos\theta - Mg - Ma = ma$$

$$ma + Ma = mg\sin\theta - \mu_k mg\cos\theta - Mg$$

$$a = \frac{mg\sin\theta - \mu_k mg\cos\theta - Mg}{m + M} = \frac{m\sin\theta - \mu_k m\cos\theta - M}{m + M}g$$

Plugging in $m = 4M$, $\theta = 45°$, and $\mu_k = 0.3$, you get

$$a = \frac{4M\sin 45° - 0.3(4M)\cos 45° - M}{4M + M}g = \frac{4\sin 45° - 0.3(4)\cos 45° - 1}{5}\left(10\text{ m/s}^2\right) = 2.0\text{ m/s}^2$$

5. (A) First, calculate the amount of time it takes for the ball to travel 20 m. Apply the first of the horizontal motion equations:

$$\Delta x = v_{0x}t \Rightarrow t = \frac{\Delta x}{v_{0x}} = \frac{20\text{ m}}{16\cos 30°\text{ m/s}} = 1.44\text{ s}$$

Next, calculate the height of the ball at $t = 2.3$ s. Apply Big Five #3 to vertical motion:

$$\Delta y = v_{0y}t - \frac{1}{2}gt^2 = (16\text{ m/s})\sin 30°(1.44\text{ s}) - \frac{1}{2}\left(10\text{ m/s}^2\right)(1.44\text{ s})^2 = 1.15\text{ m}$$

The student must swing the bat at a height of 1.15 m to hit the ball.

(B) The horizontal velocity of the ball is constant:

$$v_x = v_{0x} = v_0\cos\theta = 16\cos 30° = 13.9\text{ m/s}$$

As impact occurs when $t = 1.44$ s, the vertical velocity of the ball can be calculated using Big Five #2 to vertical motion:

$$v_y = v_{0y} - gt = 16\sin 30° - (10\text{ m/s}^2)(1.44\text{ s}) = -6.4\text{ m/s}$$

Use the Pythagorean Theorem to solve for the magnitude of the overall velocity:

$$v^2 = v_x^2 + v_y^2 \Rightarrow v = \sqrt{v_x^2 + v_y^2} = \sqrt{(13.9\text{ m/s})^2 + (-6.4\text{ m/s})^2} = 15.3\text{ m/s}$$

(C) The impulse is equal to the change in momentum of the ball. You can calculate the horizontal and vertical components of the impulse separately:

$$J_x = \Delta p_x = m\Delta v_x = m(v_{x,f} - v_{x,i}) = 2\text{ kg}(-12\text{ m/s} - 13.9\text{ m/s}) = -51.8\text{ N·s}$$

$$J_y = \Delta p_y = m\Delta v_y = m(v_{y,f} - v_{y,i}) = 2\text{ kg}(5\text{ m/s} - (-6.4\text{ m/s})) = 22.8\text{ N·s}$$

(D) You can use the equation **J** = **F**Δt to calculate the horizontal and vertical components of the average force:

$$F_x = \frac{J_x}{\Delta t} = \frac{-51.8 \text{ N} \cdot \text{s}}{0.05 \text{ s}} = -1036 \text{ N}$$

Use the Pythagorean Theorem to calculate the magnitude of the average force experienced by the ball during impact:

$$F = \sqrt{F_x^2 + F_y^2} = \sqrt{(-1036 \text{ N})^2 + (456 \text{ N})^2} = 1132 \text{ N}$$

(E) The direction of the average force can be calculated using

$$\tan\theta = \frac{|F_y|}{|F_x|} \Rightarrow \theta = \tan^{-1}\frac{|F_y|}{|F_x|} = \tan^{-1}\left(\frac{456 \text{ N}}{1036 \text{ N}}\right) = 23.8°$$

The direction of the force is 23.8° above the horizontal.

HOW TO SCORE PRACTICE TEST 4

Section I: Multiple Choice

__________ × 1.5 = __________

Number Correct (out of 50)

Weighted Section I Score (Do not round)

Section II: Free Response

Question 1:	______ (out of 12)	× 1.25 =	______ (Do not round)
Question 2:	______ (out of 12)	× 1.25 =	______ (Do not round)
Question 3:	______ (out of 7)	× 2.143 =	______ (Do not round)
Question 4:	______ (out of 7)	× 2.143 =	______ (Do not round)
Question 5:	______ (out of 7)	× 2.143 =	______ (Do not round)

Sum = ______

Weighted Section II Score (Do not round)

AP Score Conversion Chart Physics 1

Composite Score Range	AP Score
107–150	5
90–106	4
73–89	3
56–72	2
0–55	1

Composite Score

______ + ______ = ______

Weighted Section I Score

Weighted Section II Score

Composite Score (Round to nearest whole number)

Completely darken bubbles with a No. 2 pencil. If you make a mistake, be sure to erase mark completely. Erase all stray marks.

1.

YOUR NAME: (Print) ____ Last ____ First ____ M.I.

SIGNATURE: ____ DATE: __/__/__

HOME ADDRESS: (Print) ____ Number and Street

____ City ____ State ____ Zip Code

PHONE NO.: (Print) ____

IMPORTANT: Please fill in these boxes exactly as shown on the back cover of your test book.

2. TEST FORM

3. TEST CODE

0	A	J	0	0
1	B	K	1	1
2	C	L	2	2
3	D	M	3	3
4	E	N	4	4
5	F	O	5	5
6	G	P	6	6
7	H	Q	7	7
8	I	R	8	8
9			9	9

4. REGISTRATION NUMBER

0	0	0	0	0	0
1	1	1	1	1	1
2	2	2	2	2	2
3	3	3	3	3	3
4	4	4	4	4	4
5	5	5	5	5	5
6	6	6	6	6	6
7	7	7	7	7	7
8	8	8	8	8	8
9	9	9	9	9	9

5. YOUR NAME

First 4 letters of last name					FIRST INIT	MID INIT
A	A	A	A		A	A
B	B	B	B		B	B
C	C	C	C		C	C
D	D	D	D		D	D
E	E	E	E		E	E
F	F	F	F		F	F
G	G	G	G		G	G
H	H	H	H		H	H
I	I	I	I		I	I
J	J	J	J		J	J
K	K	K	K		K	K
L	L	L	L		L	L
M	M	M	M		M	M
N	N	N	N		N	N
O	O	O	O		O	O
P	P	P	P		P	P
Q	Q	Q	Q		Q	Q
R	R	R	R		R	R
S	S	S	S		S	S
T	T	T	T		T	T
U	U	U	U		U	U
V	V	V	V		V	V
W	W	W	W		W	W
X	X	X	X		X	X
Y	Y	Y	Y		Y	Y
Z	Z	Z	Z		Z	Z

6. DATE OF BIRTH

Month	Day		Year	
JAN				
FEB	0	0	0	0
MAR	1	1	1	1
APR	2	2	2	2
MAY	3	3	3	3
JUN		4	4	4
JUL		5	5	5
AUG		6	6	6
SEP		7	7	7
OCT		8	8	8
NOV		9	9	9
DEC				

The Princeton Review®

1. A B C D	14. A B C D	27. A B C D	40. A B C D
2. A B C D	15. A B C D	28. A B C D	41. A B C D
3. A B C D	16. A B C D	29. A B C D	42. A B C D
4. A B C D	17. A B C D	30. A B C D	43. A B C D
5. A B C D	18. A B C D	31. A B C D	44. A B C D
6. A B C D	19. A B C D	32. A B C D	45. A B C D
7. A B C D	20. A B C D	33. A B C D	46. A B C D
8. A B C D	21. A B C D	34. A B C D	47. A B C D
9. A B C D	22. A B C D	35. A B C D	48. A B C D
10. A B C D	23. A B C D	36. A B C D	49. A B C D
11. A B C D	24. A B C D	37. A B C D	50. A B C D
12. A B C D	25. A B C D	38. A B C D	
13. A B C D	26. A B C D	39. A B C D	

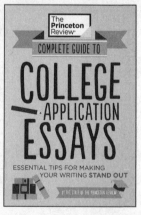

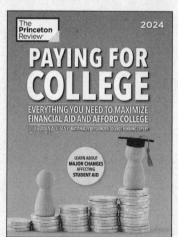